A SOURCE BOOK FOR RULE COLLECTORS

Philip E. Stanley

Photographs by the Author

Astragal Press

Copyright 2003 by Philip E. Stanley

All rights reserved. No part of this book may be reproduced or transmitted in any form or by any means, electrical or mechanical, including photocopying, recording, or by any information or retrieval system, without the written permission of the Publisher, except where permitted by law.

International Standard Book Number   1-931626-17-0
Library of Congress Control Number   2004100105

Published by
THE ASTRAGAL PRESS
5 Cold Hill Road, Suite 12
P.O. Box 239
Mendham, New Jersey 07945-0239

Cover design by Donald Kahn

Manufactured in the United States of America

Dedicated to the Memory of

KENNETH D. ROBERTS

In so many areas of tool collecting: from English tools to wooden planes; from measuring instruments to catalog reprints; from horology to tool making in New York and Connecticut; Ken Roberts was one of the first to publish in the field, doing seminal research, and blazing the way for the rest of us. His example, advice, and encouragement have helped us all.

Kenneth D. Roberts, 1916-2000

Acknowledgments

"If I have seen further (than others) it is by standing upon the shoulders of giants."
Sir Isaac Newton
Letter to Robert Hooke, February 5, 1675-1677

Newton's observation is as true in the production of a reference book as it is in scientific research. Such a book may have only a single name on the title page, but in reality it is a communal effort, the result of the efforts, past and present, of an army of collectors and researchers. This is certainly true of this book; without the aid and cooperation of many others I would not have been able to pull all this information together, and I wish to acknowledge their assistance. Thank you all.

I wish to specially thank all those individuals who have conducted research in the field of measuring instruments, and have published their results for all of our benefit. The availability of all that information is one of the things that made this book possible.

A selection of articles, in their entirety, has been reprinted in this book as being of exceptional value to collectors. For permission to reprint these articles I wish specifically to thank their authors, and the publications where they appeared:

William Baader	Bruce Babcock
Ben Blumenburg	Dale Butterworth
Clifford Fales	David Goodson
William Holden	Paul Kebabian
Dieter von Jezierski	John V. Knott
Kenneth D. Roberts (deceased)	

The Chronicle of the Early American Industries Association
The Gristmill, published by the Mid-West Tool Collectors Association
The Journal of the Oughtred Society
Stanley Tool Collector News, published by The Tool Merchant

In many cases these authors and the editors of these publications have not only permitted, but actively cooperated in helping me to reprint these articles, furnishing me with the original artwork for the illustrations, or, if it could not be found, lending me the rules themselves so I could photograph them and replicate the lost illustrations.

Some of America's major repositories of historical information are the public library local history rooms and historical societies which can be found in so many of our communities. I have used these resources extensively in preparing this book, and found the responsible individuals extremely helpful and cooperative. They often spent hours of research on my behalf, and suggested other sources when their collections had been exhausted. They have my respect and gratitude.

Specific contributions by individual collectors have also been of great value. They have made their research notes available to me, and allowed me to photograph rules in their collections. If I could not get to their homes, they sent me photographs they had taken, and in some cases the rules themselves. Other individuals have shared information with me or pointed me toward a source of information. In this context I wish to particularly thank the following:

Clarance Blanchard	George Collord
Martin J. Donnelly	Clifford Fales
Robert Glendenning	Marianne Halpern
Walter Jacob	Don Jordan
Herb Kean	Dr. Richard Knight
Scott Lynk	Douglas Orr
Philip Platt	James Price
Douglas Volk	Jared Silbersher
William Skerritt	Robert Westley
Donald and Anne Wing	

I would also like to thank my publisher, the Astragal Press, and in particular my editor Missy Staples, for their encouragement and cooperation.

Finally, I wish to thank my wife, Andrea for her encouragement and suggestions. Her unflagging patience and support has been the thing that has, more than anything else, carried me through the long and arduous task of preparing this book.

Preface

This book has its origins in a 14 page booklet entitled *Introduction to Rule Collecting* published by Ken Roberts in 1982. That booklet contained a brief history of rule making in the United States during the 19th century, together with a selection of rule lists from various makers and wholesalers. It was the first publication to actively address the information needs of rule collectors, and even today, when a great deal of other material has become available, is still widely referenced.

A large amount of information relating to measuring instruments has been published since Ken published his *Introduction* in 1982. Not all of it is readily available to collectors, however. The articles and material have appeared intermittently over 20 years in a number of different publications, some not usually seen by rule collectors, and because of this has never been available to be viewed as a connected whole.

I have long felt that it would be of great value to rule collectors and those interested in measuring instruments if the best of this available information, together with text expanding and completing it, could be gathered together and published in a single book which could be used as a basic reference in the field.

In the spring of 2000 Ken decided to discontinue the publication of the *Introduction*, which by then had been expanded into a 62 page softbound book entitled *Fundamentals of Rule Collecting*. At that time he asked if I would be interested in taking over its publication, and all of a sudden everything fell into place. I said yes, but then proposed that he allow me to expand *Fundamentals* into the reference book which I had long felt was needed. This new book would retain some of the basic structure and much of the text contained in the prototype, but would have a large amount of additional material added, material both newly written and collected from various sources. It would be an attempt to gather in one place as much information as practical which would be of value to persons interested in studying or collecting measuring instruments—a source book for rule collectors.

Ken greeted the idea with enthusiasm, and turned over to me all his originals. Since that time I have been locating, preparing, and writing the additional material.

This additional material includes:

- Copies of rule price lists from major 19th century American rulemakers. Dating from 1838 through 1867, these lists provide a capsule view of the number and types of rules offered at the time when rulemaking was in transition between the small shops of 1800-1830 and the larger manufactorys (as they were then called) which became common after the American Civil War.
- Reprints of significant articles dealing with measuring instruments which have previously appeared elsewhere, but are not readily available to collectors. These reprinted articles are particularly valuable to beginning collectors, providing them with a good overview of the field of measuring instruments, and also giving them access to material published before they began to receive tool-related periodicals.
- Reprints of the first three issues of the rule collectors' newsletter, *Mensuration*, which was published briefly in the 1980s.
- Three chapters which I wrote specifically for this expanded edition, dealing with:
  - -rule materials and construction
  - -graduations and markings
  - -rule types and applications

  Although based loosely on material which appeared initially in *Boxwood & Ivory*, these chapters are in fact totally new, incorporating a large amount of additional material, and having been completely rewritten.
- A chapter on the multiplicity of European linear measures which were in use prior to the metric system, complete with an extensive table of these measures. This table is very useful in identifying unmarked rules. Comparing graduations on an unmarked rule with the data in the table would allow its locale of use to be quickly ascertained. This chapter is the result of a study which I had done for a paper delivered at the TATHS meeting in England in October, 2000.
- A chapter on rule combination tools and accessories, aftermarket devices which could be attached to a rule to expand its functionality. This chapter is a major expansion of an article which I had written for the *Fine Tool Journal* in 1983.
- Finally, a bibliography of catalogs, books, and articles relating to measuring instruments and rule collecting.

This book, which I have named *A Source Book For Rule Collectors*, is the result of these additions.

Once this project was started it seemed to take on a life of its own, constantly growing and expanding. I kept finding new material that I wished to include; things which I had obsessively copied and filed years before suddenly became extremely useful; casual comments by other collectors suggested types of measures or areas of information which should be incorporated; actual examples encountered in collections and at antique shows and flea markets needed to be described and illustrated.

My major concern, which had initially been about having enough material for the book, rapidly became one of being able to limit the book to a reasonable size and maintain its focus without omitting anything I thought important. I hope I have succeeded.

I only wish that Ken could be here to see what his efforts and support have led to.

This is not a "coffee table" book, full of gorgeous color photographs of rare rules in mint condition. It would have been fun to do something like that, but the cost would have been prohibitive and the information content greatly reduced. Most collectors would not have been able to afford it, and its value as a reference book would have been severely compromised. Instead, the photographs are all in black and white, and have been selected primarily to support the informational value of the text.

One thing which will (regrettably) not be found in this book is a chapter detailing the history of rulemaking in the United States, tracing its origins, including its connections to rulemaking in England, and outlining its subsequent development. The chapter on rulemaking by Stanley and its competitors which appeared in *Boxwood & Ivory* could have been included, but I felt that the information was too incomplete and out of date to make it worth while. In the years since *Boxwood & Ivory* was published in 1984, Donald and Anne Wing and others have done extensive research on the history of rulemaking in America and its roots in England. The results of this research, which largely supersedes, and in some cases contradicts what I had written in 1984 is now very close to being published. Because of this I felt that the best thing I could do was to avoid recycling obsolete information and let this soon-to-be-published material be (hopefully) the final word on the subject.

I hope this book will communicate to beginning and advanced collectors not just information, but my enthusiasm for the entire field of rules and measuring instruments. They are beautiful objects to see and handle, spanning the entire breadth of 19th century manufacturing technology, and the ingenuity represented by their special scales and functions is marvelous to consider. They are truly beautiful to behold.

Philip Stanley
Worcester, Massachusetts

Table of Contents

I	INTRODUCTION TO RULE COLLECTING	1
II	EARLY RULE LISTS AND FLYERS	15
III	REPRINTED ARTICLES	23
	The English Carpenter's Rule - Notes on Its Origin	24
	Further Notes on the Early English Three Fold Ship Carpenter's Rule	28
	A Guided Tour of an 18th Century Carpenter's Rule	31
	Carpenter's and Engineer's Slide Rules (Part I: History)	36
	Carpenter's and Engineer's Slide Rules (Part II: Routledge's Rule)	41
	Carpenter's and Engineer's Slide Rules (Part III: Errors in Data Tables	44
	Joshua Routledge 1775-1829	46
	Some Notes on the History and Use of Gunter's Scale	47
	Further Notes on the Operation of Gunter's Rule	52
	William Cox on the Sector	54
	Dialling (Figuring Time)	60
	Ashton Blaine French - Scaler and Log Rule Maker	62
	William Greenleaf: Talented, Eccentric	66
	E. S. Lane: A Maine Rule Maker and Scaler	68
	A Most Unusual Rule	70
	The Stanley No. 58½ Rule	73
	How Thick is "Extra Thick"?	79
	The Lufkin Rule Co.	81
	Measuring Rules	86
	Rules	91
	Introduction to Rulemaking at Birmingham	96
	Rule Manufacturing (at Chapin-Stephens)	100
	Rulemaker's Planes	102
	Crafting Scale Rules by Hand - An Historical Note	104
IV	MENSURATION	109
	Volume I, No. 1 (Summer 1987)	110
	Volume I, No. 2 (Fall 1988)	122
	Volume I, No. 3 (Fall 1989)	130
V	MATERIALS, CONSTRUCTION, AND THE RULEMAKING PROCESS	138
	Materials	138
	Construction	141
	The Rulemaking Process	148
VI	GRADUATIONS AND MARKINGS	154
	Embellishment Lines	154
	Linear, 1:1 Graduations	155
	Linear Graduations, Not 1:1	159
	Nonlinear Graduations	163
	Tables	165
VII	SPECIAL RULE TYPES AND USES	169
	Advertising & Third Party Rules	169
	Dearborn's "Angelet"	170

Bench Rules 173
Bevels, Ship Carpenters' 173
Blacksmiths' Rules 174
"Blindman's Rules" 175
Board Measure, Board Sticks and Board Tables 175
Board & Log Canes 177
Button Gauges 178
Calculator Rule 178
Caliper Rules 178
Carpenters' (Sliding) Rules 180
Carriagemakers' Rules 181
Combination Rules 181
Coopers' Rules 187
Cordage Rules 187
Counter Measures 187
Desk or School Rules 188
Draftsmen's Rules 189
Engineers' (Sliding) Rules 192
Extension Sticks 192
Foreign Measures or Comparison Rules 194
Freight Rules or Forwarding Sticks 196
Gauging and Wantage Rods 196
Gear Rules 199
Glaziers' Rules 200
Gunter's Scales 201
Hatters' Rules 203
Horse Measures 203
Ironmongers' Rules 204
Ivory Rules 204
Lining or Accountants' Rules 205
Log Sticks, Log Canes, and Log Calipers 208
Machinists' Rules 210
Masonic Rules 212
Metric Rules 213
Milk Can Gauges 213
Multiple Slides 214
Plain Scales 216
Printers' Rules 217
Rolling Measures 218
Saddlers' Rules 220
Sealers' Rules 220
Seamstress or Workbasket Rules 223
The Sector 223
Shrinkage Rules 225
Six Fold Rules 227
Stationers' Goods 228
Steam Engineers' Rules 228
Tailors' Rules 230
Textile Industry Rules 233
Tinsmiths' Rules 236
Watch Glass Gauges 237
Woodcutters' Rules 238
Yark Sticks 238

VIII	**FOREIGN UNITS OF LENGTH**	240
	Origins	240
	Historical Measures	241
	Measures After the Renaissance	241
	European Units of Length Before the Metric System	242
IX	**RULE ACCESSORIES**	258
	Rule Fence Attachments	258
	Protractor Fence Attachments	261
	Rule Trammel Points	262
	Complex Rule Tools	264
	Other Rule Accessories	267
	Rule Accessories Yet to be Found	269
X	**A RULE BIBLIOGRAPHY**	272
	Manufacturers' and wholesalers' catalogs	272
	Books & Articles	274
	Periodicals	283
	Encyclopedias	284
	Patent Indexes	285
	Bibliographies & Directories	285

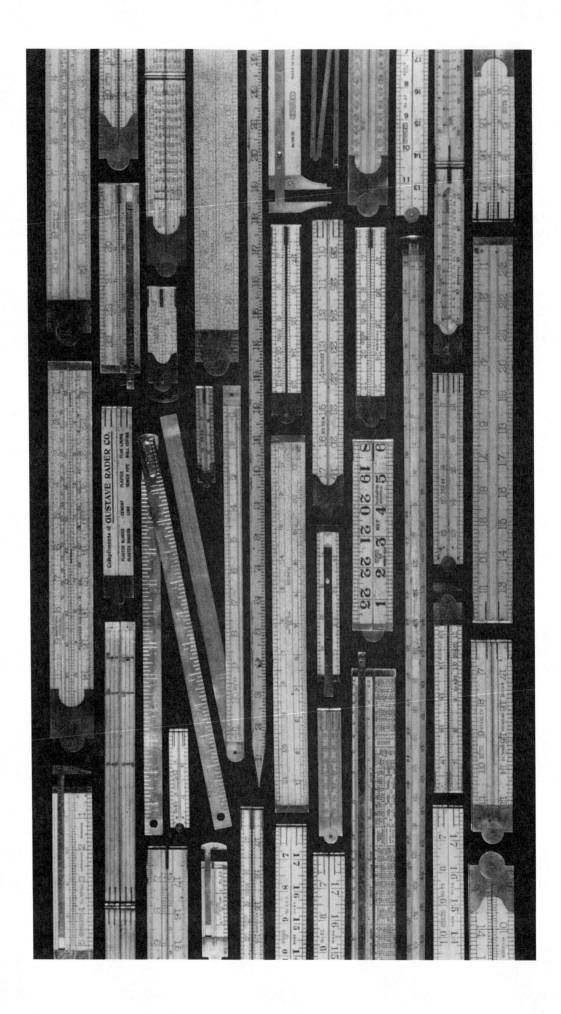

I
Introduction to Rule Collecting

Ken Roberts' interests were not limited to Connecticut clock makers, English tools, and wooden planes. In 1982 he turned his formidable talents to measuring instruments, publishing a booklet entitled *Introduction to Rule Collecting*. This booklet contained a brief history of rule making in the United States during the 19th century, together with insights into the details of rule construction and a selection of rule lists from various makers and wholesalers. It was only fourteen pages in length, but it represented a large increase in the available knowledge of rules and their makers, and created the structure for much of the research which has been done in this field since. Even today, when a great deal of other material has became available, it is still widely referenced.

WHERE IT ALL BEGAN

The impetus for *Introduction to Rule Collecting* was Ken's research focused on the activities of Hermon Chapin and his successors at Pine Meadow, Connecticut. In the company papers (in the possession of the Connecticut Historical Society) he had found a large body of material relating not only to the Chapin rulemaking activities but also to Chapin's involvement in the Rule Maker's Association. This access to the papers of that Association in turn provided valuable information on the activities of the other major American rulemakers. Drawing on that material, plus other documents and catalogs in the possession of the Society and on information in his own possession he was able to create the *Introduction* and thus lay the foundation for measuring instrument research in the United States.

During the next 15 years Ken published a number of articles on rules and measuring instruments, as well as reprints of catalogs by makers such as Lufkin, Belcher Bros., the Stanley Rule & Level Co., and others.

As a result of the interest generated by these articles and reprints, and articles written by others, in 1997 Ken determined to reissue *Introduction to Rule Collecting* in expanded form. He retained the original Introduction, but to it he added copies of the articles he (and I) had written during the last 15 years. Additionally he included an updated version of my "Concordance of Major American Rulemakers," first published in the *Fine Tool Journal* in 1985. These additions created a much larger work, a 62 page softbound book which Ken published under the title *Fundamentals of Rule Collecting*. *Fundamentals* was enthusiastically received, and only the fact that Ken chose to have it printed 20 copies at a time in a copy-center format prevented it from being as widely distributed and read as it deserved.

In the spring of 2000 I approached Ken with an idea for a further, major expansion of *Fundamentals of Rule Collecting*, proposing to add a large amount of material which had not previously been included. This new book would be an attempt to gather in one place as much information as practical which would be of value to persons interested in studying or collecting measuring instruments. Ken greeted the idea with enthusiasm, and turned over to me all his originals from *Fundamentals*, and the book you are holding, now entitled *A Source Book for Rule Collectors*, is the result.

I feel it is only fitting that we reprint *Introduction to Rule Collecting* here as the first chapter in the book. It is, after all, where it all began.

2 INTRODUCTION TO RULE COLLECTING

Stanley Rule & Level Factory in New Britain, Connecticut, ca. 1870

Chapin-Stephens Factory in Pine Meadow, Connecticut, ca. 1901

INTRODUCTION TO RULE COLLECTING

© 1982 Kenneth D. Roberts

Rules involve a fascinating study of the history of technology which up to this point seems to have been neglected among many collectors. In all crafts it was necessary to make measurements, and most certainly rules were the basis of determining sizes. Allied with rules are squares, mortise and marking gauges.

There are three basic types of rules: 1. linear measure; 2. volume measure (dry and wet); and 3. calculating. By far the majority of rules made during the 18th and 19th centuries were for linear measurements used in the woodworking trades. Among the trades for which rules were offered in the recently reprinted *1892 J. Rabone & Sons Catalogue* [Birmingham, England] were architects, builders, carpenters, coachmakers, engineers, ironmongers [hardware dealers], saddlers, shoemakers, surveyors and tailors. Additional trades using rules were: blacksmiths, machinists, optometrists, patternmakers [shrinkage rules], printers, tin smiths, and watchmakers. Rules for the woodworking trades were usually made from either boxwood or ivory; for the metal trades of brass or steel; for surveyors of fabric or metal tape. While made in a large variety of sizes, the most common were the two feet, two-fold for the tool chest and the two feet, four-fold for the pocket. The number of folds is one less than the number of hinges and equal to the number of pieces folded.

According to the studies of Ray Townsend the folding joint wood rule was invented by the Italian architect, Vincenzio Scammozi (1552-1616). This was later adopted in France by instrument makers, producing sector rules, the fore-runner of the mathematical slide rule. Subsequently this was introduced at London, England during the 16th century, then called the French joint. Plate I, reprinted from *STANLEY RULE & LEVEL Co. CATALOGUE #129*, illustrates the three common types of rule joints and the styles of middle joints and bindings. The least expensive, usually not warranted by the maker, was the round end joint, while the strongest and most desirable was the double arch, full bound. Bitted rules had a brass plate inserted on the edges of rules to protect the wood from splitting out from the closing pins. While all of these designs were developed in Great Britain, improvements were subsequently made in the United States.

The trade of rulemaking developed in England from instrument makers working at London. By the last quarter of the 18th century the center shifted to Birmingham and surrounding towns. From 1797-1900 there were 255 firms noted in Birmingham directories. (See check list in the reprint *1892 J. Rabone & Sons Catalogue.*) While a few instrument makers produced rules at such important centers as Boston, Philadelphia and New York City during the late 18th century, the first firm in the United States to develop into a large producer was the Belcher Brothers at New York City. A brief history of this firm accompanies the recent reprint *1860 Belcher Brothers & Co. Price List of Rules*.

Another New York City firm was J. & G. H. Walker, established in 1845, continuing until 1873. The senior partner of this firm, Job Walker, began there in 1843, and was also probably an apprenticed rule maker who emigrated from England as Thomas Belcher is believed to have been. Other smaller rule firms listed in New York City business directories were: 1840 — Andrew Brenmor, J. Souls; 1855 — John Gee, John Hughes, Orrin Noble; 1860 — Isaac Arm, Kirby & Davidson.

Plate II presents a line diagram showing the dates and firms involved in rule making in the United States from 1822 to the present date. There were numerous other smaller specialty firms. During the 19th century those firms associated with Hermon Chapin had the longest tenure: 65 years, plus 29 more during the 20th century. The only concerns noted on this chart presently making rules are Stanley Rule & Level Co. and Lufkin, the latter having moved to another location under different corporate ownership.

Hermon Chapin established his Union Factory at Pine Meadow, Conn. as a planemaker in 1826. In 1834 he ventured into the rule business by purchasing the stock and machinery of Franklin Bolles of Hartford. During the remainder of the 19th century rules became a very important product at this factory. Their rules were almost totally sold to the wholesale trade, such as their account with Sargent & Co. at New York City.

Lorenzo Stephens and his son, Delos, worked for H. Chapin at Pine Meadow before renting a building from this employer and forming their own rule concern in 1853. For many years thereafter they supplied H. Chapin large quantities of rules to which he added his own imprint. L.C. Stephens invented and patented a combination rule, square, level and bevel. (See CHRONICLE of EAIA, Vol. 35, No. 2, June 1982, p. 29 and back cover.) Delos Stephens developed into a mechanical genius. He designed and patented many machines for making rule joints and marking. They moved from New Hartford to Riverton, a district of Barkhamsted, in 1864. Financial difficulties led to H. Chapin's Son & Co. purchasing this business, combining it with theirs and organizing a new firm, Chapin-Stephens Co., Oct. 1, 1901.

An apprentice of Hermon Chapin at his rule factory by the name of Henry Seymour with his brother-in-law, John Churchill, removed to Bristol and formed the rule concern of Seymour and Churchill about 1849. Eventually they sold this business to a former apprentice of theirs, Thomas Conklin in 1854. This was combined with A. Stanley & Co. who formerly manufactured hardware in the adjacent town of New Britain. With the subsequent purchase of another rule firm, Seth Savage of Middletown in 1855 and merger with Hall & Knapp, level manufacturers at New Britain, A. Stanley & Co. became Stanley Rule & Level Co. on Sept. 27, 1858. This date has been erroneously reported as 1857, but the certificate of organization is filed at the Secretary of State Offices at the Capitol, Hartford.

In 1863 Stanley Rule & Level Co. acquired ownership of the E. A. Stearns Co. of Brattleboro, VT, then under the management of Charles L. Mead. Operations at Brattleboro were continued, as noted in the *1867 Stanley Rule & Level Co. Price List*. By 1870 the Brattleboro factory was closed down and the machinery moved to New Britain. Charles Mead for several years managed the Stanley Rule & Level Co. New York City sales offices at Chambers St. Mead subsequently became treasurer of the firm and upon the death of Henry Stanley in 1884 became president. (See reprint, *1872 Stanley Rule & Level Co. Catalogue.*)

Willis Thrall, a map publisher at Hartford, in 1842 purchased the former rule business of S. A. Jones of that city. In 1860 his son joined the firm, which then became Willis Thrall & Son. In addition to rules they sold bevels and gimlet screws. This remained a relatively small firm. Upon Willis Thrall's death in 1884 manufacture of rules was discontinued. His son continued the hardware business until closing about the turn of the century.

The Standard Rule Co. was established at Farmington, Conn. in 1872 at the borough of Unionville. (See *Wooden Planes in 19th Century America*, p. 216.) In 1875 when Belcher Brothers & Co. discontinued manufacturing rules, Standard Rule Co. replaced this concern in the Rule Manufacturers Association. At that date at least 95% of the rules produced in the United States were made within a 15 mile radius of Unionville. Andrew Upson, president of this firm, merged this into his other Unionville concern, the Upson Nut Co. in 1889. The firm continued to make rules under the firm name of Upson Nut Co. until 1922 when this division was sold to Stanley Rule & Level Co.

Two other concerns, R. B. Haselton (1847-1924) of Contoocook, NH and Lufkin Rule Co. (1869-1967) of Saginaw, Michigan are noted on Plate II.

The first Convention of Rule Manufacturers was held on May 11, 1869 at New York City. The firms attending were: Belcher Brothers & Co., J. & G. H. Walker, H. Chapin, L. C. Stephens and Willis Thrall. William Belcher was elected president and George Walker secretary, which proves the esteem with which these two men were then regarded by the other rule manufacturers present. At this meeting the representatives, who were for the most part the owners, agreed to adopt a uniform terminology for each type of rule, a uniform discount of 50% in consideration that 25% profit was a fair return, and also adopted a uniform price list. This price list, with a few minor exceptions, was in effect the remainder of the 19th century by all firms. However, during difficult financial times and periods when they experienced high material costs, the discounts were altered. Annual meetings of this association were held to discuss mutual problems and sales conditions.

In 1875 the members agreed to set production schedules and pooled sales according to the following allotments. These were changed in 1877.

Percentage Share of Sales for Members of Rule Manufacturers Association

	1875	1877
Stanley Rule & Level	26	33
H. Chapin's Son	22	22
Stephens & Co.	22	22
Standard Rule Co.	20	15
Willis Thrall & Son	10	8

Monthly reports of sales by each firm were made to each concern. Every six months a cash balance was made among firms according to the agreed sales quotas. Such price fixing, which was wide-spread in the United States among many other industries, was declared illegal when President Benjamin Harrison signed the Sherman Anti-Trust Act of 1890. However, indications from price lists show that the remaining firms then manufacturing rules continued to sell at the same prices as set in the 1859 Convention through 1898.

With the formation of Chapin-Stephens Co. the new firm adopted the same rule numbers as used by Stanley Rule & Level Co. Standard Rule had arranged their rule numbers the same as SR&L Co. When Upson Nut took over, these identical numbers were continued. With all three Connecticut firms having the same numbers and prices, which incidentally were reduced about 50% from those previous to 1898, it would appear that collusion continued into the 20th century in spite of the Sherman Anti-Trust Act. It is suggested that the principal reason these three ostensibly competitive firms used identical numbers for the same types of rules was to facilitate interchange of products for customers in filling orders and meeting production schedules. Also this expedited balancing accounts in semi-annual settlements among these three manufacturers in this industrial conspiracy. This practice had been prevalent among tool firms in England, as was price fixing, since the first third of the 19th century. (See Conclusions, *Some 19th Century English Woodworking Tools*, p. 477.) It had been practiced among Liverpool watchmaker artisans and London merchant watchmakers since the 16th century.

The trade of rulemaking, like planemaking, is another example of a craft that was fully developed in England and was brought to the United States by immigrants seeking better opportunities to develop their own businesses. Actually there are two distinctly different trades in this craft: brass joint makers and rule assemblers or finishers. Apprentices developed only one of these trades. The usual material was boxwood imported from Turkey or southern Russia. After proper seasoning, usually two to three years, the wood was slit into strips using circular saws. In England the hand sawn stock was planed using jointer, fore and jack planes having their irons set perpendicular to the stock. In this manner the boxwood was scraped smooth. The English rulemakers also used specially designed plows for cutting dado configurations for slides. In America this work was done with circular power driven saws and much of the finishing work with machine driven sanding drums rather than hand work. After polishing and assembling the joints, the divisions and figures were inscribed. Soon after 1850 machine methods were employed, whereas previously such marking was done by hand.

The advantages of boxwood as a material were its close-grained, dense structure, making it moisture resistant; dimensional stability; hardness and potential for being polished to an attractive finish. Ivory was more expensive, costing three to four times the amount for the same size as boxwood, but was more attractive and easier to read. However, disadvantages were that it was not as dimensionally stable; the ink wore off from handling or was bleached by the sun and it became rather brittle, frequently breaking off at the joints. Ivory rules have been known to shrink 10% in size. (See p. 754., *Chas. A. Strenglinger & Co. 1897 Catalogue, Wood Working Tools.*)

The three forms of finished rules were: unbound; half-bound [outside edges with brass]; and full bound [all edges bound with brass]. Bitted unbound and half-bound rules usually had a small brass plate set in the inside edge containing the closing pins. About 1850 the alloy known as "German Silver" [commonly 66% copper, 17% zinc and 17% nickel (but containing no silver)] was employed for joints and binding ivory rules, thus offering a more attractive product, appearing to be silver. This practice started in England and was soon followed in the United States.

Plates III-VII, reprinted from the 1901 *Sargent & Co. Hardware Catalogue*, illustrate and note the diversified varieties and types of rules. These rules were all made by H. Chapin's Son & Co. and were inscribed with this manufacturer's numbers. As a matter of interest the 1869 *Sargent & Co. Catalogue* noted "Agents for H. Chapin's Son's RULES, GAUGES, HAND SCREWS, BENCH SCREWS, PLUMBS, LEVELS, &c." The cuts illustrating these rules were the same in both firms' catalogues and are the same as shown on Plates III - VII in this study. The prices of these rules were the same as offered by the Chapin firm. However, these were retail list prices and were subject to commercial wholesale discounts. As previously explained the Rule Manufacturers' Association agreed to fix uniform prices during the period of 1859-1901. The Comparative List of Rules, at the bottom section of Plate VII, shows the equivalent numbers of rules followed by Stanley Rule & Level Co., Stephens & Co., and H. Chapin's Son & Co. As previously noted, after the consolidation of Chapin-Stephens Co. this new firm also adopted the identical rule number of Stanley Rule & Level Co. Accordingly, after that date Sargent & Co. sold rules with the SR&L Co. numbers. The 1902 *Sargent & Co. Catalogue* tabulated a list showing the relationship between the old and new rule numbers brought about by this change. This is shown as Plate VIII.

Among the specialty purpose rules are caliper (two and four fold, Nos. 70-73, Plate III) for measuring thickness; architects (four fold, No. 17½, Plate IV) with beveled inside edges with various scales; and carpenter's slide rules (two feet, two fold, Nos. 46-49, Plate VI) with Gunter scales for calculating matematical problems.

The common form of calculating rules are exemplified by the engineer's and carpenter's slide rules with Gunter scales. The former was introduced in England by J. Routledge at Bolton, c. 1811; the latter also in England about 1840. Both styles were made in America from 1840 through the early part of the 20th century. The engineer's rules permitted weight determination by calculating the volume and using a gauge point inscribed on the rule which incorporated the density of the material. Also calculations of proportion, pumping engines, etc., could be determined by manipulating the logarithmic Gunter scales. The carpenter's slide rule permitted the same calculations by using different gauge points from a separate printed table. Rather than gauge points, proportion scales were incorporated on the leg adjacent to the Gunter's scales. The opposite side had a scale arrangement for shaping timber into octagonal masts. A book of instructions, *The Carpenter's Slide Rule, History and Instructions*, originally published by J. Rabone & Sons, 1880, has been recently reprinted and is available from this Publisher.

Rules from which volume can be calculated are of two types: dry and liquid. Timber measure is the best example of the former, enabling the contents in board feet of a plank or log to be determined, as noted on Plate VI. There are numerous special designs of such rules, as there are also with liquid measure. The latter involves dip sticks placed through the bung hole of casks or barrels. These were called wantage or gauging rules.

There is very little literature concerning the speciality of rules and their manufacture. John Rabone, Jr. wrote an article in 1866, which has been reprinted recently with the *1892 J. Rabone & Sons Catalogue of Rules, etc.* Delos Stephens, proprietor of Stephens & Co., wrote an article on the subject which was published in *The Great Industries of United States* [Hartford, CT, 1872, pp. 739-744]. In my forthcoming Vol. II, *Wooden Planes in 19th Century America*, it is my intention to publish a chapter with historical and technical data concerning rulemaking as practiced by H. Chapin and successor firms and also Stephens & Co.

Philip Stanley is now preparing a book manuscript concerning rules made by Stanley Rule & Level Co. Ray and Jim Hill are presently working on a future book publication concerning linear measurements. At present the best manner to become acquainted with the wide diversity in rules is to study trade catalogues. A recommended list of these accompanies this essay.

With the purchase of Chapin-Stephens & Co. by Stanley Rule and Level in 1929 and its immediate closing, the production of folding boxwood and ivory rules was soon after ended. The advances in technology after the turn of the century brought better materials, mechanized production and new designs. *SR&L Co. Catalogue No. 26*, 1900 noted "Stanley's ZIG-ZAG Rules" in lengths of 2,3,4,5,6, and 8 feet. These folding rules were made of flexible hardwood with blued steel joints. The one inch divisions were graduated in units of 1/16, clearly marked. These could be produced in large quantities and sold at lower prices. The advancement of the two foot carpenter's framing square with tables and the availability of graduated scales on try, mitre and combination squares decreased the necessity of rules. Finally the introduction of the flexible retractable steel tape measure introduced in 1922 as the Farrand Rapid Rule resulted in a rapid decline of the folding boxwood and ivory rules.

A peculiar difference between American and English folding rules is that the former read from right to left; whereas the latter read from left to right. No known authoritative explanation has yet been found to account for this difference. It is suggested that it was simply a matter of custom, similar to driving on different sides of the road. Yet it seems strange that such difference occurred, since American practice in reading rules should have followed the English system.

In cleaning boxwood rules a waterless hand cleaner containing lanolin is effective [Boraxo brand is recommended]. This may be applied with a paper towel, rubbed until cleaned and then wiped away with a fresh paper towel. If there is paint to be removed, a mild paste paint and varnish remover is similarly applied [5f5 Brand, made by Sterling-Clark-Lurton Corp., Malden, MA 02148 is recommended]. After this has been wiped away, an application of the waterless cleaner should be followed.

BOXWOOD RULES

Points of Superiority of STANLEY Rules—

1. The selection and seasoning of the wood.
2. The weight of the metal in the joints and trimmings.
3. The accuracy of the graduations.
4. The use of brass to prevent rusting, in all joints, plates, bindings etc.
5. The careful finishing.

The Distinguishing Feature of all Boxwood Rules is the Main Joint

THE ROUND JOINT

Is used in the less expensive Rules. In this form, one plate is embedded in each leg of the rule. The leg and plate are pinned together.

THE ARCH JOINT

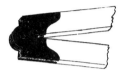

Is practically of the same construction as the Square Joint. The plates, however, are larger, more graceful, and cover more of the surface of the wood, which adds to the life of the rule.

THE SQUARE JOINT

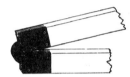

Has two plates to each leg, one on each outside face of the wood. These are held securely together by rivets which go through the two plates and leg. This is a much stronger joint than the Round Joint.

THE DOUBLE ARCH JOINT

Is the same as the Arch Joint, but is repeated at the folding joint as well as the central joint, which again adds to the strength of the rule.

Other important Features of the Boxwood Rule are the Middle Joint and the Binding.

THE MIDDLE PLATE JOINT

The Middle Plates are set in the center of the wood and pinned.

HALF BOUND

A Half Bound Rule has a protective brass binding pinned along the outside edges of the legs only.

THE EDGE PLATE JOINT

The Edge Plates are fastened to the outer edges by rivets which go through both the wood and plates holding them firmly together. This makes a stronger joint than the Middle Plates.

FULL BOUND

A Full Bound Rule has a protective brass binding pinned to both the inside and outside edges of each leg.

"ENGLISH" AND METRIC GRADUATIONS

Boxwood Rules can be furnished with "English" graduations, i.e.: numbers reading from left to right. They can also be furnished with the "English and Metric" graduations, i.e.: metric graduations on one side and inches (from left to right) on the other side. When ordering rules with "English" markings add "E" to the class number of the rules, i.e., 61E. When ordering rules with "English and Metric" markings add "EM" i.e.; 61EM.

DRAFTING SCALES

Are used for laying out or reading drawings where a scale of ¼, ½, etc., to the foot is used.

PLATE I - Joints of Boxwood Rules

8 INTRODUCTION TO RULE COLLECTING

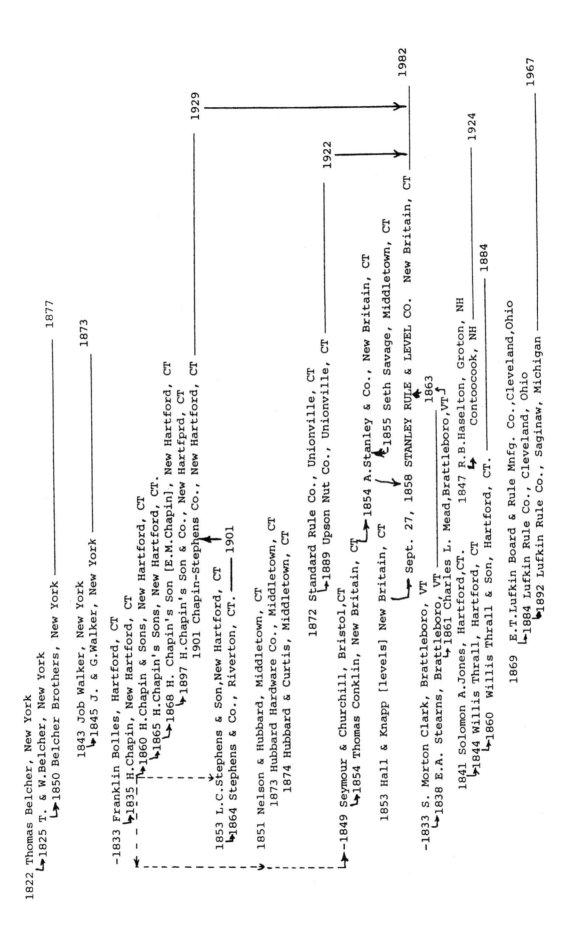

PLATE II - Boxwood and Ivory Rule Manufacturers in the United States

1123

Boxwood Rules.

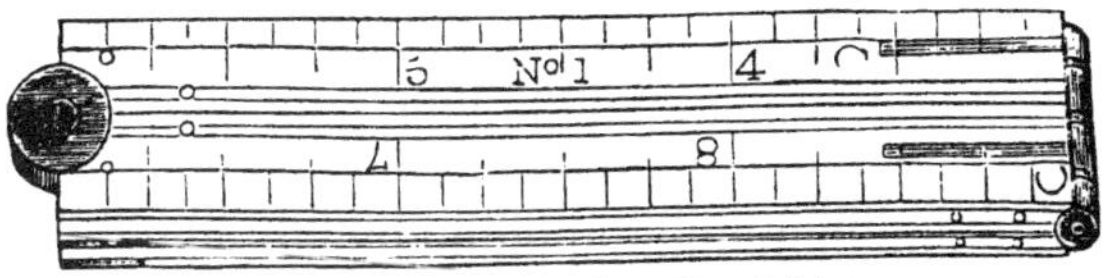

Middle Plates. Cut of No. 1, Round Joint.

One Foot, Four Fold.

No. **1**, Round Joint, Middle Plates, 8ths and 16ths, ⅝ inch wide . . . . per dozen, $3 00

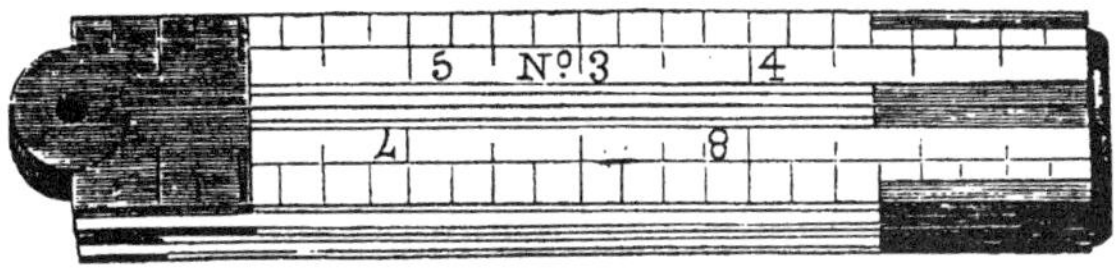

Edge Plates. Cut of No. 3, Square Joint.

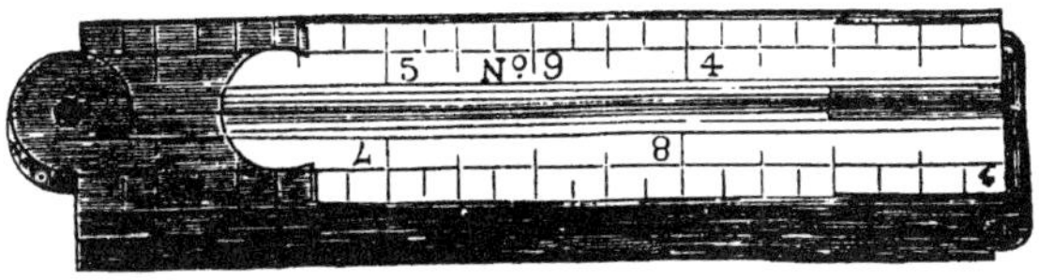

Full Bound. Cut of No. 9, Arch Joint.

One Foot, Four Fold.

Square Joint, 5-8 Inch Wide.		Arch Joint, 5-8 Inch Wide.	
No. **2**, Middle Plates, 8ths and 16ths . . . per dozen, $3 50		No. **6**, Middle Plates, 8ths and 16ths . . . per dozen, $4 00	
No. **3**, Edge " " " . . . " 5 00		No. **7**, Edge " " " . . . " 6 00	
No. **4**, Half Bound " " " . . . " 9 00		No. **8**, Half Bound " " " . . . " 10 00	
No. **5**, Full " " " " . . . " 11 00		No. **9**, Full " " " " . . . " 12 00	

Nos. 4, 5, 8 and 9, half dozen in a box; other numbers, one dozen.

Boxwood Rules—Calliper.

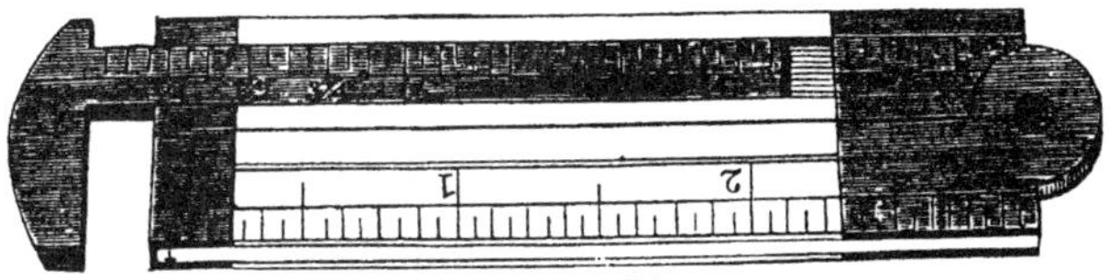

No. 70, Square Joint.

6 Inch, Two Fold.

No. **70**, Square Joint, 8ths and 16ths. ⅞ inch wide . . . . per dozen, $7 00
No. **71**, " " " " Brass Case, ⅞ inch wide . . . . " 12 00
No. **71½**, " " " " " Full Bound. ⅞ " " . . . . " 12 00

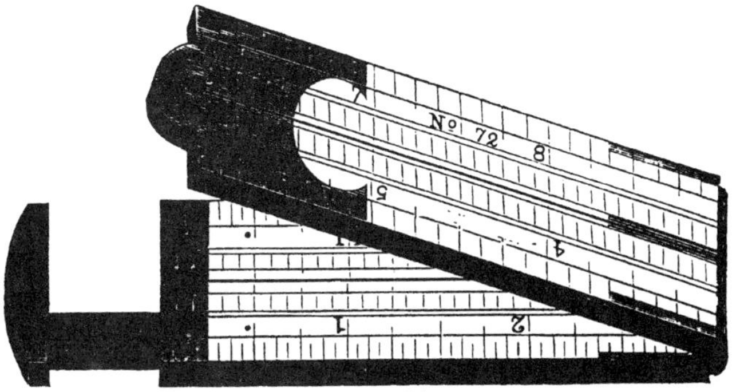

No. 72, Arch Joint.

One Foot, Two Fold.

No. **70½**, Square Joint, 1⅜ inch wide, 8ths and 16ths . . . . per dozen, $12 00

One Foot, Four Fold.

No. **72**, Arch Joint, Edge Plates, ⅞ inch wide, 8ths and 16ths . . . . per dozen, $12 00
No. **73**, " " Full Bound, ⅞ " " " . . . . " 20 00

Nos. 72 and 73, half dozen in a box; other numbers, one dozen.

PLATE III - Chapin Rule Listings in a 1901 Sargent & Co. Catalogue (First of Five Pages)

1124

Boxwood Rules—Narrow.

Two Feet, Four Fold.

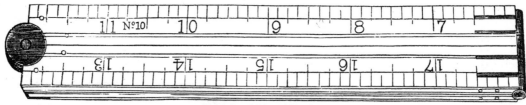

No. 10, Round Joint, Middle Plates.

No. **10**, Round Joint, Two Feet, Middle Plates, 8ths and 16ths, 1 inch wide . . . . per dozen, $4 00

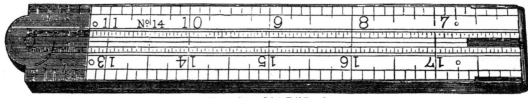

No. 14, Square Joint, Half Bound.

Square Joint, 3-4 Inch Wide.		Square Joint, 1 Inch Wide.	
No. **11½**, Middle Plates, 8ths and 16ths . . per dozen, $5 50		No. **11**, Middle Plates, 8ths and 16ths . . . per dozen, $5 00	
No. **13**, Edge " 8ths, 16ths, 10ths and 12ths " 8 00		No. **12**, Edge " 8ths, 16ths, 10ths, 12ths and Scales " 7 00	
No. **15½**, Full Bound, " " " " " 15 00		No. **14**, Half Bound, " " " " " " 12 00	
		No. **15**, Full " " " " " " 15 00	

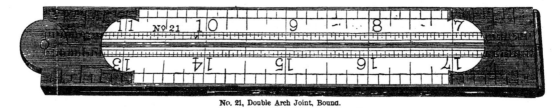

No. 21, Double Arch Joint, Bound.

Arch Joint, 1 Inch Wide.	Per dozen	Double Arch Joint, 1 Inch Wide.	Per dozen
No. **16**, Middle Plates, 8ths, 16ths, 10ths, 12ths and Scales	$6 00	No. **20**, 8ths, 16ths, 10ths, 12ths and Scales .	$9 00
No. **17**, Edge " " " " "	8 00	No. **21**, Full Bound, " " " " " .	21 00
No. **18**, Half Bound " " " " "	13 00		
No. **19**, Full " " " " "	16 00		

Architect's Rule.

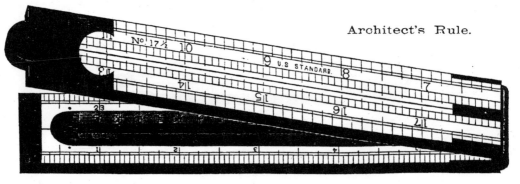

No. **17½**, Arch Joint, Edge Plates, 8ths, 10ths, 12ths and 16ths, Inside Edges Beveled, Drafting Scale for Architects' Use, 1 in. wide, per doz., $15 00

Nos. 14, 15, 15½, 17½, 18, 19 and 21, half dozen in a box; other numbers, one dozen.

PLATE IV - Chapin Rule Listings in a 1901 Sargent & Co. Catalogue (Second of Five Pages)

Boxwood Rules—Broad.

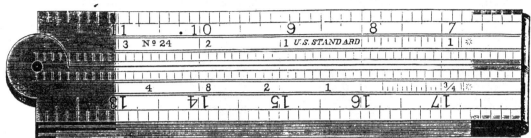

No. 24, Square Joint, Edge Plates.

Two Feet, Four Fold.

1 3-8 Inches Wide.

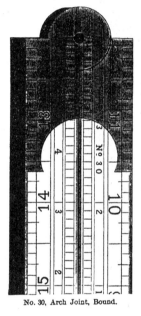

No. 30, Arch Joint, Bound.

No. **22**,	Round Joint, Middle Plates, 8ths and 16ths . . . . per dozen,	$5 00
No. **23**,	Square " " " " . . . . "	7 00
No. **24**,	" " Edge Plates, 8ths, 16ths, 10ths and Scales . . . . . "	9 00
No. **25**,	" " Half Bound, " " " " "	14 00
No. **26**,	" " Full Bound, " " " " "	18 00
No. **27**,	Arch " Middle Plates, " " " "	9 00
No. **28**,	" " Edge Plates, " " " "	11 00
No, **29**,	" " Half Bound, " " " "	16 00
No. **30**,	" " Full Bound, " " " "	20 00
No. **31**,	Double Arch Joint. " " " "	12 00
No. **32**,	" " " Full Bound. " " " "	24 00
No. **32½**,	" " " Half Bound. " " " "	20 00
No. **33**,	Arch Joint, Edge Plates, with Slide, 8ths, 16ths, 10ths and Scales, . "	14 00

Two Feet, Four Fold—Board Measure.

1 3-8 Inches Wide.

No. **34½**, Square Joint, Edge Plates, Board Stick Table, 9 lines, 10 to 18 feet, . per dozen,	$11 00
No. **35**, " " Full Bound, 8ths, 16ths, 10ths and Scales . . . "	20 00
No. **36**, Arch Joint, Edge Plates, " " " " . "	13 00
No. **37**, " " Full Bound, " " " " . "	22 00

Nos. 22, 23, 24, 27 and 28, one dozen in a box; other numbers, half dozen.

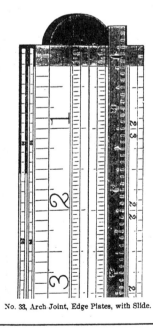

No. 33, Arch Joint, Edge Plates, with Slide.

Two Feet, Two Fold.

1 1-2 Inches Wide.

No. **38**, Round Joint, 8ths and 16ths . . . . per dozen,	$3 50
No. **39**, Square " " " . . . . "	5 00
No. **41**, Arch Joint. 8ths, 16ths, 10ths and Scales . "	7 00
No. **44**, " " Extra Scales, Thin. " " " " . "	10 00
No. **45**, " " " " Full Bound. " " " "	16 00

No. 44, one dozen in a box; other numbers, half dozen.

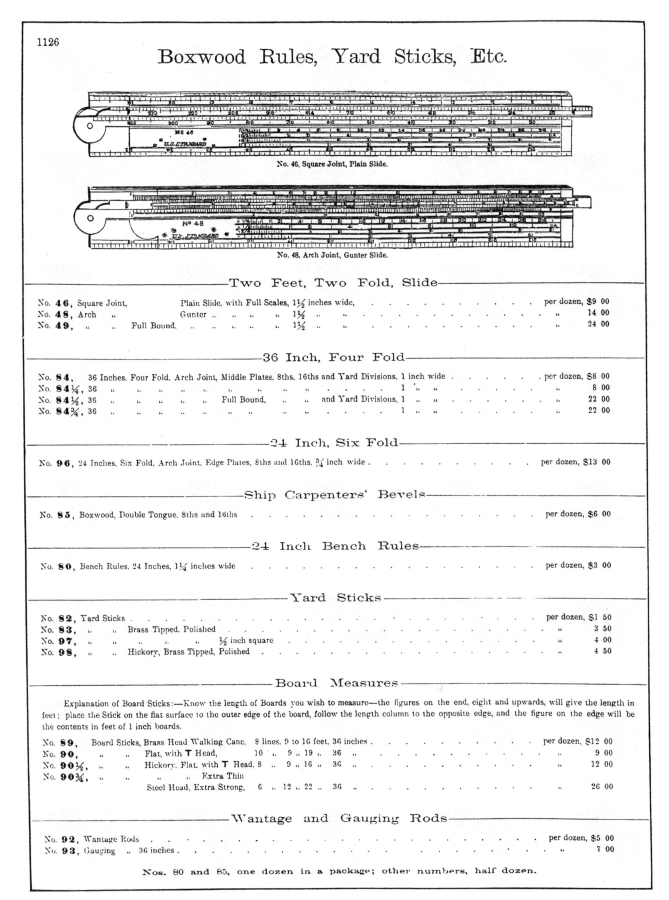

PLATE VI - Chapin Rule Listings in a 1901 Sargent & Co. Catalogue (Fourth of Five Pages)

1127

Ivory Rules.

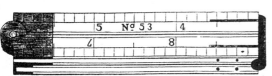

No. 53, Square Joint, Middle Plates.

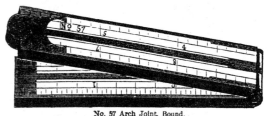

No. 57 Arch Joint, Bound.

One Foot, Four Fold.

No. **52**, Round Joint, Middle Plates, Brass,	8ths and 16ths,	½ inch wide	. . . .	per dozen,	$10 00
No. **53**, Square " " " "	" " "	½ "	. . . .	"	12 00
No. **54**, " " " German Silver "	" " "	½ "	. . . .	"	14 00
No. **55**, " " Edge " "	" " "	⅝ "	. . . .	"	17 00
No. **57**, Arch " Full Bound, "	" " "	½ to ⅝ "	. . . .	"	32 00
No. **58**, Square " " "	" " "	½ to ⅝ "	. . . .	"	28 00

Two Feet, Four Fold.

No. **59**, Square Joint, Edge Plates, German Silver,	8ths and 16ths, with Full Scales.	⅞ inch wide	. . . .	per dozen,	$54 00
No. **60**, Arch " " " "	" " " "	1 "	. . . .	"	64 00
No. **61**, " " Full Bound, " "	" " " "	1 "	. . . .	"	80 00
No. **62**, Double Arch Joint, Full Bound, German Silver,	" " " "	1 "	. . . .	"	92 00

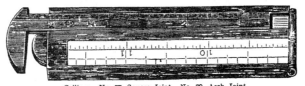

Calliper. No. 77, Square Joint. No. 79, Arch Joint.

6 Inch, Calliper, Two Fold.

No. **74**, Square Joint,	German Silver, 8ths and 16ths, ⅞ inch wide	. . . . . .	per dozen,	$15 00
No. **74½**, " " Full Bound, " " "	" " " ⅞ "	. . . . . .	"	30 00

One Foot, Calliper, Four Fold.

No. **76**, Square Joint, Edge Plates, German Silver, 8ths and 16ths,	⅞ inch wide	. . . . . .	per dozen,	$38 00
No. **77**, " " " " " "	½ to ⅝ "	. . . . . .	"	44 00
No. **78**, Arch " " " " "	½ to ⅝ "	. . . . . .	"	48 00
No. **79**, " " " " " "	⅞ "	. . . . . .	"	48 00

Rules with Spanish or Metric Measure furnished to order.

Comparative List of Rules.

Stanley	Stephens	Chapin	Stanley	Stephens	Chapin	Stanley	Stephens	Chapin	Stanley	Stephens	Chapin	Stanley	Stephens	Chapin
1	13	41	39	98½	76	54	49	19	65	71	2	78	. . .	38½
4	15	44	40	99¼	77	55	73	6	65½	72½	5	78½	63	32
5	17	45	41	111	83	56	74	7	66	32	84	81	66	36
12	14	48	42	31	85	57	75	9	66½	. . .	84¼	82	67	37
15	27	49	43½	. . .	90½	58	33	96	67	53	22	83	. . .	33
18	2	39	44	. . .	92	59	50	20	68	41	10	84	42¼	14
26	9	46	45	. . .	93	60	52	21	69	70	1	85	77	59
29	1	38	48	. . .	89	61	42	11	70	54	23	86	83	60
32	98	72	49	. . .	90¾	61½	44	11½	72	56	24	87	84	61
32½	99	73	50	112	98	62	42½	15	72½	54½	26	88	94	57
34	124	80	51	46	16	62½	42¾	15½	73	57	27	89	86	62
36	95	70	52	49¼	18	63	45	12	75	59	28	90	89	52
36½	100	70½	53	48	17	63½	44½	13	76	60	30	92	91	55
38	95½	74	53½	39	17½	64	72	3	77	61	31	92½	90½	54

Comparative List of Rules, Etc.

Old Numbers Compared with the New.

Old Nos.	New Nos.	Old Nos.	New Nos.	Old Nos.	New Nos.	Old Nos.	New Nos.	Old Nos.	New Nos.	Old Nos.	New Nos.	Old Nos.	New Nos.	Old Nos.	New Nos.	Old Nos.	New Nos.
1	69	15½	62½	27	73	37	82	50	6	63	94½	74	38	83½	67¼	90	43¼
2	65	16	51	28	75	38	29	51	16	64	95	74½	40¼	84	66	90½	43½
3	64	17	53	28½	75½	39	18	52	90	65	95½	75	40¾	84¼	66½	90¾	49
5	65½	17½	53½	29	76½	40	22	53	90½	66	95¾	76	39	84½	66¼	91	48¾
6	55	18	52	30	76	41	1	54	92½	67	97	77	40	84¾	66¾	91½	48¼
7	56	19	54	31	77	42	2	55	92	68	98½	77½	40½	85	42	92	44
9	57	20	59	32	78½	43	23	56	88½	69	99½	78	99¾	85¼	73¼	93	45
10	68	21	60	32½	78	44	3½	57	88	70	36	79	39½	85½	43	94	45¼
11	61	22	67	33	83	44½	4	58	91½	70¼	13½	80	34	85¾	77¼	94½	45½
11½	61½	23	70	33½	83½	45	5	59	85	70½	36½	81	35	86	46	95	58¾
12	63	23¼	61¼	34	79	46	26	59½	85½	70¾	13	81½	30	87	46½	96	58
13	63½	23½	70¼	34½	79½	47	27	60	86	71	14	82	33½	88	47	96¼	58½
13½	53¾	24	72	34¾	79¾	48	12	60½	86½	71½	14½	82½	33		47½	96½	94
14	84	25	72¼	35	79¼	49	15	61	87	72	32	83	41	89	48	97	41½
15	62	26	72½	36	81	49½	49½	62	89	73	32½	83¼	68¼	89½	48½	98	50

PLATE VIII - Comparative Table of New versus Old Rule Numbers From a 1902 Sargent & Co. Catalogue

II
Early Rule Lists and Flyers

The earliest advertising by American rulemakers was in the form of one- or two-page flyers listing their product lines, prices and trade discounts. These inexpensive lists were primarily for the use of wholesalers and retailers, who had to decide which, if any, rules by a given maker they would choose to order and stock. They were not usually illustrated, or at most had only one or two line cuts of rules. (This paucity of illustrations was typical of printed advertising materials before about 1860; the extensive use of line cuts to illustrate catalogs and flyers did not become common until later in the century.)

The descriptions in these lists were also somewhat cryptic. Abbreviations were extensively used, and ditto marks, and the expression "do." (ditto) was frequently used. Since they were intended primarily for the use of dealers and retailers who, presumably, were familiar with the different types of rules, it was probably not regarded as necessary to give a point-by-point description of each rule. Just enough information was included to differentiate the different rules for ordering purposes.

Few of these early rule lists have survived. They were intended to be valid for only a year or two, with new editions being issued frequently. Almost all of them were discarded as soon as the new ones were received by dealers, and the ones which have come down to us have survived mostly by happenstance. Some survived by being attached to a more durable object; a circa 1848 E. A. Stearns list, for instance, was once found pasted into one of the company's early account books. Others survived by being mixed in with other papers; the Willis Thrall list was found in the archives of a defunct Connecticut company which had been donated to the Connecticut Historical Society. Some were printed in catalogs, and thus survived as long as the catalog did; the Asa Richardson list was printed in the 1838 catalog of Carr & Co., of Philadelphia.

Seven surviving examples of such lists are reproduced here, as being of possible interest to collectors. The oldest is that of of Asa Richardson & Son, of Middletown, Connecticut, dating from 1838, the newest is that of Willis Thrall & Son(*), of Hartford, Connecticut, dating from 1867. The five others are the lists of Belcher Brothers & Co.(*), of New York/New Jersey, dating from 1853, A. Stanley & Co., of New Britain, Connecticut, dating from 1855, Anthony Gifford & Son, of Westport, Massachusetts, dating from approximately 1859, Stephens & Co., of Riverton, Connecticut, dating from approximately 1860, and E. A. Stearns & Co.(*), of Brattleboro, Vermont, dating from 1861.

(*) The Belcher Bros. and Willis Thrall price lists are copied from originals in the manuscript collection of the library of the Connecticut Historical Society, to whom grateful acknowledgment is made for the use of these documents.

The E. A. Stearns price list is a second generation reprint, having been copied from an earlier one published by Antique Crafts & Tools of Vermont, that reprint having been made from an original then in the possession of J. Lee Murray, Jr.

ASA RICHARDSON'S
JOINERS' CARPENTERS' AND ENGINEERS' RULES AND BEVILS.

624	*Arch,*	No. 1, Joint Slide Rule.	655	*Satinwood,* 32, do Gunter, two foot.
625	"	" 2, do do	656	" " 33, Plotting do do
626	*Square,*	" 3, do do	657	" " 34, Lumber table rule.
627	"	" 4, do do	658	*Gauging,* " 35, four foot rods.
628	"	" 5, do do	659	" " 36, 3½ do do
629	*Broad,*	No. 6, four fold, extra arch.	660	" " 37, Box Wantage Rods.
630	"	" 7, do side plates.	661	" " 38, Satin do do
631	"	" 8, do do	662	" " 39, S. makers size sticks.
632	*Square,*	" 9, joint 4 fold top plate.	663	" " 40, Ship Joiners' Bevils.
633	"	" 10, do do do	664	*Boxwood* " 41, Plain Gauge.
634	"	" 11, do do do	665	" " 42, Marked do
635	*Thin,*	" 12, ½ narrow do	666	" " 34, Mortising do
636	"	" 13, " do do	667	" " 44, Round Yard Sticks.
637	"	" 14, narrow, 4 fold.	668	" " 45, Square do
638	"	" 15, do do	669	*Ivory,* " 46, Arch Joint Slide.
639	*Thick,*	" 16, do do	670	" " 47, Two foot, four fold.
640	"	" 17, do do	671	*Ivory,* No. 48, square joint 4 fold.
641	"	" 18, do do	672	" " 49, ½ narrow, do
642	"	" 19, 36 inch, 4 fold.	673	" " 50, narrow, do
643	*Arch,*	" 20, joint, 2 fold thin.	674	" " 51, ½ do 2 fold one foot.
644	"	" 21, do do common.	675	" " 52, narrow, do do
645	"	" 22, do do do	676	" " 53, one foot, 4 fold.
646	*Square,*	" 23, do do thin	677	" " 54, do do
647	"	" 24, do do common.	678	" " 55, ½ foot fold.
648	"	" 25, do do do		
649	"	" 26, do do do		
650	*Straight,*	" 27, two foot Rule Plane.		
651	"	" 28, Broad M. Table.		
652	"	" 29, one foot four fold.		
653	*Gunter,*	" 30, scales, two foot.		
654	"	" 31, do one "		

Asa Richardson Price List, 1838

PRICE LIST OF BOX-WOOD AND IVORY RULES,

MANUFACTURED BY

BELCHER BROTHERS & CO., DEPOT NO. 221 PEARL-STREET, NEW-YORK.—WM. BELCHER, Agent.

DOUBLE FOLD BOX-WOOD RULES.

ONE FOOT.

DESCRIPTION OF JOINTS, &c.	½ inch wide. Medium.	¾ inch wide. Medium.	⅞ inch wide. Medium.
Common Joint	0—$3 25	10—$3 25	20—$4 00
Square do. do. edge plates	1— 3 75	11— 3 00	21— 5 00
Do. do. half bound	2— 5 00	12— 3 00	22— 5 00
Do. do. bound	3— 8 00	13— 8 00	23— 7 50
Arch Joint	4—10 00	14—10 00	24—12 00
Do. do. edge plates	5— 4 50	15— 4 50	25— 5 00
Do. do. half bound	6— 5 50	16— 5 50	26— 7 00
Do. do. bound	7— 8 50	17— 8 50	27— 7 30
Both Joints Arch	8—11 00	18—11 00	28—13 00
Do. do. edge plates	9— 7 50	19— 7 30	29— 9 00
	E9— 9 30	E19— 9 30	E29—11 00

TWO FEET.

DESCRIPTION OF JOINTS, &c.	¾ inch wide. Thin.	One inch wide. Medium.	1¼ inch wide. Thick.	One inch wide. Medium.	1¼ inch wide. Medium.
Common Joint	40—$5 00	30—$4 50		50—$3 50	70—$7 00
Square do. do. edge plates	41— 6 00	31— 5 00		51— 4 00	71— 8 00
Do. do. half bound	42— 7 00	32— 6 00		52— 6 00	72— 9 00
Do. do. bound	43—12 00	33— 9 30		53—12 00	73—14 00
Arch Joint	44—14 00	34—12 00		54—16 00	74—18 00
Do. do. edge plates	45— 7 00	35— 6 50		55— 7 00	75—10 00
Do. do. half bound	46— 8 00	36— 7 00		56— 8 00	76—10 00
Do. do. bound	47—13 00	37—12 50		57—13 00	77—15 00
Both Joints Arch	48—17 00	38—16 50		58—17 00	78—19 00
Do. do. edge plates	49—10 00	39— 9 50		59— 9 00	79—10 00
	E49—12 00	E39—11 50		E59—12 00	E79—14 00

SINGLE FOLD BOX-WOOD RULES.

TWO FEET.

DESCRIPTION OF JOINTS, &c.	1½ inch wide. Plain.	1½ inch wide. Slide.
Square Joint (no name on)	80—$3 25	90—$7 00
Do. do. brass pin holes	81— 4 00	91— 8 00
Do. do. brass pin holes	82— 5 00	92— 10 00
Do. do. thin	83— 6 00	(Engineers) 93—14 00
Do. do. bound	84—14 00	91—20 00
Arch Joint	85— 5 00	95— 10 00
Do. do. brass pin holes	86— 6 00	96—12 00
Do. do. thin	87— 7 00	(Engineers) 97—16 00
Do. do. bound	88—15 00	98—22 00
Common Joint. do. Calliper. SINGLEFOLD SIX INCH RULES.		
Square do. do.	89— 1 75	Slide. 99— 3 00

DOUBLE FOLD IVORY RULES.

ONE FOOT.

DESCRIPTION OF JOINTS, &c.	¾ inch wide.	⅞ inch wide.
Common Joint	100—$8 00	110—$9 00
Square Joint	101— 9 00	111—10 00
Do. do. edge plates	102—10 30	112—11 30
Do. do. bound	104—15 00	114—19 00
Arch Joint	105—10 50	115—11 50
Do. do. edge plates	106—12 00	116—13 00
Do. do. half bound	108—20 00	118—21 00
Both Joints Arch	109—16 00	119—16 00
Do. do. edge plates	E109—20 00	E119—21 00

TWO FEET.

DESCRIPTION OF JOINTS, &c.	¾ inch wide.	One inch wide.	1¼ inch wide.
Common Joint	140—$24 00	160—$30 00	170—$36 00
Square Joint	141—27 00	161—33 00	171—40 00
Do. do. edge plates	142—30 00	162—36 00	172—44 00
Do. do. edge plates	144—40 00	164—46 00	174—54 00
Do. bound	145—30 00	165—36 00	175—44 00
Arch Joint	146—33 00	166—40 00	176—50 00
Do. do. edge plates	148—14 00	168—50 00	178—40 00
Do. do. bound	149—36 00	169—42 00	179—52 00
Both Joints Arch	E149—40 00	E169—46 00	E179—56 00

DOUBLE-FOLD IVORY RULES, WITH GERMAN SILVER JOINTS, &c.

ONE FOOT.

DESCRIPTION OF JOINTS, &c.	¾ inch wide.	⅞ inch wide.
Common Joint	200—$9 00	220—$10 00
Square Joint	201—12 00	221—13 00
Do. do. edge plates	202—15 00	222—16 00
Do. do. bound	204—23 00	224—24 00
Arch Joint	205—15 00	225—16 00
Do. do. edge plates	206—19 00	226—20 00
Do. do. bound	208—26 00	228—27 00
Both Joints Arch	209—22 00	229—23 00
Do. do. edge plates	E209—26 00	E229—27 00

TWO FEET.

	¾ inch wide.	One inch wide.	1¼ inch wide.
	240—$26 00	260—$34 00	270—$42 00
	241—28 00	261—39 00	271—46 00
	242—34 00	262—42 00	272—52 00
	244—48 00	264—56 00	274—66 00
	245—34 00	265—42 00	275—52 00
	246—40 00	266—46 00	276—56 00
	248—52 00	268—60 00	278—70 00
	249—42 00	269—50 00	279—60 00
	E249—48 00	E269—56 00	E279—66 00

MISCELLANEOUS ARTICLES.

No.	Prices.		No.	Prices.
Ivory Six-inch Rules, single-fold	160—$3 50	Ship-joiners' Bevels, two blades	901—$4 50	
Do. Calliper rules	199—12 00	Ship-joiners' do. do. inches	902— 5 00	
Double-fold Three Feet Rule, style of No. 31	351— 7 00	Two Feet Rules for the Bench or Glaziers' use	911— 4 00	
Do. Four Feet Rules, do. do.	476—30 00	Do. do. box-wood	913—6 30	
Do. Board Measure, du.	71 571—10 00	Do. Board Measure, octagonal	921— 3 00	
Single-fold do. do.	do. 86 586—12 00	Three Feet do. do. walking canes	928—12 00	
Ship-joiners' Bevels, one blade	900— 4 30	Yard Sticks, common. Nos. 932 at $6, 933 at $7, and 934—$8 00	930—3 00	

Shoe Size-sticks, Gauging Rods, Gunter Scales, &c., at List Prices.

BELCHER BROTHERS & CO., in adopting the above new List of Numbers and Prices per dozen, which they have arranged with much care, believe that, although a temporary inconvenience will be caused by the change, yet, as the classification of the articles and the system of numbers are at once so simple, definite and comprehensive, there will probably be no need of further change.

It will be noticed in the List of Box-wood Rules, that the unit columns of the numbers are all alike, consequently the first column is a Key as respects the mountings of all of them, excepting only a portion of the class of Single-fold Rules. The figures in the columns of tens distinguish the ten different sizes or classes into which the whole are divided. The Ivory Rules are arranged on the same plan, but numbered exactly one hundred higher; those mounted with German silver, two hundred higher; consequently the figures not in the hundred columns are, in each instance, the identical number of the corresponding Rule made of box-wood.

B. B. & Co. hereby assure dealers, that their Rules will continue to be made of the same uniform good quality which has obtained for them a preference in the market ever since their introduction in 1821.

N. B.—The Rules numbered 20, 50, 70, 80 and 90, are not stamped Belcher Brothers & Co.

New-York, May 1st, 1853.

Belcher Brothers & Co. Price List, 1853

PRICE LIST
OF
BOXWOOD AND IVORY RULES,
MANUFACTURED BY
A. Stanley & Co.,
NEW BRITAIN, CONN.

BOXWOOD RULES.
TWO FOLD.

Price pr. doz.

No.					
1.	Arch Joint,			2 ft.	$6.00
2.	do.	with Bits,		"	7.00
3.	do.	Bound Outside,		"	15.00
5.	do.	Full Bound,		"	19.00
7.	do.	Bits, Board Measure,		"	11.00
12.	do.	Slide, Bits, Extra Scales,		"	13.00
14.	do.	do.	Bound Outside,	"	21.00
15.	do.	do.	Full Bound,	"	28.00
16.	do.	do.	do. Engineering,	"	30.00
18.	Square Joint,			"	4.50
19.	do.	Extra Scales,		"	7.00
22.	do.	Bits, Board Measure,		"	9.50
26.	do.	Slide, Bits,		"	10.00
27.	do.	do. Extra Scales,		"	12.00
28.	do.	Full Bound,		"	17.00
29.	English Joint,			"	4.00
34.	Bench Rules,			"	4.00
35.	do.	Board Measure,		"	8.00

CALIPER RULES.

No. 36.	Boxwood Caliper Rules, 2 fold, 6 in.	9.00
36½	do. do. do. 2 " 12 in.	12.00
37.	Ivory do. Brass, 6 in.	12.00
38.	do. do. Ger. Silver, 6 in.	15.00
39.	do. do. do. 4 fold, 12 in.	32.00
40.	do. do. do. do. do. Bound,	40.00

No. 41.	Yard Sticks,	3 ft.	4.50
41½	Wood Measure,	5 ft.	8.00
42.	Ship Carpenters' Bevels,	12 in.	6.50
42½	do. do. Rosewood,	14 "	9.00
44.	Board Sticks, Square,	24 "	8.00
45.	do. do. do.	36 "	12.00
46.	do. do. Octagon,	24 "	10.00
47.	do. do. do.	36 "	14.00

NARROW FOUR FOLD.

No. 51.	Arch Joint,	2 ft.	8.00
53.	do. Edge Plates,	2 "	9.50
54.	do. Full Bound,	2 "	22.00
55.	do.	1 "	5.50
56.	do. Edge Plates,	1 "	7.50
57.	do. Full Bound,	1 "	15.00
59.	Double Arch Joint, Bits,	2 "	12.00
60.	do. do. Full Bound,	2 "	28.00
61.	Square Joint,	2 "	6.00
61½	do	2 "	5.50

Discount on BOXWOOD RULES, per cent.

NARROW FOUR FOLD, Continued.

Price pr. doz.

No. 62.	Square Joint, Full Bound,	2 ft.	$20.00
63.	do. Edge Plates,	2 "	8.00
64.	do. do.	1 "	6.00
65.	do.	1 "	5.00
65½	do. Full Bound,	1 "	13.00
66.	do. Yard Measure,	3 "	8.50
67.	English Joint, Broad,	2 "	7.00
68.	do. Narrow,	2 "	4.50
69.	do.	1 "	3.75

BROAD FOUR FOLD.

No. 70.	Square Joint,	2 ft.	9.50
71.	do. Bits,	"	10.00
72.	do. Edge Plates,	"	11.50
72½	do. Full Bound,	"	24.00
73.	Arch Joint,	"	11.00
74.	do. Bits,	"	11.50
75.	do. Edge Plates,	"	13.00
76.	do. Full Bound,	"	26.00
77.	Double Arch Joint, Bits,	"	16.00
78.	do. do. Bound Outside,	"	23.00
78½	do. do. Full Bound	"	30.00

With Board Measure Table.

No. 79.	Square Joint, Edge Plates,	2 ft.	13.50
80.	do. Full Bound,	"	26.00
81.	Arch Joint, Edge Plates,	"	15.00
82.	do. Full Bound,	"	30.00

IVORY RULES.
NARROW FOUR FOLD.

			Brass.	Ger. Silv.
No. 85.	Square Joint, Edge Plates,	2 ft.	$28.00	$32.00
86.	Arch do. do.	2 "	30.00	34.00
87.	do. Full Bound,	2 "	45.00	50.00
88.	do. do.	1 "	25.00	30.00
88½	do. Edge Plates.	1 "	16.00	20.00
89.	Double Arch Joint, Bound,	2 "	60.00	68.00
90.	English joint,	1 "	8.50	10.00
90½	do. Edge Plates,	1 "	10.00	11.00
91.	Square joint, do. Broad,	1 "	15.00	18.00
92.	do. do. Narrow,	1 "	11.00	13.00
93.	Eng. joint, 2 Fold, do.	6 in.	4.00	5.00

BROAD FOUR FOLD.

No. 94.	Arch joint, Edge Plates,	2 ft.	45.00	52.00
95.	do. Full Bound,	2 "	56.00	64.00
96.	Dble Arch joint, Edge Plates,	2 "	52.00	60.00
97.	do. do. Full Bound,	2 "	64.00	72.00

Discount on IVORY RULES, per cent.

A. Stanley & Co. Price List, 1855

ANTHONY GIFFORD,

MANUFACTURER OF

Boxwood and Ivory Rules and Ship Bevels.

TERMS:—SIX MONTHS, OR FIVE PER CENT. DISCOUNT FOR CASH.

TWO FOLD BOXWOOD RULES.

Per Dozen.

No. 1	Arch joint, slide lined and bound,	$20 00
2	" " " bound,	13 00
3	" " " "	8 00
4	Square joint slide,	7 00
5	Arch " plate, extra thin,	6 00
6	" " "	5 00
7	Square " "	4 00
33	English, " "	2 75

BROAD FOUR FOLD BOXWOOD RULES.

No. 8	Arch joint, slide lined and bound,	25 00
9	" " " "	21 00
10	" " bound,	14 00
11	" " Arch back, side plates,	14 00
12	" " " inside plates,	13 00
13	" " side plates,	9 00
14	" " "	8 00
15	Square, "	7 00

For Rules marked with Board Measure will be charged the additional price of $2.

IVORY RULES.

No 37	Calliper brass mounted,	6 inch,	12 00	
38	" Ger. Silver,	6 "	15 00	
39	" " four fold, 12 "		32 00	
40	" " " 12 " bound,		40 00	

NARROW FOUR FOLD.

		Brass.	Ger. Silver.
88 Arch joint full bound, 12 inch,		25 00	30 00
88½ " " Edge Plates, 12 "		16 00	20 00
90 English joint, 12 "		8 00	9 50
90½ " " Edge Plates, 12 "		10 00	11 00
91 Square " Edge pl't, broad 12 "		15 00	18 00
92 " " 12 "		11 00	13 00
93 English " 2 fold, 6 "		3 00	4 00

Discount on Boxwood Rules   per cent.

NARROW FOUR FOLD BOXWOOD RULES.

No. 16	Arch joint, lined and bound,	18 00
17	" " bound,	12 00
18	" " Arch back, side plates,	12 00
19	" " " inside "	11 00
20	" " extra thin, side "	9 00
21	" " side plates,	8 00
22	" " "	7 00
23	Square joint, back plates,	8 00
24	" " side "	7 00
25	" " "	5 00
26	English joint,	2 75

MISCELLANEOUS BOXWOOD RULES.

27 Square joint, broad 12 in.		5 00
28 English " narrow 12 in.		3 00
29 Square " lined and bound Calliper, 6 in.		10 00
30 " " Calliper, 6 "		6 00
31 Arch " " 6 "		7 00
32 Ship Carpenters' Bevels, brass tongue, 12 in.		4 00

TRY SQUARES,

18 inch ROSEWOOD, with Rests,		15 50
15 do do do		12 50
12 do do		9 00
10 do do		7 50
9 do do		6 75
7½ do do		6 00
6 do do		5 25
4½ do do		4 00
3 do do		3 00

SLIDING T BEVELS.

14 inch ROSEWOOD,		7 50
12 do do		7 00
10 do do		6 50
8 do do		6 00
6 do do		5 50

Discount on Ivory Rules   per cent.   Discount on Ship Bevels   per cent.

Anthony Gifford Price List, ca. 1859

PRICE LIST.
STEPHENS & COMPANY,

MANUFACTURERS OF

U. S. Standard Boxwood and Ivory Rules;

ALSO EXCLUSIVE MANUFACTURERS OF

L. C. STEPHENS' PATENT COMBINATION RULE.

HITCHCOCKVILLE, CONN.

TERMS:		WAREHOUSES:
40 Per Cent. Discount from the entire List, and 5 Per Cent. off for Cash on delivery of Goods.	☞ Orders sent to the Factory will receive Prompt attention.	77 John St., N. Y. W. A. DODGE, Agent. 415 Commerce St., Phila., CHARLES A. MILLER, Agent.

BOXWOOD RULES.
Two Foot Two Fold.

	No.	Price pr doz
Round Joint,	1	5 50
Square Joint,	2	7 50
Square Joint Slide,	9	13 50
Arch Joint,	13	12 00
Arch Joint Extra Thin,	15	15 00
Arch Joint Bound,	17	24 00
Arch Joint Board Measure,	18	15 00
Arch Joint Board M. Bound,	22	24 00
Arch Joint Slide,	23	15 00
Arch Joint Slide Bound,	27	30 00

MISCELLANEOUS RULES.

	No.	Price pr doz
Ship Carpenters' Bevels,	31	9 00
Yard Rules Four Fold,	32	10 50
Two Fd. Ivory Rules 6 in. Brass,	33	7 00
Two Fold Ivory, 6 in G. Slvr.	34	9 00

PATENT COMBINATION RULES.

Boxwood Brass Bound,		60 00

BOXWOOD RULES.
Two Foot Narrow Four Fold.

	No.	Price pr doz
Round Joint,	41	5 00
Square Joint,	42	7 00
Square Joint Bound,	42½	21 00
Square Joint Extra Narrow,	44	8 00
Sq. Jt. Ex. Nar Edge Plates,	44½	12 00
Square Joint Edge Plates,	45	10 00
Arch Joint,	46	9 00
Arch Joint Edge Plates,	48	12 00
Arch Joint Bound,	49	24 00
Double Arch Joint,	50	13 50
Double Arch Joint Bound,	52	32 00

BOXWOOD RULES.
Two Foot Broad Four Fold.

	No.	Price pr doz
Round Joint,	53	7 50
Square Joint,	54	10 50
Square Joint Bound,	54½	27 00
Square Joint Edge Plates,	56	13 50
Arch Joint,	57	18 00
Arch Joint Edge Plates,	59	16 00
Arch Joint Bound,	60	30 00
Double Arch Joint,	61	18 00
Double Arch Joint Bound,	63	36 00

BOXWOOD RULES.
Two Ft. Broad Four Fold Board Measure.

	No.	Price pr doz
Arch Joint Edge Plates,	66	19 00
Arch Joint Bound,	67	33 00

BOXWOOD RULES.
One Foot—Pocket.

	No.	Price pr doz
Round Joint,	70	4 00
Square Joint,	71	5 50
Square Joint Edge Plates,	72	7 50
Square Joint Bound,	72½	16 50
Arch Joint,	73	6 00
Arch Joint Edge Plates,	74	9 00
Arch Joint Bound,	75	18 00

IVORY RULES.
Two Foot Narrow Four Fold.

	No.	Price pr doz
Square Jt. Edge Plates G Slvr	77	76 00
Square Joint Bound G Silver,	78	108 00

IVORY RULES.
Two Foot Broad Four Fold.

	No.	Price pr doz
Arch Jt. Edge Plates G Slvr,	83	96 00
Arch Joint Bound G Silver,	84	116 00
Double Arch Jt. Bd. G Silver,	86	136 00

IVORY RULES.
One Foot—Pocket.

	No.	Price pr doz
Round Joint Brass,	89	16 00
Round Joint G Silver,	89½	18 00
Square Joint Brass,	90	20 00
Square Joint G Silver,	90½	22 00
Sq. Jt. Edge Plates G Silver,	91	28 00
Square Joint Bound G Silver,	92	48 00
Arch Jt. Edge Plates G Silver,	93	32 00
Arch Joint Bound G Silver,	94	54 00

CALIPER RULES BOXWOOD.

	No.	Price pr doz
Two Fold 6 inch Plain,	95	10 50
Two Fold 6 inch Brass Case,	96	12 00
Two Fold 6 inch Bound,	97	18 00
Four Fold 12 inch Plain,	98	18 00
Four Fold 12 inch Bound,	99	30 00

CALIPER RULES IVORY.

	No.	Price pr doz
Two Fold 6 inch G Silver Plain,	95½	26 00
Two Fold 6 inch Brass Case,	96½	24 00
Two Fold 6 in. Bound G Silver,	97½	48 00
Four Fold 12 inch Plain,	98½	60 00
Four Fd. 12 in. Bound G Silver,	99½	76 00

Stephens & Co. Price List, ca. 1860

PRICE LIST
OF
BOX WOOD AND IVORY RULES,
MANUFACTURED BY
CHARLES L. MEAD,
SUCCESSOR TO
E. A. STEARNS & CO.,
BRATTLEBORO, VT.

Two Fold Rules.

No.				PRICE Per Doz.
1	Broad Arch Joint, Slide, Engineering,	24 Inch.		24 00
2	Broad Arch Joint, Slide, Engineering,	"	"	16 00
3	Broad Arch Joint, Slide, Full Scale, Bound,	"	"	20 00
4	Broad Arch Joint, Slide, Full Scale,	"	"	13 00
5	Broad Arch Joint, Slide, Scales,	"	"	12 00
6	Broad Arch Joint, Full Scale,	"	"	8 00
6½	Broad Arch Joint, Thin, Full Scale,	"	"	9 00
7	Broad Arch Joint, Scales,	"	"	6 00
8	Broad Square Joint, Slide, Full Scale,	"	"	11 00
9	Broad Square Joint, Slide, Scales,	"	"	8 00
10	Broad Square Joint, Scales,	"	"	5 00
11	Broad Square Joint,	"	"	4 00
12	Medium Iron Mongers, Caliper, Brass Leg,	6 Inch		10 00
13	Medium Iron Mongers, Caliper,	"	"	7 00
13½	Narrow Iron Mongers, Caliper,	"	"	6 00

Four Fold Rules.

No.				
14	Broad Arch Joint, Board Measure, Bound,	24 Inch.		24 00
15	Broad Arch Joint, Full Scale, Bound,	"	"	23 00
16	Broad Arch Joint, Arch Back, Board Measure,	"	"	17 00
17	Broad Arch Joint, Arch Back, Full Scale,	"	"	16 00
18	Broad Arch Joint, Board Measure,	"	"	13 00
19	Broad Arch Joint, Full Scale, Slide,	"	"	15 00
20	Broad Arch Joint, Full Scale,	"	"	12 00
21	Broad Arch Joint, Scales,	"	"	9 00
22	Medium Arch Joint, Board Measure, Bound,	"	"	21 00
23	Medium Arch Joint, Full Scale, Bound,	"	"	20 00
24	Medium Arch Joint, Arch Back, Board Measure,	"	"	15 00
25	Medium Arch Joint, Arch Back, Full Scale,	"	"	14 00
26	Medium Arch Joint, Board Measure,	"	"	11 00
27	Medium Arch Joint, Scales,	"	"	8 00
28	Medium Arch Joint, Yard Sticks,	36 Inch.		10 00
28½	Narrow Arch Joint, Full, Six Fold,	24 Inch.		12 00
29	Narrow Arch Joint, Bound, Full,	12 Inch.		12 00
29½	Narrow Arch Joint, Caliper, Bound, Full,	"	"	16 00
30	Narrow Arch Joint, Full,	"	"	8 00
30½	Narrow Arch Joint, Caliper, Full,	"	"	12 00
31	Broad Square Joint, Full Scale,	24 Inch.		11 00
32	Broad Square Joint, Scales,	"	"	8 00
32½	Broad Square Joint, Board Measure,	"	"	11 00
33	Medium Round Joint,	"	"	5 00
33½	Medium Square Joint, Board Measure,	"	"	9 00
34½	Medium Square Joint, Scales,	"	"	7 00
34	Narrow Square Joint, Back Plates,	"	"	9 00
35	Narrow Square Joint, Scales, Bound,	"	"	16 00
36	Narrow Square Joint, Edge Plates,	"	"	7 00
37	Narrow Square Joint, Edge Plates,	"	"	6 00
38	Narrow Square Joint, M'dle Plates,	"	"	5 00
41	Narrow Round Joint, M'dle Plates,	"	"	3 00
39	Narrow Square Joint, Edge Plates,	12 Inch.		5 50
40	Narrow Round Joint, M'dle Plates,	"	"	3 00
42	Round Joint Pocket,	"	"	3 00
43	Round Joint Pocket,	"	"	2 00
44	Round Joint Pocket, Two Fold,	6 "		1 50

Ivory Rules.

No.				PRICE PER DOZEN. Unbound.	Bound
45	Broad Arch Joint, Slide, Eng'rng, Two F'ld	24 In.		72 00	92 00
46	Broad Square Joint, Slide, " Two Fold,	"	"	66 00	84 00
47	Broad Arch Joint, Four Fold, Full Scale,	"	"	66 00	84 00
48	Medium Arch Joint, Four Fold, Full Scale	"	"	52 00	70 00
49	Broad Square Joint, Four Fold, Full Scale,	"	"	60 00	78 00
50	Narrow Square Joint, Four Fold, Full Scale,	"	"	40 00	56 00
50½	Narrow Arch Joint, Six Fold, Full,	"	"	52 00	
51	Narrow Arch Joint, Four Fold, Full,	12 In.		28 00	40 00
52	Narrow Square Joint, Four Fold, Full,	"	"	25 00	37 00
53	Narrow Arch Joint, Four Fold, Caliper, Full,	"	"	36 00	48 00
54	Narrow Square Joint, Four Fold, Caliper, F'll,	"	"	32 00	44 00
55	Narrow Square Joint, Two Fold, Caliper,	6 In.		18 00	26 00
55½	Narrow Square Joint, Silver Leg, Caliper,	"	"	24 00	
56	Narrow Square Joint, Brass, Caliper,	"	"	15 00	22 00
57	Narrow Square Joint, Pocket, Four Fold,	12 In.		14 00	24 00
58	Narrow Square Joint, Pocket, do. Brass,	"	"	11 00	
59	Round Joint Pocket, Four Fold, Brass,	"	"	8 00	
60	Round Joint Pocket, Four Fold, Brass,	"	"	7 00	
61	Round Joint Pocket, Two Fold, Brass,	6 "		3 00	

Miscellaneous Articles.

No.				
62	Bench Rule, Board Measure, Box Wood,	24 Inch.		12 00
62½	Bench Rule, Board Measure and Slide, Box Wood,	"	"	15 00
63	Bench Rule, Board Measure, Satin Wood,	"	"	8 00
63½	Bench Rule, Boxwood, Bound,	"	"	16 00
64	Bench Rule, Plain,	"	"	4 00
65	Board Measure, Square, Cap'd,	"	"	8 00
66	Board Measure, Square,	"	"	7 00
67	Board Measure, Square, Cap'd,	36	"	11 00
67½	Board Measure, Octagon, Cap'd,	"	"	12 00
68	Board Measure, Square, Cap'd,	48 Inch		14 00
69	Board Measure, Octagon, Cap'd,	24	"	9 00
70	Board Measure, Octagon,	"	"	8 00
71	Yard Sticks, Cap'd,	36 Inch.		5 00
72	Yard Sticks,	"	"	4 00
73	Ship Bevels, Prime Box Wood, Db'l or S'gl Tongue,	12 Inch		6 00
74	Ship Bevels, Single Tongue,	"	"	5 00
77	Ship Bevels, Br. Box Wood, Db'l Tn'gd and Tip'd	"	"	7 00
78	Gauge Rod, Maple,	36		10 00
79	Gauge Rod, Maple, Wantage Tables,	43		16 00

N. B. From this list the Joints and Bindings of all Ivory Rules will be made of good quality GERMAN SILVER, excepting those numbers marked BRASS. Improved Machinery is used in manufacturing and graduating the Rules, and a prime quality of TURKEY BOX WOOD worked, strict attention being given to the selecting and seasoning.—The uniform quality of the Goods for many years, has obtained for them a preference in market, and the rates at which they are now afforded, recommend them to the attention of Dealers. The Rules will continue to bear the original stamp,

E. A. STEARNS & CO. MAKERS, BRATTLEBORO, VT.

☞ Rules designated FULL SCALE, are graduated 8ths, 10ths, 12ths, and 16ths of inches, with Drafting Scales (and Octagonal Scales on Broad Rules,) also 100 parts to the foot on edges of Unbound Rules. Those marked Scales, bear Drafting and Octagonal Scales, with 8ths and 16ths of inches. FULL, signifies graduations of 10ths, 12ths, and 16ths of inches, with 100 parts to the foot. All other Rules are graduated 8ths and 16ths of inches. SLIDE RULES (except No. 9, which has simply a slide graduated 8ths of inches,) are designed for Instrumental Calculations in Engineering and Mechanics. THE SLIDE RULE GUIDE, a book explaining in full their use, will be sent to order. Price 50 cts., with discount to dealers.

Brattleboro, Jan. 1, 1861.

E. A. Stearns Price List, 1861

Please Destroy all Former Lists.

WILLIS THRALL & SON,

MANUFACTURERS OF

BOX WOOD AND IVORY RULES.

HARTFORD, CONN.

WARRANTED ACCURATE, and made of well seasoned and selected Turkey Box Wood.

1867

LIST OF PRICES.

TWO FOLD RULES.

			Per Doz.
No. 1,	ARCH JOINT,	Full Bound Slide,	$24.00
2,	"	Bound Outside Slide,	20.00
3,	"	Four Fold Slide,	16.00
4,	"	Slide,	14.00
5,	SQUARE JOINT,	Slide,	12.00
6,	"	Cheap Slide.	9.00
7,	ARCH JOINT, Thin,		10.00
8,	"	Extra Scales.	8.00
10,	"		7.00
14,	SQUARE JOINT,		5.00
15,	ROUND, "		3.50

BROAD FOUR FOLD.

No. 16.	ARCH JOINT, Board Measure,		13.00
18,	" " Full Bound,		20.00
19,	" " Double Arch,		12.00
20,	" " Edge Plate,		11.00
21,	" "		9.00

NARROW FOUR FOLD.

No. 22,	ARCH JOINT, Bound Outside,		13.00
23,	" " Full Bound,		16.00
24,	" " Edge Plate,		8.00
25,	" " Double Arch Ex. Scales,		9.00
26,	" "		6.00
27,	" " Edge Plate, Light,		8.00
28,	" "		7.00

BROAD FOUR FOLD.

No. 29,	SQUARE JOINT, Edge Plate,		$9.00
30,	" "		7.00
67,	ROUND JOINT,		5.00

NARROW FOUR FOLD.

			Per Doz.
No. 31,	SQUARE JOINT, Extra Divisions,		$7.00
31½	" " Black Plate,		10.00
33,	" "		4.50
34,	ROUND JOINT,		3.50
35,	SQUARE JOINT, Edge Plate, Light,		8.00
36,	" "		5.50
37,	" " Pocket, 12 inch,		3.50
37½	" " Edge Plate,		5.00
37¾	" " Pocket, 12 in., brass bd.		13.50
38,	ROUND JOINT, 12 "		2.50
38½	SQUARE JOINT,		3.50
39,	ROUND JOINT Ivory, 12 "		10.00
39½	" " " 12 " G. S. Edge Plate,		17.00
39¾	SQUARE JOINT, Ivory, 12 inch, G. S. Bound,		32.00
40,	ROUND JOINT, " 6 inch,		4.50
41,	" " Boxwood 6 "		1.75
42,	SQUARE " Calipers, 6 "		7.00
43,	SQUARE JOINT, Edge Plate, 3 foot,		12.00

MISCELLANEOUS ARTICLES.

Two Feet Board Measures,	9.00
Two Feet Bench Rules,	6.00
Shoemakers' Size Sticks,	3.50
Yard Sticks,	6.00
Wantage Rods,	5.00
Gauging Rods, 4 feet long,	18.00
" " 3 "	7.00
Ship-Carpenter's Bevels, single and double tong,	6.00

No. 1, BEST PLATED TRY SQUARES.

2½	3	3½	4½	6	7½	9	12	15	18 inch
$5.00	5.50	6.00	6.50	8.00	9.50	11.00	16.00	20.00	24.00 doz.

NO. 2 PLATED TRY SQUARES.

2½	3	3½	4½	6	7½	8 inch
$3.00	3.00	3.50	3.75	5.00	5.75	6.00 dz.

9	10	12	15	18 inch.
$6.50	7.00	8.50	12.50	15.50 doz.

No. 1 BEST SLIDING T BEVELS.

4	6	8	10	12	14	16	18 inch
$7.50	8.00	8.50	9.00	9.50	10.00	10.50	11.00 doz.

No. 2 SLIDING T BEVELS.

4	6	8	10	12	14 inch.
$5.00	5.50	6.00	6.50	7.00	7.50 per doz

SCREW MORTISE GAUGE, 10.00 "
COMMON MARKING GAUGE, 1.00 "

GIMLET POINT SCREWS.

Stamped (HARTFORD.)

We have only the following Sizes.

¼ In. 2, 3, 7, 8, 9; ⅝ in. 3, 9, 10; ¾ in. 4, 5; ⅞ in. 11, 13, 14, 15, 16, 17; 1 in. 5, 6, 13, 15; 1¼ in. 16, 17, 18, 19, 20; 1½ in. 15, 16, 17, 18, 19, 20; 2 in. 17, 18, 19, 20, 21; 2¼ in. 11, 15, 16, 17, 18. 19, 20; 2½ in. 11, 12, 13, 14, 15, 16, 17, 19, 20; 2¾ in. 11, 12, 13, 14, 15, 20; 3 in. 14, 15, 16, 17, 19, 20; 3½ in. 14, 15.

The price of Goods, on orders received by us through the mail, is fixed on the day of their receipt, whether any change appears on our list or not.

CASH DISCOUNTS,

ON RULES, TRY SQUARES, BEVELS, ETC.

Discount on Rules, 25 and 5 per cent.
" on Try Squares, Nos. 1 and 2, 15 & 5 pr. ct.
" on Sliding T Bevels, Nos. 1 & 2, 15 & 5 "
" on Miscellaneous Articles, 25 & 5 "
" on Gauges, 5 per cent.
" on Gimlet Pointed Screws, 15 & 5 "

Willis Thrall price list, 1867

III
Reprinted Articles

This chapter is an attempt to bring together and reprint a selection of outstanding and worthwhile writings on rules and measuring instruments where they will be more accessible to collectors and students. These articles are drawn from a variety of sources, some having appeared relatively recently in tool collector publications, others in less familiar journals, and still others in books no longer in print. All are felt to have exceptional merit as sources of information and insight into the subject of measuring instruments, and deserve to be made available where collectors may read (or reread) them.

THE IMPORTANCE OF PRESERVING ARTICLES RELATED TO LINEAR MEASURES

During the last 30 years a large number of articles have been published dealing with linear measures and related artifacts. Many of these have appeared in tool-collector publications such as the *Chronicle of the Early American Industries Association* or *The Gristmill*. Others have appeared in publications not ordinarily received by tool collectors, such as the *Journal of the Oughtred Society* (a publication for slide rule collectors) and *Rittenhouse* (for those interested in American scientific instruments). In addition to these relatively recent offerings, several important articles have appeared as chapters in books published over the last 150-200 years, books long out of print.

The author feels that it is important to reprint articles such as these, gathering them together and bringing them freshly before the collecting public. It makes them fully accessible to all collectors (important in the case of older articles), and, in cases where several articles bear on the same subject, provides greater insight and understanding. Older articles are not readily available to new collectors who do not have a file of old *Chronicles* or *Gristmills* to study. Even long time recipients of these publications sometimes choose to retain only the last couple of years. Also, collectors may be entirely unaware of articles which have appeared in other publications. Few rule collectors subscribe to the *Journal of the Oughtred Society* or to *Rittenhouse*.

The choice of which articles to include in this *Source Book* was not an easy one. Many more articles of near equal worth have been omitted due to space constraints. One has only to review the bibliography in Chapter X to realize this.

Selection was based on several factors: the quality and depth of an article's intended coverage; the availability elsewhere of the information it contains; the likelihood that a collector would be aware of and have access to it; and (to a certain extent) my personal interests.

If any authors feel that their articles have been wrongly omitted, I can only reply that I had to make choices. To attempt to reprint all articles of worth would result in a book of unmanageable size, and even then some articles would probably have to be omitted in order to allow publication in a single volume.

Each of the articles reprinted here has been reprinted with the permission of the author(s) and of the publication where it originally appeared (where possible).

Please note that in all cases, the copyright remains the property of the indicated author(s), and no part thereof may be reproduced by any means without written permission of these copyright holder(s).

The English Carpenter's Rule
Notes on Its Origin
by Paul. B. Kebabian

■ *Author, amasser of a noted tool collection and past-president of EAIA, Paul Kebabian can be depended on for solidly researched examinations into forgotten tool history.*

This limited study of the carpenter's rule was prompted by the acquisition of two English boxwood rules with tables and graduations that were unfamiliar to the writer. In an effort to determine an approximate date for the rules, and to understand the graduations, the investigation started with a 16th century source.

William Goodman, in his *History of Woodworking Tools* and Raphael Salaman, in the *Dictionary of Tools Used in the Woodworking and Allied Trades*, place the start of industrial manufacturing of modern boxwood rules in the early decades of the 19th century. This was the period during which the Rabone firm was established in Birmingham, England, and Belcher Bros. (1821), S. Morton Clark (ca. 1833), Hermon Chapin (1834), and E.A. Stearns (1838) were introducing factory production of accurately graduated boxwood and ivory rules in the United States. Prior to this time, rule making both in England and America had been largely in the hands of mathematical instrument makers or individuals who made their own — often imprecise — rules for linear and other measurement. Measuring devices for performing mathematical computation, for cask gauging, &c were also the product of the mathematical instrument maker.

Some two hundred and seventy-five years before the period of factory production, Leonard Digges, "Gentleman," published a significant contribution to the subject of mensuration entitled *A Boke Named Tectonicon.*[1] Digges, who died about 1571, studied at Oxford and subsequently became an accomplished architect, a student of optics, mathematician, surveyor, and author of several books on meteorology, military science, and on linear measurement and the measurement of solids. The *Tectonicon* had an unusual publishing history: sixteen editions or printings were made between 1556 and 1637. Digges explained that his purpose was to provide the Landemeater (surveyor), Carpenter, or Mason with a way to obtain "the true measurynge and readye accompte of all maner of Lande, Timber, Stoone, Borde, Glasse, Pavement. &c." Portions of the *Tectonicon* particularly relevant to the measurement of timber and boards, and the rule, are "A table to finde the just Radix or Square of any Tymber," and chapters on the table of timber measure, on superficial or board measure, and on the description and construction of the carpenter's rule.

The emphasis placed by Digges — and later by Richard More[2] — on the need to determine cubic measurement of timber by one lineal foot units, and also to make superficial measurement of boards by the square foot, was based on the fact that wood sold to the carpenter was then, as today, priced by per foot measurement. Thus the carpenter needed to know, for example, what length of a timber of certain dimensions of width constituted a cubic foot. His rule, by means of specific tables and graduations, could provide the answer if the timber were in cross-section an equal-sided square, if his rule was accurate, and if he knew how to use it. But if the sides were of unequal dimension, Digges explains, the untutored were accustomed simply to add the width and depth dimensions and halve the result, thereby obtaining an incorrect figure for applying to the rule to find what length would give one cubic foot.[3] With a square timber of 8 inches on each side, Digges' table for timber measure reads 2 feet 3 inches as the length of one cubic foot. But if the timber were of unequal widths — as 4 inches and 12 inches — and if one added these and divided by two the resulting 8 inch figure would be grossly incorrect for determining cubic measurement by means of the carpenter's rule. For one cubic foot of the latter dimensions, the carpenter should get a piece some 3 feet long, rather than 2 feet 3 inches.

Digges provided a "Tabula Radicum" to give ready response to the question: What figure can be used with the table of timber measure to determine the length of a cubic foot of timber of specified, *unequal dimensions of width*. The coordinate of 4 and 12 and his Tabula, for instance, gives the figure "7." Opposite 7 on the table of timber measure is the answer: 2 feet, 11 2/7 inches. The single "pointe adjoyned" to the figure 7 of the Tabula Radicum was not a typographical aberration, but the introduction by Digges of one of the earliest examples of a qualifying mathematical symbol.[4] He explains that by one "prycke" beside a figure he means a *little quantity less than*, and by two points (:) *a fractional amount more than*. The square root of 4 times 12 is 6.928 or not quite 7, the figure to be used with the table of timber measure. It would not be possible to accommodate a series of whole and fractional decimal figures on a table, and of necessity the figures of Digges' tables are in certain instances but close approximations.

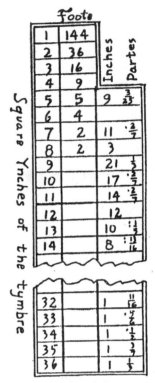

Fig. 1 Copy of part of the timber measure from The Tectonicon. For "Square Ynches," read "Inches Square."

Following the tables and discussion of timber measurement, Digges provides a square foot board measure table and instructions for its use. The columns are headed "Fo" and "Yn" for foot and inches, for widths of board, glass, &c from 1/4 to 12 inches by quarters of an inch, and "Yn" and "Par" for inches and parts of inches for material from 12 1/4 to 36 inches in width. The number of feet, inches, and fractions of inches in length that make a superficial square foot is given for each width dimension. Thus, opposite 6 inch width is the figure 2 feet, and opposite 25 inch width, the answer 5 3/4: inches (or slightly more than that length) per board foot.

Starting the chapter on the Carpenter's Ruler, Digges writes: "Because the effect of this Ruler is above declared by Tables, an Instrument also wel knowen and commune amonge good Artifycers: I will not spende many woordes, in opening it. Behold the fyguers, and learne by them howe ye ought to make, and commonly to decke youre Ruler, bothe with Tymber and bourde measure."[5] We note that Digges uses the term *board measure*. Strictly speaking, a

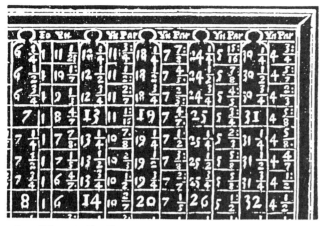

Fig. 2 A section of Digges' table of board measure.

board foot is a measurement including thickness. That is, a piece of wood 12 inches by 36 inches represents 3 superficial square feet; if one inch thick, it is 3 board feet (or if two inches thick, 6 board feet &c). Popular usage through the centuries has given to the superficial measurement of lumber the term "board" measure. That Digges uses "bourde measure" in 1562 would indicate that 16th century measurement of sawn boards was then, as now, based on boards of 1 inch thickness. Edmund Gunter, however, writing in 1624, preferred the term "broad measure."[6]

Digges describes the ruler as "wel playned, twelve Inches longe, a quarter of an Inche thicke, and two Inches yn breadth. Truly yt were more commodious if it hath two foote in length." The 12 inch ruler he illustrates is scaled in inches, and half, quarter, and eighths of the inch. The timber measure is then added to the rule by writing or graving the proper figures, taken from his table of timber measure: near 1 inch on the rule is marked 144 (feet); 2 inches, 36 (feet); 3 inches, 16 (feet); &c to 12 inches, 12 (inches). The second figure represents the running length to make one cubic foot from 1 inch, 2 inch, 3 inch &c of square timber. Instructions are given for continuing the square timber scale on the rule for material up to 36 inches on a side, by inserting the numbers 13 to 36 on the 1/4 inch thick side of the rule at their appropriate places according to the figures of the timber table. In a similar manner the figures for board measure are to be marked on the face of the rule, and continued along the second or opposite 1/4 inch wide edge.

Digges does not stop at this point, but makes of his carpenter's rule a rather sophisticated instrument. He gives directions for marking on the back a "Quadrant Geometricall" with a 90° scale, for mounting a tiny plummet on a fine thread to hang from the quadrant, and attaching level sights "wel bored . . . made of wode, or rather metall," to be fastened on one edge of the rule when it is to be used for sighting.

Forty-six years following the first edition of the *Tectonicon* Richard More of London published *The Carpenter's Rule*. This 1602 work is addressed to "The Worshipful, The Master, Wardens, And Assistants of the Companie of Carpenters of the Citie of London . . ." The author, himself a member of the Company, affirms that gross errors are constantly being made in the measurement of timber. The most common resulting in loss to the buyer are purchasing "waynie" or twisted-growth timber and measuring it as square, and secondly "taking halfe the breadth and thickness of a peece being added together, for the square thereof." More gives full credit at more than one point to Digges: "The ordinarie Rule which Carpenters, Shipwrights, and others doe use to measure Timber withall," he notes, "was invented and published by Paster Leonard Digges." However, he finds that most rules are "very false" and that "those divisions or strikes which are set on them for measuring of Boord and Timber, are not in their right places."[7]

More describes a variety of ways to measure boards and planks, and makes a case for using an arithmetical method. He provides instructions for measuring square, rectangular, round, oval, waynie, and taper growth timber. In chapter 12 of part 2, he includes tables for timber and board measure, with illustrations of the tables which are based essentially on those of Digges. More does not describe or illustrate a graduated carpenter's rule, but his references to Digges and his timber and board measure tables suggest that the construction of the instrument and its graduations and other markings had undergone no significant change from 1556.

The expansion of knowledge in the field of mathematics which took place in the 17th century, however, was to result in new scales being added to the carpenter's rule. John Napier published his work on logarithms in a Latin edition of 1614, followed by a translation in English in 1616.[8] Based on the work of Napier and other contemporary mathematicians, logarithmic lines of numbers were developed by Edmund Gunter, and in his 1624 work on the sector he described "The use of the line of Numbers in broade measure, such as boord, glasse, and the like," in "solid" or cubic measure, and in "Gaugeing of vessels." The invention of the rectilinear slide rule, using two juxtaposed rules with Gunter's lines, is generally credited to William Oughtred.[9] It is this invention that subsequently became an essential part of the carpenter's slide rule.

Coggeshall's *Timber-Measure by a Line of More ease . . .* (i.e., a modified Gunter's line) was made public in this 1677 pamphlet,[10] and Thomas Everard's *Stereometry*, on cask gauging by use of his slide rule, was first published in 1684. Coggeshall's sliding rule is carefully described in the 1728 first edition of Chambers' *Cyclopaedia*[11] and in subsequent editions at least through the 7th of 1751-52. It is this rule, in a two foot, two fold form with Gunter's line as modified by Coggeshall that persisted into the 20th century in both England and America.

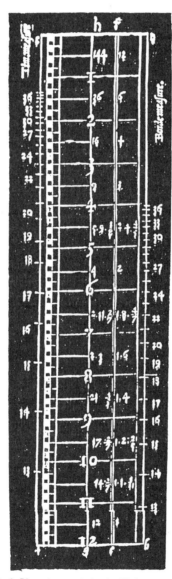

Fig. 3 Digges' carpenter's rule. Timber measure graduations are to the right of the line g h, and continue on the left edge of the rule. Board measure graduations are on the line f e, and on the right edge b d.

The 1728 Chambers description of Coggeshall's sliding rule reads substantially as follows:[12] it is principally used in measuring the superfices and solidity of timber, &c. It consists of two rulers, each a foot long, which are framed or put together in various ways: sometimes they are made to slide by one another, like glaziers rules: sometimes a groove is made in the side of a common two foot joint rule, and a thin sliding piece put in, and Coggeshall's lines added on that side (the method used in 19th century American manufacture): the most usual way is to have one of the rulers slide in a groove made along the middle of the other. Coggeshall's carpenter's slide rule utilized Gunter's lines for the A, B, and C scales but the fourth scale, "called the girt-line, and noted D, whose radius is equal to two radius's of any of the other lines, is broken for easier measurement of timber . . .." The D line is graduated from 4 to 40, and is so illustrated on Coggeshall's rule in the plates of Chambers' *Cyclopaedia*.

The two English boxwood rules here illustrated in Figs. 4 and 5 bear graduations for board and timber measurement consistent with the tables developed by Digges. They also carry Gunter's line (two lines figured 1 to 10 as described by him in 1624 rather than Coggeshall's A, B, C, & D scales). To perform arithmetical calculations it is necessary to use dividers.[13]

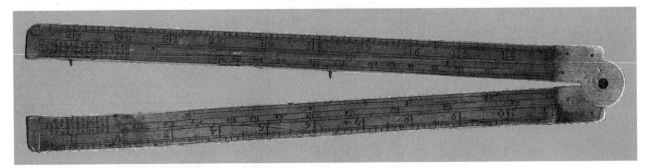

Fig. 4   The 2 foot, 2 fold boxwood carpenter's rule.

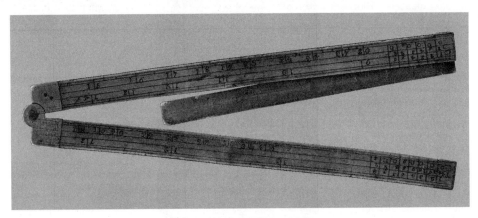

Fig. 5   The 2 foot, 3 fold rule of two hinged 9 inch legs and a 6 inch brass hinged extension.

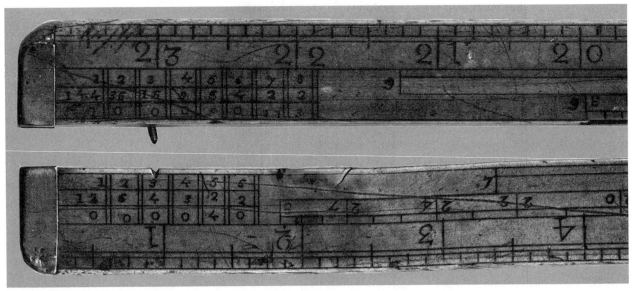

Fig. 6   Tables of timber (above), and board measure (below) on the 2 fold rule.

The tables below, which represent those of the two rules, provide for superficial or board measure and for timber or cubic measure of wood up to 6 inches wide, and 8 inches wide, respectively. Both of the rules also include *lines of board and timber measure* which extend the tabular data for dimensions beyond 6 and ;8 inches in width. As Digges and More explain in their books, the figures 1 to 6 of the board measure table represent inches of breadth, the second and third lines the feet and inches of length that make a square foot for the corresponding width figure on the top line. Thus, a board 1 inch wide requires 12 running feet to make 1 board foot; a board 5 inches wide requires 2 feet and four inches to make one board foot.

The line of board measure (graduated for 7 to 36 inch widths on one rule and 9 to 36 on the second) is used as follows: a figure for width, such as 11 inches, is located on the line. The rule is turned over, and on the 24 inch scale of the back side, and opposite the location of 11 on the first side, is found 13 1/16 inches. Thus a board 11 inches wide measures approximately 13 1/16 inches in length per board foot. That rule makers considered it necessary to provide for measurement of boards 30 inches and more in width speaks to the dimensions of trees being cut for timber (or perhaps being imported for sale) in England in the 16th and 17th centuries.

The top line of the table of timber measure reads 1 to 8, representing figures for square timber of 1 to 8 inches width on each side; the second and third lines are the number of running feet and inches required for one cubic foot. Thus a 4 inch square of timber requires a piece nine feet long to yield one cubic foot; or if a 6 inch square, a length of 4 feet for one cubic foot. For square timber more than 8 inches on the sides, one uses the line of timber measure (graduated for 9 to 30 inch widths on the first rule pictured, and 11 to 30 inch widths on the second). In using this line one measures — using a second rule to perform the measurement — from the figure for width per side, to the end of the rule: for example, from 11 on the line of timber measure to the end of the rule measures approximately 14 1/4 inches. That is to say, an 11 inch square sided timber (or a timber of unequal width sides, the square root of whose product equals 11) contains 1 cubic foot for every 1 foot 2 1/4 inches of length.

These applications of the tables and lines are also explained in the editions of Chambers *Cyclopaedia* from the 1st edition of 1728 through the 7th in 1751-52 (and possibly later editions not available to the writer). Curiously, there is a substantial error in the example given for use of the board measure table which persists through all the editions noted, viz, "If a surface be one inch broad, how many inches long will make a superficial foot? Look in the upper row of figures for one inch, and under it in the second row, is 12 inches, the answer to the question." Unfortunately, the example errs not only by giving the wrong answer, but by asking the wrong question. The second row of the board measure table gives measurements in feet and not inches, and the question should ask how many feet make a superficial foot for a surface 1 inch wide, and the answer from the table is 12 feet. It is apparent that Chambers' work was considered a most authoritative source (as it was). *The Builders Dictionary* of 1734, borrowing virtually word for word from Chambers, contains the same error.[14]

That the two foot, two fold rule was in general use by 1728 is apparent from reading the descriptions of Coggeshall's sliding rule in Chambers, where this rule with Gunter's line is explained. Making use of the "common" 2 foot joint rule by inserting the slide with logarithmic scale is stated as one method of constructing it. One could conclude that between Digges and More (1556/1602), and the first edition of Chambers (1728) the 2 foot, 2 fold carpenter's rule had become a standard form of measuring instrument. Its illustration with tables and lines of board and timber measure in the 7th edition of Chambers is very similar to the rule in Fig. 4 above. In 1758 Edmund Stone's translation of Bion's book on mathematical instruments was published in a 2nd edition with Stone's "Additions of English Instruments." Stone's first example (Book I, p. 14) describes the construction and use of the "Carpenter's Joint-Rule." Both in textual description and illustration in the plates, the rule is again similar to that of Fig. 4. Both rules are professionally made, and neither bears a maker's name. It would seem reasonable to suggest that these two English rules are of late 17th to mid-18th century manufacture, and that the 3 fold rule with 6 inch brass leg is probably somewhat later than the 2 foot example.

NOTES

1. Digges, Leonard. A Boke Named Tectonicon. London: Thomas Gemini, 1562. It is of interest that in his imprint Gemini notes that his location is within Black Friars and that he is "there ready exactly to make all the Instrumentes apperteynynge to this Booke." The first edition published by Digges was dated 1556.
2. More, Richard. The Carpenters Rule, Or, A Booke Shewing Many plaine waies, truly to measure ordinarie Timber . . . London, 1602. Reprint. New York: Da Capo Press, 1970.
3. Digges. leaf 11 recto.
4. Cajori, Florian. A History of Mathematical Notations. 2 v. Chicago: Open Court Pub. Co. (1928-29) v. 2, p. 197.
5. Digges. leaf 17 verso.
6. Gunter, Edmund. The Description and use of the Sector. The Crosse-staffe and: other instruments. London, 1624. Reprint. New York: Da Capo Press, 1971. The first Booke of The Crosse-Staffe, p. 31-50.
7. More. p. 2.
8. Napier, John. A Description of the Admirable Table oe (sic) Logarithmes. London, 1616. Reprint. New York: Da Capo Press, 1969.
9. Cajori, Florian. William Oughtred. Chicago: Open Court Pub. Co., 1916. p. 47. See also: Nicolas Bion. The Construction and Principal Uses of Mathematical Instruments. Translated from the French . . . by Edmund Stone. 2nd ed., London, 1758. Reprint. London: Holland Press, 1972. Book I, p. 16.
10. Coggeshall, Henry. Timber-Measure by a Line of more ease, dispatch and exactness, then any other way now in use, by a double scale . . . London: Printed for the Author, 1677.
11. Chambers, Ephraim. Cyclopaedia: or, An Universal Dictionary of the Arts And Sciences . . . 2 v. London: Printed for J. and J. Knapton (et al), 1728.
12. Chambers. Entry under: Sliding rule - - Coggeshall's Sliding rule.
13. For an explanation of using Gunter's line see: The Carpenter's Slide Rule, its History and Use. Published by John Rabone & Sons. 3rd ed. Birmingham, 1880. Reprint. Fitzwilliam, NH: Ken Roberts, 1982. p. 8-10.
14. The Builder's Dictionary: or, Gentleman and Architect's Companion. 2 v. London, 1734. Reprint. Washington: Association for Preservation Technology, 1981. Entry under: Rule.

Figure 1 is a slightly altered, reduced copy.

Figures 2 and 3 are reductions from negative microfilm.

Further Notes on the Early English Three Fold Ship Carpenter's Rule

by Paul B. Kebabian

■ *This piece and Past President Paul Kebabian's seminal article,* The English Carpenter's Rule, *June 1988, demonstrate what fertile ground your magazine provides for the development of a topic.*

The two foot rule illustrated in Fig. 1 was described in some detail in an earlier article in the *Chronicle*.[1] Its two boxwood nine inch legs, a six inch brass hinged extension, 24 inch graduations, tables and lines of timber and board measure, and its Gunter's lines for mathematical calculations were identified. Two additional sets of graduations (on the sides reverse to those bearing the board and timber measures) were not, however, discussed. The reasons for these omissions were that the article was primarily concerned with the board and cubic measures of the two fold and three fold rules; that the writer was ignorant of the functions of the two sector lines on the rule; and that he overlooked the line of figures which constitute an early introduction of an octagon scale. It is these latter series of graduations — the sectors with nautical designations and the octagon scale — that clearly bring the tool from the general description of *carpenter's rule* to the more specific, i.e., *ship carpenter's rule*.

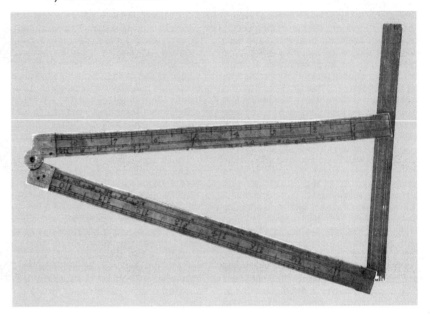

Fig. 1 - The two foot, three fold shipwright's rule with sectors and octagon scale.

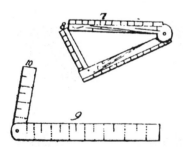

Fig. 2 - Rules illustrated on plate K by Rålamb in 1691. 7 & 8 is the three fold "English" rule with sector, and 9 & 10 a "Dutch" bevel.

One of the earliest illustrations of a three fold rule of this type (Fig. 2) is in the 1691 Swedish work on shipbuilding by Rålamb.[2] Although his representation of the rule, no. 7 & 8 of plate K, is not sufficiently detailed to show accurate graduations, it does clearly reveal two sector lines. The rule is described as 7 "Een Tumståk," a *(folding) rule*, and 8 "Tungan nti en Engelst Tumståk af 1½ fot lang," *Tongue of an English rule of 1½ foot length.* The description of length seems ambiguous (1½ rather than 2 feet), but it is likely that the author identifies at no. 8 the six inch tongue leg which is hinged to no. 7, a two fold part of 18 inches length, rather than suggesting that the rule is 1½ feet long overall. The "English" attribution of this rule is of interest in the light of rule 9 & 10 (Fig. 2) which Rålamb refers to as a "Dutch" rule (Hållenst Tumståk) and which appears to be a ship carpenter's bevel with graduated body and tongue. Nicolaes Witsen's Dutch shipbuilding text, *Architectura Navalis et Regimen Nauticum*, 2nd ed.,

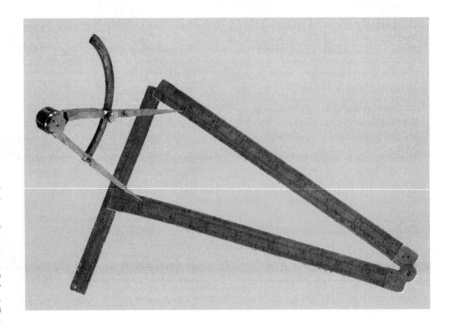

Fig. 3 - Making a measurement on the sector.

1690, contemporary with Rålamb, shows no three fold rule in the illustrations of shipwright's tools.

One end of one of the boxwood legs of the three fold rule is laterally slotted to a depth of about five sixteenths of an inch, into which the six inch brass extension can be snugly fitted. (see Fig. 1) This gives the instrument rigidity so that dividers can be used to take off accurate measurements from the sector lines. The inner sector lines on each of the boxwood legs are graduated *P, 3Q, 2Q, 1Q, and MH,* representing Partners, third, second, and first Quarters, and Masthead. The function of this sector is to provide a series of diameter measurements for shaping the taper of a mast. Below decks mast dimensions are not taken into account by the sector measurements. The partners are the heavy timbers forming a supporting framework around the opening for a mast at the ship's main deck

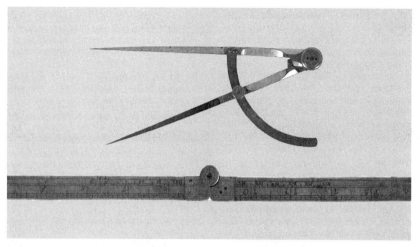

Fig. 5 - The octagon scale, to left and right of the rule joint, scaled 0 to 28.

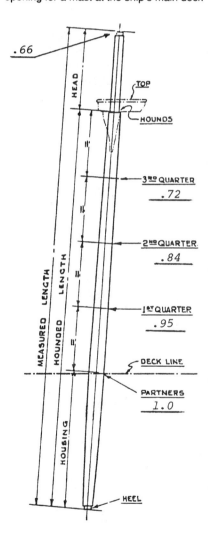

Fig. 4 - Mast diagram from Masting and Rigging *by Underhill. The figures for diameter proportions obtained from the mast sector of the rule, on an arithmetical base of 1 at the partners, have been added at partners, quarters, and masthead.*

level. The mast head is here understood to be the top section of the lower mast (from which additional masts — topmast, topgallant, etc. — could be successively raised as required by the architecture and rigging of a particular vessel). The sector may be opened to an angle so that the distance between point P on each leg, using a pair of dividers, forms a convenient, arbitrary, transverse measurement, such as four inches. This transverse dimension, and the distances between the succeeding points Q3, Q2, Q1, and MH may then be measured with the dividers and laid off on a plan. (see Fig. 3)

Anderson states that the manuscripts of Keltridge and Battine on shipbuilding, written in 1675 and 1684 respectively, agree that the diameter at the head of a mainmast or foremast should be two thirds of the diameter of the mast at the partners.[3] Duhamel du Monceau, writing some eighty years later, gives the same 2/3:1 proportion for diameters of the mast.[4] It is of note that the figures obtained by use of the sector on the rule give a masthead dimension that is precisely two thirds of the diameter at the partners. (see Fig. 4)

The second, or outer set of the two sector lines are designated S, 3Q, 2Q, 1Q and YA, for Slings, third, second, and first Quarters, and Yardarm. The slings are the middle part of a yard, where it is suspended by rope or chain from the mast; the yard's function is to support a square or other shaped sail. Yardarm is the term given the end sections of a yard; yardarms are of somewhat arbitrary length, extending from the quarters to the two extremities of the yard, and in lateral dimension they may include the outer quarter. The second set of sector lines is used to determine diameter measurements at various points on the length of either *half* of a yard, from slings to yardarm end. The diameter of a yard at its ends should be one third of the diameter at its slings, according to Keltridge.[5] Again, the results of measurement by the sector are quite consistent with Keltridge's 1675 dimensions. For example, a three inch setting on the rule at the sector points for S gives a transverse measurement of only slightly more than one inch at the Y graduations. A work dealing with more modern sailing ships by Underhill includes a table of diameter dimensions for timber yards that are proportionately of somewhat greater size at the yardarms than the Keltridge figures.[6] For example, a yard of twenty-one inch diameter at the slings measures nine inches at the yardarm, or about 3 to 1.28, rather than 3 to 1.

It is probable that in shaping a spar, 20th century shipwrights carry on long-standing, traditional techniques. A log is squared by the broadaxe, further reduced to an octagonal spar by hewing, and rounded by the mast drawing knife and spar plane.[7] Anderson, in describing Lord Pembroke's ship model of 1692, the 1701 model of the *St. George*, and others of the early years of the 18th century, reports that the lower yards had an octagonal center section at the slings.[8] An *octagon scale* is included on the three fold rule. (Fig. 5) It would have been an obvious convenience for the shipwright to measure and lay out construction lines to reduce the square of a yard to an octagon, while portions beyond the center were rounded and tapered to dimension. The eight square line of this three fold rule is graduated with figures 0 to 28, and is of the type referred to as an "M" or middle scale on 19th century rules.[9] A dimension taken from the scale would be measured to right and left from the mid-points of the square sides of a yard, rather than from the corners, to mark the layout lines.

To review the data of the three fold rule of Fig. 1, it bears tables and lines of board and timber measure developed during the 17th century. Two Gunter's logarithmic lines on the outer narrow edges of the rule are consistent with that mathematician's descriptions published as early as 1624. On the other hand, the rule does not incorporate the more convenient *sliding lines of numbers* which were developed in the latter part of the 17th century by Oughtred, Coggeshall (1677), and Everard (1684). It incorporates sectors for the taper of masts and yards, the octagon scale, and two foot scale graduated in inches for linear measurement.

Further utility of the rule as a shipwright's tool has been suggested in that two simultaneous angle measurements can be made for compound bevels by use of the two boxwood legs and the third, brass leg. Taken together, the scales and graduations of this versatile member of the measuring instrument family suggest that it was probably a product of the latter part of the 17th century or the early decades of the 18th century.

The author wishes to thank Carl Bopp of Audubon, New Jersey, and Philip Walker of Bungay, Suffolk, England, for helpful information. Particular thanks are due to Richard Knight of Birmingham, England. His explanation of the letters and figures of the sectors was essential for preparation of this article.

NOTES

1. "The English Carpenter's Rule; Notes on its Origin," by Paul B. Kebabian. *Chronicle of the Early American Industries Association*, v. 41, no. 2, June 1988, pp. 24-27.

2. Rålamb, Åke Classon. *Skeps Byggerij eller Adelig Ofnings*. Stockholm, 1691. Facsimile reprint, 1943. Plate K.

3. Anderson, R. C. *Seventeenth-Century Rigging; a Handbook for Model-Makers*. London: Percival Marshall, 1955. p. 9.

4. *Duhamel du Monceau, Henri Louis. Elemens de l'Architecture Navale; Traite Pratique de la Construction des Vaisseaux.* Paris: C-A. Jombert, 1758. p. 135.

5. Anderson. p. 29.

6. Underhill, Harold A. *Masting and Rigging the Clipper Ship & Ocean Carrier*. Glasgow: Brown, Son & Ferguson, 1949. p. 258.

7. Story, Dana A. *The Building of a Wooden Ship*. Barre, Mass.: Barre Publishers, 1971. (*Two unnumbered pages describe and illustrate shipwright Arthur Gates making masts*).

8. Anderson. p. 30.

9. The octagon, or eight square lines are commonly found on 19th and 20th century two foot, two fold carpenters' rules of both England and American manufacture. A middle line (M), and edge (E) scales are frequently provided.

A Guided Tour of an 18th Century Carpenter's Rule

Bruce E. Babcock

Last summer, on the last day of our vacation, I rather casually purchased a two-foot, two-fold rule with a Gunter slide in a small, second-story antique shop not far from the coast of Maine. What attracted my attention to this rule was the fact that the slide was made of wood rather than of brass. This was different from any other two-fold rule that I had seen with a Gunter slide. After I returned home and had time to study the rule more closely I began to suspect that this rule might be older than I had originally assumed.

The rule is a two-foot, two-fold Coggeshall type with one leg incorporating a slide and the other bearing a table of timber measures. When folded, the rule is 12 inches long, 1-7/16 wide and 1/4 inch thick. There is no indication on the rule of who made it or when it was made.

The reverse side of both legs of the rule contain inch measures and primitive architect's scales. This rule differs from many other Coggeshall rules in at least two ways. First, the slide on the rule is made of wood rather than of brass, and secondly, the joint is made partly of wood by incorporating an extension of one leg. A drawing of the rule is shown in Figure 6.1. In the following pages I will try to elaborate on these and other details and provide some background information relevant to this rule and its history.

Some History of Carpenter's Rules

The Coggeshall carpenter's rule got its name from Henry Coggeshall, an Englishman who, according to Cajori [4], first described a version of the rule in 1677. In 1722 the 3rd edition of the Coggeshall book, *The Art of Practical Measuring Easily Performed by a Two Foot Rule that Slides to a Foot* [5] included a drawing of an early version of his slide rule. In this drawing it appears that the slide rule consisted simply of two separate pieces that were not attached in any way. See Figure 6.2. In a later edition of the book published in 1729 [6] the rule seems to have evolved to the form that existed, with variations, for almost 200 years. A drawing of this rule that appeared in the 1729 edition is shown in Figure 6.3.

The history of carpenter's rules, however, begins at least 100 years before Coggeshall added the logarithmic slide to a two-foot, two-fold rule. It appears that this history may have begun as early as 1556 when an architect named Leonard Digges developed a table of timber values and placed them on a rule for convenient reference (Kebabian, [10]). In the intervening years other carpenter's rules were made. For example, under the heading of 'slide rules' Baxandall and Pugh [1] list (#431, page 79) a "carpenter's joint rule (pearwood) by Richard Assheton 1659." No other details are given. Kebabian [11] describes a two-foot, two-fold carpenter's rule that utilizes a Gunter's scale (requiring the use of dividers to perform the mathematical calculations) that may have also predated Coggeshall's design.

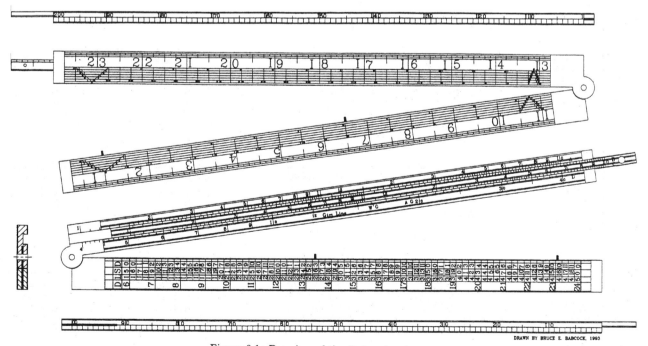

Figure 6.1: Drawing of the Babcock rule.

From the *Journal of the Oughtred Society*, Vol. 3, No. 1 (March, 1994), by permission

Figure 6.2: Coggeshall rule of 1722.

Arrangement of the Scales

One of the features that identifies a Coggeshall carpenter's rule is the arrangement of the scales on the slide rule. In Coggeshall's arrangement the A, B, and C scales are identical, and are basically the same as the two-cycle A and B scales on 20th century slide rules. The D scale, however, was a single line of numbers running from 4 or the left to 40 on the right. This D scale Coggeshall labeled the ``Girt Line." This label does not appear in the 1722 drawing or the 1729 drawing but can be seen in Figure 6.1. According to Baxandall and Pugh [1], it was in 1685 that Coggeshall added the D scale and the term girt line ``chiefly for the use of carpenters, for measuring the superfices and bulk of timber."

Construction of the Joint

The joints on these rules appear to have evolved from the joints that were originally used on sectors. There were two basic types of joints used on these rules. These were described by Rabone [13]:

> "The joints of the earliest made rules were cut from the wooden pieces which formed the legs, and the brass plates were attached to the outside to strengthen the otherwise weak wooden joints. Afterward the joints were composed of metal, and attached to the wooden sides, instead of being made out of them"

The joint on the author's rule is constructed in a manner similar to the early one described by Rabone. Jay Gaynor [8] shows a picture of a similar joint on a rule that is now at Williamsburg. He states that this type of joint dates back at least as far as the 1670's. The rule that Gaynor describes was made by a C. Stedman in London around 1750 and Gaynor considers it to be a typical joint-rule of that period. However, as mentioned above, this rule does not include a Gunter slide.

Gauge Points on the Rule

Coggeshall's carpenter's rules sometimes include gauge points to assist in simplifying calculations. On the author's rule, the letters WG and AG appear on the D scale to mark the gauge points corresponding to the diameter of containers that would hold one gallon for each inch of depth. There are two separate gauge points because prior to January 1826 there were two definitions of the gallon in common use (Heather, [9]). One was called the ale gallon (AG) and the other the wine gallon (WG). Between the letters W and G there is a line on the rule corresponding to the value of 17.15 inches for the wine gallon. However, there is no line to correspond with 18.95 inches for the ale gallon, possibly because it falls so close to the line for 19 but more likely an inadvertent omission by the maker. Heather observes that "These marks are consequently found upon rules constructed prior to January, 1826." After that date the gauge point was located at 18.79 to correspond with the imperial gallon.

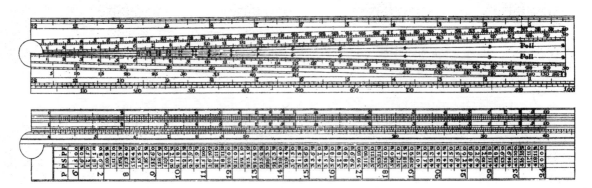

Figure 6.3: Coggeshall rule of 1729.

The Table of Timber Measure

As mentioned earlier, carpenter's rules often included tables of timber measures. The illustration of a carpenter's rule in the 1729 edition of Coggeshall's book shows one example of this type of table (see Figure 6.3). It appears that the headings in this table—P, P, S, P, and F—represent Pence, Pounds, Shillings, Pence and Farthings.

Similar tables are mentioned by several authors.[1]

Turner [20] briefly describes a two-foot jointed rule from the 18th or 19th century with a brass Gunter slide and a table that gives the price in pounds, shilling and pence of various units of timber.

Baxandall and Pugh [1], in their catalog of the slide rules in the Science Museum in London, describe a three-foot rule made in 1742 by Voster that contains "... a table giving the cost of a load of timber at any price from 6d to 2s per foot, reckoning 50 ft to the load." (The author assumes that 6d and 2s represent the same units as 6D and 2S.)

John Bonnycastle [3], mentions that the carpenter's folding rule contains "... a table of the value of a load of timber, at all prices, from 6d. to 2s. a foot." Taylor [19], however, indicates that this book was first printed in England in 1782.[2] Thus 1782 is the more likely date for the table since it would seem that inflation between 1742 and 1844 would have made the 6d to 2s values obsolete. (These values are explained below.)

The table on the author's rule (see Figure 6.1) appears to fit the descriptions of these tables. The leg of the rule adjacent to the one with the Gunter slide contains a table with columns headed D, £, S, & D with numbers from 6 to 24 down the left side. The first column "D" is assumed to be the price in pence per foot of timber. The table starts with 6D at the top and proceeds to 24D at the bottom. With one shilling (S) being worth 12 pence (D) the range of values in this table is the same as given by Baxandall and Pugh and by Bonnycastle, 6d to 2s. Next, if the pounds (£), shillings (S) and pence (D) in each row of the table are converted to pence and added together, the sum is 50 times the value of D in the first column. Thus, if the price per foot is 13 pence(D) the value of a 50-foot load will be (from the table) 2£ 14S 2D which equals 650 pence.

2£	@ 240D/£	=	480 D
14S	@ 12D/S	=	168 D
2D		=	2 D
50 feet/load	@ 13D/foot	=	650 D/load

The Architect's Scales

The reverse side of the author's rule is marked off in inches from 1 to 24 and also contains scales of ¼", ½", ¾", and 1" to the foot divided into 12 parts. This appears to be the forerunner of the modern architect's scale. It is interesting to note that none of the divisions of these scales or the inch or tenths of a foot scale extend to the edge of the rule as on modern rules and scales. The reason this was not necessary was that measurements were not generally taken directly with the rule, as they are today, but were transferred to or from the rule with the use of a set of dividers (more commonly known then as compasses). This seems to be a continuation from the practice of making calculations on a Gunter's scale with the use of dividers. The only divisions on the rule that extend to the edge of the rule are the inch and fractions of an inch divisions on the back of the slide.

Alignment of the Markings

The lines for the inches and eighths of inches on the architect's scales are not quite perpendicular to the edges of the rule, but are consistently angled at approximately 0.9 degrees clockwise. In contrast, the lines on the side of the rule with Gunter's slide show no observable deviation. Gaynor notes that the lines on the Stedman rule mentioned above are also somewhat askew. He attributes this to "... some idiosyncratic aspect of the cutter's working method, his sloppiness or ineptitude, or an unsquare marking guide." Judging by the precision of the markings on the Gunter slide and the consistency of the angle, none of these explanations seems to be entirely satisfactory, with the possible exception of the idea of an idiosyncrasy.

Right-to-Left Numbering

When the rule is unfolded, it can be observed that the inch divisions on the rule are numbered starting with 1 on the right and progress toward 24 on the left.[3]

Ken Roberts [15] noted that there seems to be a correlation between the country where a rule was manufactured and the direction of the numbering of the inches. Looking at the 13 other folding rules with Gunter's slides in the author's collection it was noted that the six that were made in England have the 1 at the left end and are numbered toward the right. The seven that were made in the U.S. have the 1 at the right end and are therefore numbered toward the left. Also, the rule mentioned above that was made by Stedman, around 1750 in London, is numbered from left to right. All of this would seem to strongly support the possibility that this rule was made in the U.S. or in the American Colonies.

Materials of Construction

As mentioned earlier, the author's rule is made almost

---

1. The author assumes that the references to a foot of timber refer to a board foot of 144 cubic inches.
2. It appears that Bonneycastle died in 1821.
3. Actually 23 is the last number, 24 would fall on the brass end piece and was therefore omitted.

entirely of wood (probably boxwood). There is no brass bitting or binding on the rule. However, there are two brass pins to align the two legs when they are folded together. The ends of the rule are capped with brass and there is a 0.15-inch-diameter brass stud in the right hand end of the wooden slide to facilitate movement of the slide. The only other brass pieces are the center portion of the joint and the two pieces that reinforce the wooden half of the joint.

In the texts from 1722 and 1729 there was no mention made of materials of construction. However, in Figure 6.3, which is taken from the 1729 book, the markings continue to the ends of the rule and there is no line to indicate an interface between two materials in the area of either the joint or the extreme ends of the rule. If the drawing is accurate to this degree of detail, it would seem that the 1729 rule must have been made entirely of either wood or metal. The most likely case seems to be that the rule was made entirely of brass, much like the one shown on page 42 of the August, 1992 issue of the *Journal of the Oughtred Society* (Otnes, [12]).

Accuracy of the Rule

The accuracy of the inch markings on the rule seems to vary with the location on the rule. The one-inch mark is only 0.975 inches from the right end of the rule. The 23-inch mark is 0.990 inches from the left end. The greatest error occurs at the joint, with the distance across the joint from the eleven-inch mark to the thirteen-inch mark being 2.075 inches. The distances from one inch mark to another, with the above exception, appear to be very close to one inch. The overall length of the unfolded rule measures 24-1/16 inches.

Some Other Details

The edge of the author's rule is divided into 10ths and 100ths of a foot. According to Spang [16], this is an arrangement that was devised around 1620 by Edmund Gunter, the creator of Gunter's scale.

The back side of the slide is marked off in inches and 1/8ths of an inch. Turner [20], in describing an example of Coggeshall's slide rule, speculates that this is to allow the rule, with the slide extended, to measure a yard.

When the slide on the author's rule is moved to the left the number ``7'' is visible in the bottom of the groove.

Conclusions

It appears that the following conclusions can be drawn safely:

1. The arrangement of the girt line and the table of timber values clearly indicate that this rule was designed for use by carpenters.

2. The rule was made prior to 1826, as indicated by the gauge points for ale gallons and wine gallons.

3. The rule was made after 1685, because that is the date of Coggeshall's invention.

4. The similarity of the wooden joint to the one from around 1750 described by Gaynor and the values of a 50-foot load of timber used in the table being the same as on the 1742 rule described by Baxandall and Pugh [1] might indicate that the rule was made as early as the 1740's or 1750's.

5. The numbering of the inch scale gives some reason to believe that this rule was made in the U.S. or in the American Colonies. However, according to Phillip Stanley, right and left numbering is not an entirely reliable way of determining the origin of a rule.

I would like to acknowledge the advice and assistance provided by Henry Aldinger in preparing this article.

References

1. Baxandall, D. and Pugh, Jane. *Catalogue of the collections in the Science Museum; Calculating Machines and Instruments.* London: The Science Museum, 1975.
2. Belcher Brothers and Co. *Price List of Boxwood and Ivory Rules*, 1860. (reprinted by Ken Roberts) Fitzwilliam NH: The Ken Roberts Publishing Co., 1982.
3. Bonneycstle, John. *Introduction to Mensuration & Practical Geometry.* Philadelphia: Kimber & Sharpless, 1844.
4. Cajori, Florian, *A History of the Logarithmic Slide Rule.* New York: The Engineering News Publishing Co, 1909. (reprinted by the Astragal Press) Morristown NJ: The Astragal Press, 1994
5. Coggeshall, Henry, *The Art of Practical Measuring Easily Perform'd by a Two Foot Rule Which Slides to a Foot; etc.* 3rd edition. London: Printed by H. Clark at the Prince's Arms for Richard King, 1722.
6. Coggeshall, Henry, *The Art of Practical Measuring Easily Perform'd by a Two Foot Rule Which Slides to a Foot; etc.* 4th edition, revised by John Ham. London: Printed by Prince's Arms for Richard King, 1729.
7. Cordingly, J. L. Et al. *Instructions for the Sliding Rule.* Cincinnati OH: A. Derrough, 1841.
8. Gaynor, Jay. "Mr. Hewlett's Tool Chest - Part II." *Selections from the Chronicle; The Fascinating World of Early Tools and Trades.* Emil and Marty Pollak, editors. Morristown NJ: The Astragal Press, 1991.
9. Heather, J. F. *A Treatise on Mathematical Instruments: Their Construction, Adjustment, and use Concisely Explained.* 13th Edition. London: Crosby Lockwood & Co., 1878

10. Kebabian, Paul B. "The English Carpenter's Rule, Notes on It's Origin." *The Chronicle of the Early American Industries Association*, 41:2 (June 1988).
11. Kebabian, Paul B., "Further Notes on the Early English Three Fold Ship Carpenter's Rule," *The Chronicle of the Early American Industries Association*, 42:1 (March 1989).
12. Otnes, Bob. "The Market Report." *The Journal of the Oughtred Society* 1:2 (August 1992).
13. Rabone, John, & Sons. *Instructions for the Use of the Practical Engineers' and Mechanics' Improved Slide Rule, as Arranged by J. Routledge, Engineer*, 1867. (reprinted by Ken Roberts) Fitzwilliam NH: The Ken Roberts Publishing Co., 1983.
14. Rabone, John, & Sons. *Catalog of Rules Tapes Steel Straight Edges, Steel Band Chains, Spirit Levels, Etc.*, 1892. (reprinted by Ken Roberts) Fitzwilliam NH: The Ken Roberts Publishing Co., 1982.
15. Roberts, Ken. *An Introduction to Rule Collecting*. Fitzwilliam NH: The Ken Roberts Publishing Co., 1982.
16. Spang, David A. "All About Those `Gunters." *Selections from the Chronicle; The Fascinating World of Early Tools and Trades*. Emil and Marty Pollak, editors. Morristown NJ: The Astragal Press, 1991.
17. Stanley, Phillip E. "Carpenter's and Engineer's Slide Rules: Routledge's Rule" *Selections from the Chronicle; The Fascinating World of Early Tools and Trades*. Emil and Marty Pollak, editors. Morristown NJ: The Astragal Press, 1991.
18. Stanley, Phillip E. Personal communication, 1994.
19. Taylor, E.G.R. *The Mathematical Practitioners of Hanoverian England, 1714-1840*. Cambridge: The University Press, 1966. (Reprinted by the Gemmary). Redondo Beach CA: The Gemmary, 1989
20. Turner, Gerard L'Estrange. Antique Scientific Instruments. Poole: Blandford Press, 1980.

CARPENTER'S AND ENGINEER'S SLIDE RULES
Part I - History
by Kenneth D. Roberts, P.E.

Fig. 1 The factory of Belcher Bros. & Co., at Camptown, NJ, a village near Newark; Camptown does not appear on modern maps. They were located there as early as 1834; their business location was at various addresses in New York City, 1822-79. (Information from Alexander Farnham).

The usual form of the carpenter's and engineer's slide rule is an arch joint, two-foot, two fold rule, having a brass slide and scales arranged in logarithmic progression (Gunter's scale) on one leg. Examples are shown in Plate 1. One side, folded out, has a conventional 24-inch scale in divisions of 1/16 or 1/8 increments. One leg of the opposite side has four Gunter scales: A, B, C & D, approximately 10 inches in length. The A & D scales are inscribed in the boxwood frame, but the B and C scales are inscribed on a brass slide, 1/4 inch wide by 1/16 inch thick which may be moved between the A & D scales. The A, B & C scales are all identical double repeating scales, progressing from 1 to 10. In the engineer's rule (Routledge) the D scale runs from 1 to 10, but is twice the length of each scale on the A, B & C. In this manner when the C & D scales are placed adjacent with the 1 division at the left ends coinciding, the square of any number (product of the number by itself) on the D scale is read directly above on the C scale; conversely the square root of a number on the C scale may be read directly on the D scale. A more detailed explanation of the construction of this rule appears in Plate II.

On the carpenter's slide rule the D scale is so arranged that the figure 12, which is frequently used in board feet calculations, is at the middle. In this manner the logarithmic scale begins at 4 at the left and runs to 40 at the extreme right. See Plate III for more detailed information about this design.

From numerous examples observed it is apparent that both carpenter's and engineer's rules were extensively manufactured at Birmingham, London and elsewhere in England from 1865 - 1900. The 1892 J. Rabone & Sons Rule Catalogue (reprinted in 1982 by Roberts) noted the availability of books of instructions for the Slide Rule (Carpenter's), Routledge's, Carrett's and Hawthorn's designs. Examples of these rules are known to exist in ivory with nickel silver slides. J. Rabone & Sons also listed a two-foot, 4 fold Routledge rule in their 1892 Catalogue. Carpenter's and engineer's rules were made in three styles: bitted, half-bound and full bound. In America such rules were made by Belcher Brothers, New York, 1821 - 1876; H. Chapin and successor firms, Pine Meadow, CT, 1834-1914; E. A. Stearns, Brattleboro, VT, 1838 - 1863; A. Stanley & Co., New Britain, CT 1854 - 1858, Stanley Rule & Level Co., 1858 - 1916, New Britain, CT; Stephens & Co., Riverton, CT, 1864 - 1901; and Standard Rule Co., Unionville, CT, 1872 - 1876.

The invention of the slide rule followed the discovery of logarithms by John Napier, Baron of Merchiston, published in 1614. Both occured in Great Britain. A predecessor calculating device was the sector compass invented by Galileo, c. 1597. Edward Gunter invented the logarithmic scale in 1620. William Oughtred is generally credited with initiating a form of slide rule as early as 1621, but his work, *The Circles of Proportion and the Horizontal Instrument*, was not published until 1632. As early as 1662 special application slide rules for measuring timber were being made. The fifth edition of *Stereometry or the Art of Gauging Made Easie by Help of a New Sliding Rule* by Thos. Everard was published in London in 1705 (See *Active Scrap Book*, No. 30, July 1979, "The Engineer's Rule" by Paul Kebabian). Sir Isaac Newton employed a slide rule in his calculations, suggesting the use of an index runner (cursor) as early as 1675. However, this

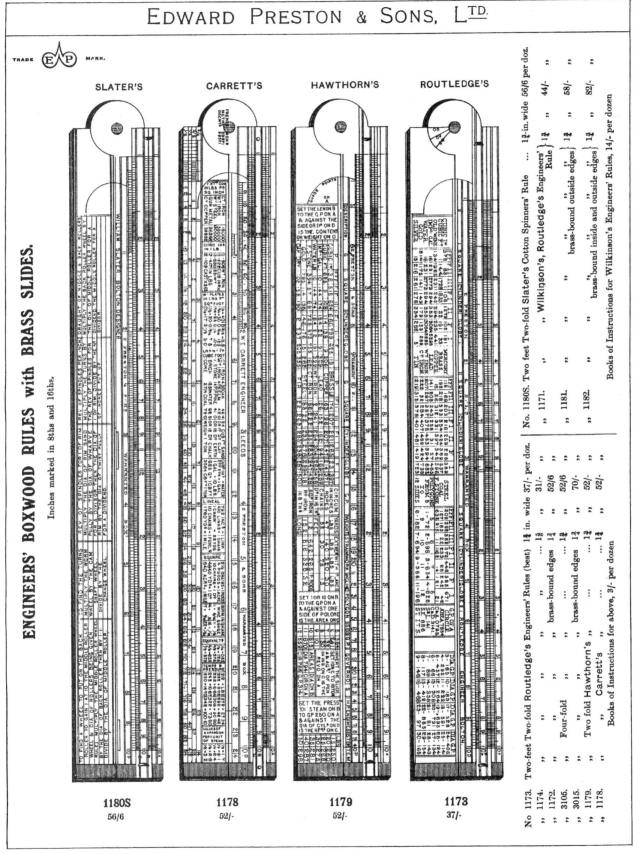

PLATE I

INSTRUCTIONS

FOR THE USE OF THE

PRACTICAL ENGINEERS' & MECHANICS'

IMPROVED SLIDE RULE,

AS ARRANGED BY

J. ROUTLEDGE,

ENGINEER,

Containing a full description of the various lines upon it, with copious directions for its use, and its applicability to Multiplication, Division, the Rule of Three, the Extraction of the Square and Cube Roots, Timber and Land Measuring, Cask and Malt Gauging, the Mensuration of Superficies and Solid Bodies, the Weighing of Metals and other Bodies, with much other matter extremely useful and valuable to the Practical Engineer and Mechanics generally; comprising also THE WHOLE OF THE TABLES FROM ROUTLEDGE'S RULE, now first presented in an intelligible form; and other Tables of Reference; the mode of Construction and Measurement of Angles; forming with the Improved Rule, a true, ready and convenient means of making all measurements and calculations, which no Practical Engineer or other Mechanic should be without.

LATEST EDITION, WITH ALTERATIONS AND CORRECTIONS.

JOHN RABONE AND SON,

Wholesale Manufacturers by Steam Machinery of every description of warranted

BOXWOOD, IVORY & other RULES,

SPIRIT LEVELS, MEASURING TAPES, &c.

ALSO MAKERS OF

ENGINE-DIVIDED STEEL RULES,

OF FINEST QUALITY FOR MACHINISTS.

HOCKLEY ABBEY WORKS, BIRMINGHAM.

The oldest House in the Trade. Established 1784.

INSTRUCTIONS, &c.

DESCRIPTION OF THE RULE.

The Engineer's Rule, as planned by Routledge, and which for ninety years past has been so extensively used by Engineers and Mechanics, is now manufactured by J. RABONE & SON, of unique and superior description. The rules are made by steam machinery, and are marked by a new process, which is used exclusively by themselves; and which—besides presenting uniformity and regularity of appearance—*ensures perfect truthfulness and unvarying accuracy in the various tables, &c.*, which cannot be guaranteed by any other makers. The advantages of such Rules being marked by one unvarying process, must be obvious to all. The Rule is made of good box wood or ivory, and is 24 inches long when open. One side of it is marked with inches and drawing scales, which serve all the purposes of the ordinary 2-ft. Rule. The edges are marked with decimals of a foot and inches divided into 10ths and 12ths. On the other side of the Rule are the lines of numbers or the working slide, and a table of the gauge points required for measuring and weighing various bodies in squares, cylinders, or globes; a table of gauge points required for finding the diameters of steam engine cylinders, that will work pumps from 3 to 30 inches diameter—the one with a pressure of 10lbs., and the other with a pressure of 7lbs. to the square inch; a table of gauge points for estimating the contents of regular polygons of not more than 12 sides; and another table of gauge points for measuring the circumferences, diameters, and areas of circles, with the relative value of such circles to squares and triangles.

On the joint will be found the relative value of the French metre or standard measure to English measure; also its subdivisions into centimetres and millimetres, likewise expressed in English inches.

EXPLANATION OF THE LINES OF NUMBERS.

There are four lines, marked A, B, C, D. The first three lines, A, B, C, are all exactly alike, consisting of two radiuses, and numbered from the left to the right hand with the figures, 1, 2, 3, 4, 5, 6, 7, 8, 9—1, 2, 3, 4, 5, 6, 7, 8, 9, 10. The line D is a single radius, double the length of the others, and numbered from left to right with 1, 2, 3, 4, 5, 6, 7, 8, 9, 10; the lines B and C slide between the other two, and by this operation are all questions answered upon the rule the same as by figures.

OF NUMERATION.

Numeration is the first thing to be learned upon this instrument, for when once that is perfectly understood, everything else will be rendered quite easy; in order that this may be made as plain as possible, let it be first

PLATE II - Title Page and Description from INSTRUCTIONS ... ENGINEER'S ... IMPROVED SLIDE RULE ... BY J. ROUTLEDGE ... [J. Rabone & Son, Birmingham, c. 1867] (Author's Collection)

was not added until 1778, attributed to John Robertson at London.

A drawback or difficulty in the acceptance of the slide rule was that the accuracy of calculations are limited to two or three significant figures. Thus for many applications it provides only an approximation or check of more accurate calculations. In 1850 Lt. Amedee Mannheim, a French artillery officer, supplemented the four scales with those for sines, tangents and logarithms and the slide rule's use was greatly advanced.

James Watt was apprenticed in London as a rule and instrument maker. Later while working with Matthew Boulton at Birmingham he is credited with the first application of a slide rule designed for an engineering purpose. This was in connection with making calculations for steam engines. From the *Treatise on the Steam Engine* (London, 1827, John Farey, p 531.)

"the early slide rule was crudely and inaccurately constructed, but since 1775 Watt and Boulton in their shop located at Soho, near Birmingham, used a slide rule of higher type designed especially for engineers for the computation in the design of steam engines. James Watt, himself, is reported to have used the instrument. Who the manufacturer of these rule was is uncertain, but it has been surmised to have been William James, a very skilled mechanic of the times"

The first practical combination measuring and calculating slide rule for engineer's use is believed to have been conceived by Joshua Routledge (1773 - 1829) while working at Bolton, c. 1811. Very little is known about Joshua Routledge's career. He had an ironmonger (hardware) business at Bolton, 1814 - 1820. In 1818 he was granted Patent No. 4232 for "Improvement upon the rotary Steam Engine". He died at Warsaw, Poland while working there on a consulting engineering contract. (*Bolton Journal*, Jan. 21, 1888)

The earliest known reference to Routledge's rule is his fourth edition of *Instructions for the Engineer's Improved Sliding Rule*, 1813, published at London. This noted that at that date the rule was exclusively made by John Jones, Crown Court, Soho, London, (See Plate IV). It is significant to note that the *Instructions for the Engineer's Improved Sliding Rule*, published by Hermon Chapin in 1858 (See Plate V) is an almost verbatim copy of Routledge's sixth edition, which was originally published in London In 1823. Also the book of instructions offered by J. Rabone & Son c. 1867 (Plate II) was an almost verbatim copy of the instructions in the Routledge 6th edition.

The 1862 Stanley Rule & Level Co. *Catalogue* stated:

THIRD EDITION.

THE CARPENTER'S SLIDE RULE, ITS HISTORY AND USE;

CONTAINING INSTRUCTIONS

For the Measurement of all kinds of Boards and Planks, Timber in the Round or Square, Glaziers' Work and Painting, Brickwork, Paviours' Work, Tiling and Slating, the Measurement of Vessels of various shapes, the Wedge, Inclined Planes, Wheels and Axles, Levers, the Weighing and Measurement of Metals and all Solid Bodies, Cylinders, Cones, Globes, SHIP CARPENTERS' EIGHT SQUARE LINES, the Measurement of Circles, and a Comparison of French and English Measures, with much other information, useful to Builders, Carpenters, Bricklayers, Glaziers, Paviours, Slaters and other Mechanics, now first printed, in a complete form, at a moderate price.

PUBLISHED BY

JOHN RABONE & SONS,

MANUFACTURERS, BY STEAM MACHINERY,

OF

BOXWOOD, IVORY & METAL RULES;

ENGINE DIVIDED STEEL STRAIGHT EDGES,

WITH VARIOUS GRADUATIONS;

Measuring Tapes, with Steel, Metallic Wire or Linen Tapes,

SPIRIT LEVELS, ETC.

HOCKLEY ABBEY WORKS, BIRMINGHAM.

ESTABLISHED 1784.—THE OLDEST HOUSE IN THE TRADE.

[ENTERED AT STATIONERS' HALL.]

1880.

PRICE THREEPENCE.

15

on looking at the Carpenters' Slide Rule that there are four lines of logarithmic numbers upon it, marked respectively at the right hand end, A, B, C, and D. The lines A, B, and C are alike and consist of two radii, as was described on page 9, Fig. 4. The reason of the slide having two lines upon it is that the upper one B may be easily read off against the A line, and the C against the D line, or *vice versâ.* The line D is formed of but one radius, and that is double the length of the others. It is the same in both the Engineer's and Carpenter's Rule, but in the first case it runs from 1 to 10, and on the Carpenter's Rule from 4 to 40; or the figures may be read as 40 to 400, &c. On the Carpenter's Rule this D line is called the "girt line," and is so arranged as to bring the number 12, which is a constant gauge point in timber measuring, near the centre of the Rule, that examples may the more readily be worked out by the lines C and D, as in using the Engineer's Rule for this purpose, the parts required frequently lie near the end of the lines, which is not so convenient. The D line is used in all questions in which cubic or solid measure is required to be estimated, and is sometimes called the line of squares, because twice the log. of any number is equal to the log. of its square.

MULTIPLICATION.

Rule.—To either of the given numbers on A set 1 upon B; then against the other number on B is the product on A.

To take the same example as given on page 9: to multiply 15 by 4. Set one upon B to 15 on A, and against 4 on B is 60, the answer on A.

A	15	60	Answer.
B	1	4	

The Rule, as now set, is really a rule of three instrument, on which three terms are given to find a fourth, or a scale of proportionals, on which any figure on A has the same ratio to the figure opposite to it on B as 15 has to 1, or as 60 to 4, as shewn below.

A	15	45	60	120	300	&c.
B	1	3	4	8	20	&c.

and so on of every other point in the lines A and B.

Multiply 75 by 36.

A	75	2700	Answer.
B	1	36	

Multiply 13½ by 5¾.

A	13.5	77.62	Answer.
B	1	5.75	

PLATE III - Title Page and Description of Carpenter's Slide Rule THE CARPENTER'S SLIDE RULE, Its History and Use [J. Rabone & Sons, Birmingham, England; 1880] From the collection of Rabone-Chesterman Ltd.

"We have prepared a Treatise on the above Gunter's Slide and Engineer's Rules, showing their utility and use; with full and complete instructions, enabling mechanics to make their own calculations. It is also particularly adapted to the use of those having charge of cotton or woolen machinery; 200 pages, bound in Cloth, each net 0.75"

The price of this publication advanced to $1.00 (see page 7, SR&L 1867 Cat.) The last listing of this booklet, offered at the same price, was in SR&L Co. Cat. 1902.

H. Chapin offered both the "Engineer's and Carpenter's Rules" as early as 1839. He and his successor firms continued to make these rules through 1914, but they were not listed in their Cat. #122 of 1922. His *Book of Instructions* was first offered in July 1859 at $1.50, increased in price to $2.00 in 1874 and finally was last offered in 1905 at 25 cents.

A. Stanley & Co. offered four rules: three Carpenter's (No. 12, 15 and 27) and one Engineer's (No. 16 full bound) in his 1855 Catalogue. The earliest extant SR&L Co. Cat. (1859) shows these four rules to have been continued and the No. 6 (Engineer's, bitted) to have been added. These five rules were continued through 1902 which was the last listing for the two Engineer's rules. The three above listed Carpenter's Rules continued to be listed in Catalogue #34 until 1915.

To what extent carpenters and engineers actually used these rules for calculations is not known. A significant commentary regarding their use was published in *Historic Instruments for the Advancement of Science* (Oxford, England, 1925)

"The saying that a prophet has no honour in his own country is well exemplified in the case of the Slide Rule and its invention. For many years this useful instrument was greatly undervalued in England, the country in which it was invented, and even as late as 1850 it was very little known. Yet as De Morgan has aptly put it "for a few shillings most persons might put as much power of calculation in their pockets many hundred times as contained in their heads and the use of this instrument is attainable without any knowledge of the properties of logarithms on which principle it depends."

Very little is presently known concerning the inventors of the three other rules shown in Plate I.

Slater's Rule for calculating winding speeds and yarn patterns in the textile trade was designed by William Slater of

Carrett's Rule was the work of William Carrett, engineer at Leeds. In addition to similar calculation as made on the Routledge rule this had four drafting scales.

Hawthorn's Rule is believed to have been designed by Robert Hawthorn the locomotive engineer. This had the D scale arranged from 4 to 40, similar to the carpenter's rule, but had gauge

INSTRUCTIONS
FOR THE
ENGINEER'S
IMPROVED
SLIDING RULE;
WITH
A Description
OF THE SEVERAL LINES UPON IT,
AND DIRECTIONS HOW TO FIND ANY NUMBER THEREON:
TOGETHER WITH
The Application of those LINES to MULTIPLICATION, DIVISION, the RULE OF THREE, &c. &c.

The Mensuration of Superfices and Solids are likewise made perfectly easy; it is also particularly useful in weighing all kinds of Metals and other Bodies.

THE SIXTH EDITION,
IMPROVED AND ENLARGED,
By J. ROUTLEDGE, ENGINEER.
BOLTON.

London:
PRINTED BY GOLD AND WALTON, 24, WARDOUR ST
FOR JOHN JONES, CROWN COURT, SOHO.
(Successor to the late Mr. A. Wellington.)
1823.
Entered at Stationers' Hall.

ADVERTISEMENT.

The IMPROVED ENGINEER'S RULES, are made and sold only by JOHN JONES, (Successor to the late Mr. A. Wellington,) OPTICIAN, CROWN COURT, Princes Street, Soho London, where Merchants and Shopkeepers may be supplied with the best and most accurate Instruments in the MATHEMATICAL and OPTICAL Line, on the most liberal Terms.

PLATE IV - Title Page and Advertisement from Instructions for the Engineer's Improved SLIDING RULE [J. Routledge, London, 1823] from the collection of the Bolton Library courtesy of Michael Holmes, Blackburn, England.

factors for making engineering calculations.

Additional information concerning these four other rules may result from further research which I hope to conduct on my next trip to England.

Acknowledgement is made of assistance in data preparing this article to Michael Holmes of Blackburn, England for supplying information regarding J. Routledge from the Bolton Library; to Paul Kebabian for use of his 1858 H. Chapin *Book of Instructions for the Engineer's Improved Slide Rule*; and to Philip Stanley for constructive criticism.

An article by Philip Stanley on the application, uses, and errors of the slide rule, is now in preparation.

INSTRUCTIONS.

DESCRIPTION OF THE RULE.

This instrument is made of good box or ivory; it has a joint in the middle, and is 24 inches long when opened out; one face of the rule is marked with inches and a drawing scale, which answers every purpose of a common two foot rule. One of the edges is marked with the decimals of a foot, and the other with inches divided into 10ths and 12ths. On the other face of the rule are the lines of numbers and a table of guage-points for square, cylinder and globe; to which is now added a table of guage-points for pumping engines; another for regular polygons; and a third for the properties of the circle, squares, and triangle.

EXPLANATION OF THE LINES OF NUMBERS.

There are four lines marked A, B, C, D; the first three lines, A, B, C, are all exactly alike, consisting of two radii, and numbered from the left to the right hand with the figures, 1, 2, 3, 4, 5, 6, 7, 8, 9;—1, 2, 3, 4, 5, 6, 7, 8, 9, 10. The line D, is a single radius double the length of the other, and numbered from left to right with 1, 2, 3, 4, 5, 6, 7, 8, 9, 10; the lines B and C slide between the other two; and by this operation are all questions answered upon the rule the same as by figures.

NUMERATION.

Numeration is the first thing to be learned upon this instrument; for when once that is perfectly understood, everything else will be rendered quite easy.

Plate V. Page 1 from the Instruction Book Published by Hermon Chapin, 1858. [Collection of Paul B. Kebabian]. The Explanation and Numeration in this book are the same verbatim as those in the J. Rabone book, 1867. See Plate II.

Carpenters' and Engineers' Slide Rules
Routledges' Rule

Philip E. Stanley

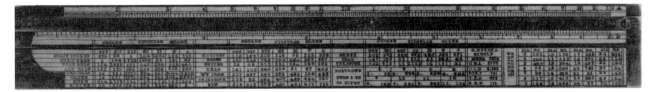

Figure 1: Routledge's Engineer's Rule

■*Phil Stanley has just completed a book, BOXWOOD AND IVORY: STANLEY TRADITIONAL RULES, 1855-1975. An electrical engineer, Phil holds 15 patents in computer design but his real love is obviously old tools, particularly those with numbers stamped on them.*

BACKGROUND

Probably the most elaborately marked and graduated boxwood rules ever made were the so-called "Engineers" rules manufactured between the early 1800's and the first decades of the 20th century. These rules were made in a number of different types, each type being known by its inventor's name (Armstrong's, Hawthorne's, Routledge's, etc.) and intended for a particular branch of engineering (see Roberts, Ref. 1).

Of all the engineer's rules, the one most popular, and most frequently encountered today, is Routledges' engineers' rule, invented ca. 1811 by Joshua Routledge of Bolton, England. Routledge's rule was an attempt to provide the user with not only the means for performing rapid calculations (the Gunter's slide), but also much of the physical data required to work out the common engineering problems of that day. With Routledge's rule, a user could perform volume conversions, weight calculations, geometric analysis, and steam engine/pump computations (this last was a common type of problem 150 years ago; from its inception, the steam engine had been used to drive mine drainage pumps, and was still more frequently used for that than any other purpose).

Physically, Routledge's rule was a 2-foot, 2-fold rule with slide, similar in construction to the more familiar carpenter's sliding rule, but differing from its counterpart in two significant respects: the drafting scales of the carpenter's rule were omitted, to be replaced by tables of physical and geometrical data and "Gauge Points", and the D scale of the Gunter's slide on the engineer's rule was arranged differently.

Using the data on the rule, and the Gunter's slide, the skilled user of the engineer's rule could solve whole classes of problems in mensuration, engineering, and mining without recourse to any other aids or tables.

THE TABLES

The tables on the engineer's rule were 5 in number, and contained reference data organized as follows:

TABLE 1: Conversion factors for relating the volumes of various geometric solids (SQUARE [rectangular parallelopipeds], CIRCULAR [cylinders], and GLOBE [spheres]), to the various units of volume (CUBIC INCHES, WINE GALLONS, etc.). The relationships are expressed in reciprocal form with the decimal point omitted (e.g.: a one foot cube (FFF) contains 1728 cubic inches; 1/1728=0.0005787. . . ; hence the gauge point in the CUBIC INCHES row for the FFF column is 578).

TABLE 2: Conversion factors for relating the volumes of the same geometric solids as in Table 1 to their weight in pounds for various materials. Again, the relationship is expressed in reciprocal form with the decimal point omitted (e.g.: a one-foot long, one inch diameter cylinder (FI) of water weighs 0.3403 pounds . . . ; 1/0.3403. . .=2.938. . ; hence the gauge point in the WATER row for the FI column is 294).

		SQUARE			CIRCULAR		GLOBE	
		FFF	FII	III	FI	II	F	I
TABLE (1)	Cubic Inches	578	83	1	106	1273	1105	191
	Cubic Feet	1	144	1728	1833	22	191	33
	Wine Gals	134	1925	231	245	294	255	441
	Ale Gals	163	235	282	299	359	312	538
	Imp. Gals	16	231	2773	294	353	3064	5295
TABLE (2)	Water	16	231	2773	294	353	3064	5295
	Gold	814	1175	141	149	179	155	269
*	Silver	15	216	261	276	334	286	5
	Mercury	118	169	203	216	258	225	389
	Brass	193	278	333	354	424	369	637
	Copper	18	26	312	331	394	344	596
	Lead	141	203	243	258	31	27	465
	Wt. Iron	207	297	357	378	453	394	682
	Cp. Ir & Zinc	222	32	384	407	489	424	733
*	Tin	219	315	378	401	481	419	723
	Steel	202	292	352	372	448	385	671
	Coal	127	183	22	233	280	242	42
	Fre Stone	632	915	11	1162	14	121	21

C.P. IR: Cupola (Cast) Iron
Fre Stone: Freestone; an easily cut variety of sandstone

Figure 2

From *The Chronicle of the Early American Industries Association*, Vol. 37, No. 2 (June, 1984), by permission

TABLE (3)	Polygons From 5 To 12 Sides	5	6	7	8
		1.72	2.598	3.634	4.828
		9	10	11	12
		6.182	7.694	9.366	11.196

Figure 3: Table 3

G. PTS. OF A CIRCLE AREA 7854 C & A 0795 C & D 3.141 SQR.I 141 S.E.A. 886 S.E.T. 115	PUMPING ENGINES	DIA	G.P.	DIA	G.P.	DIA	G.P.	DIA	G.P.
		3	165	10	183	17	528	24	106
		4	292	11	222	18	591	25	114
		5	457	12	264	19	661	26	124
		6	66	13	308	20	731	27	134
		7	89	14	358	21	81	28	143
		8	117	15	412	22	885	29	154
		9	148	16	468	23	97	30	165

Figure 4: Tables 4 & 5

TABLE 3: A table of the areas (in square units) of regular polygons (pentagon, hexagon, etc.) of unit side, for polygons from 5 to 12 sides. (e.g.: an octagon with one-foot sides has an area of 4.828. . . square feet).

Table 4: Gauge Points of a Circle. A table of numbers relating the diameter, area, and circumference of circles, and further relating them to the dimensions of their inscribed squares and triangles.

The signficance of the various terms is as follows:

AREA: The area of a circle of unit diameter (Actual value: 0.078539)

C & A: The area of a circle of unit circumference (Actual value: 0.7958)

C & D: The circumference of a circle of unit diameter (This is Pi; actual value: 3.1416 . . .)

SQR. I: The diameter of a circle within which is inscribed a square of unit side (Actual value: 1.414 . . .)

S.E.A.: The side of a square equal in area to a circle of unit diameter (Actual value: 0.8862 . . .)

S.E.T.: The diameter of a circle within which is inscribed an equilateral triangle of unit side (Actual value: 1.155 . . .)

Table 5: Gauge Points (G.P.) For Pumping Engines. A table of values which permit the computation of the required diameter of steam engine cylinder necessary to drive a pump of known diameter which is raising water a known height. This table presupposes steam pressure to operate the engine of 7 pounds per square inch, and that the cranks on the engine and the pump have the same swing.

The gauge points are calculated for a series of pump diameters (DIA), such that the square root of the product of the gauge point and the height is the diameter required for the pumping engine cylinder. (e.g.: In order to drive a 6-inch pump which raises water 4½ yards, a pumping engine run by 7 psi steam would be required to have a cylinder diameter of D= 6.60x4.5 = 29.7 =5.45 inches).

In order to fit these tables onto one leg of the rule (an area of about 10-1/2 Inch by 3/4 Inch) some rearrangement of tables 1, 2, and 3 was required. These tables were run together vertically, as shown in Figure 2, and then arbitrarily divided (at the points indicated by the asterick (*)) into three pieces of uniform height. These pieces were then formatted onto the rule beginning at the left, and tables 4 & 5 placed to their right, as shown in Figure 5.

THE GUNTER'S SLIDE

The D scale on the Gunter's Slide of the Routledge's rule was slightly different from the D scale used on the 2-fold sliding carpenter's rule.

On the carpenter's rule, the D scale was "folded" at the value 4; that is, instead of beginning at 1 on the left and progressing through 2, 3, etc. to 1 again on the right, the D scale began at 4 on the left, and progressed through 5, 6, etc. to 1 near the middle, and thence through 2, 3, etc. to 4 again on the right.

The purpose of this "folding" to the D scale on the slide of the carpenter's rule was to reduce the number of manipulations required in common operations. Folding accomplished this in two ways. First, a folded scale introduces the folding point as a multiplying or dividing factor when transferring points between the folded and unfolded scales during computations; a scale with the proper folding point can thus in many cases reduce the number of slide

Figure 5: Arrangement of Tables on Routledge's Engineer's Rule

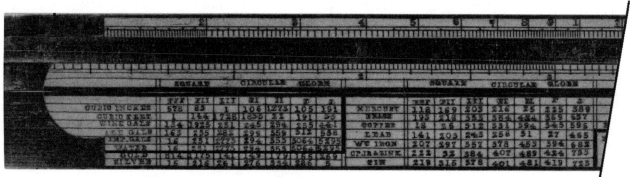

settings by one or more. The factor 4 appears frequently in many engineering problems; hence its choice as a folding point for the D scale on the carpenter's rule. Second, folding a scale has the effect of moving numbers near 1 & 10 towards the physical center of the scale. The physical location of the numbers involved in a slide rule calculation will often influence the number of steps required to perform that calculation; by moving the number 12 to the near-center of the D scale certain calculations, particularly those related to timber measurement, were thus made easier.

The D scale on Routledge's rule could not be folded, however. When Routledge computed the values for the various tables while designing the rule, he included in the gauge points all factors (such as 4 or Pi) other than the specific geometric dimensions of the shapes; the use of these values with a folded D scale would have introduced an unwanted factor and rendered the results useless. The table values could have been computed to allow for the use of a folded scale, but this would have effectively obscured their physical significance and recognizability.

INSTRUCTIONS FOR USE

Most of the large makers of Routledge's engineer's rule published instruction books on its theory and use. The earliest of these is the set of instructions written by Routledge himself, published about 1811, shortly after he invented the rule. These instructions were reissued several times over the next 25 years, the 4th edition appearing in 1813, and the 6th in 1823 (Ref. 2). Other manufacturers who published such instructions were Hermon Chapin (1858, Ref. 3), John Rabone & Sons (1867, Ref. 4, through 1892, Ref. 7), The Stanley Rule & Level Co. (1867, Ref. 5, through 1892, through 1902, Ref. 6), Edward Preston & Sons, Ltd. (Ref. 8), and the Chapin-Stephens Co. (1914, Ref. 9). It is interesting to note that the Chapin and Rabone instructions were exact copies of the Routledge 6th edition, even copying any typographical errors in the original.

These instructions begin with a description of the rule and an explanation of the theory and use of the Gunter's slide, beginning with multiplication and division, and then discussing "the rule of three" (proportions) and square roots. After this they then illustrate the uses of the rule by means of a series of sample problems and solutions. There are sections on mensuration (calculating the areas and volume of solids), land measuring, liquid measure and cask gauging, the weighing of metals, gear and machinery problems, and pump/engine calculaions.

The sample problems/solutions have a slightly archaic ring to those used to the jargon of modern engineering. A couple of examples from the Chapin instructions (Ref. 3) clearly illustrate this:

"If a cask of the third variety is 20 inches at the head, 26 at the bung, and 29 inches long, what will be its contents in old ale and imperial gallons?

Set 29 upon B to the separate circular gauge-points for old ale and imperial measure (say 359 for old ale and 353 for imperial) upon A, and against 23.3, the mean diameter upon D, are 43.8 old, and 44.6 imperial gallons, the answer, upon C."

"What will be the diameter of a cylinder to work a pump, 12 inches in diameter, at 70 yards deep, and loaded with 7 lbs. on the inch?

In the . . . table (of gauge points for pumping engines) for a 12-inch pump is 264. Set 1 upon B to 264 upon A, and against 70 yards upon C, is 43 inches upon D, the answer."

Routledge's, Chapin's, and Rabone's instructions are all approximately 30-40 pages in length. The Stanley instructions are described in their catalogue as "200 pages long"; this would indicate that they had been completely rewritten, but until a copy of this edition is found and studied this can only be a surmise.

ACKNOWLEDGEMENTS

The author wishes to acknowledge the assistance of Paul Kebabian and Ken Roberts, for allowing him to examine their copies of the Routledge, Chapin, and Rabone instruction books, and of Ken Roberts and Jim Hill for helpful criticism.

REFERENCES

1. CARPENTER'S AND ENGINEER'S SLIDE RULES (PART 1, HISTORY); Kenneth D. Roberts, P.E.; The Chronicle of the Early American Industries Association, Vol. 36, No. 1 (March 1983), pp. 1-5
2. INSTRUCTIONS FOR THE ENGINEER'S IMPROVED SLIDING RULE, WITH A DESCRIPTION OF THE SEVERAL LINES UPON IT, AND DIRECTIONS HOW TO FIND ANY NUMBER THEREON (6th Edition); J. Routledge, Engineer, Bolton; 1823
3. INSTRUCTIONS FOR THE ENGINEER'S IMPROVED SLIDING RULE, WITH EXAMPLES OF ITS APPLICATION; Published by Hermon Chapin, Pine Meadow, Conn.; 1858
4. INSTRUCTIONS FOR THE USE OF THE PRACTICAL ENGINEER'S & MECHANIC'S IMPROVED SLIDE RULE, AS ARRANGED BY J. ROUTLEDGE, ENGINEER; Published by John Rabone and Son, Birmingham; Ca. 1867 (Reprinted by the Ken Roberts Publishing Co.)
5. PRICE LIST OF U.S. STANDARD BOXWOOD AND IVORY RULES, LEVELS, TRY SQUARES, GAUGES, HANDLES, MALLETS, HAND SCREWS, &c., Stanley Rule and Level, Co., January 1, 1867 (Reprinted by the Ken Roberts Publishing Co.)
6. STANLEY RULE & LEVEL CO., CATALOGUE NO. 28, Stanley Rule & Level Co., January 1902.
7. CATALOGUE OF MEASURING RULES, TAPES, STRAIGHT EDGES, AND STEEL BAND CHAINS; SPIRIT LEVELS, & C., John Rabone & Sons, July 1892 (Reprinted by the Ken Roberts Publishing Co.)
8. ILLUSTRATED CATALOGUE OF RULES, LEVELS, PLUMBS & LEVELS, THERMOMETERS, MEASURING TAPES, PLANES, IMPROVED WOODWORKER'S AND MECHANIC'S TOOLS, & C. (6TH EDITION), Edward Preston & Sons Ltd., July 1901 (Reprinted by the Ken Roberts Publishing Co.)
9. RULES, PLANES, GAUGES, PLUMBS AND LEVELS, HAND SCREWS, HANDLES, SPOKE SHAVES, BOX SCRAPERS, ETC., CATALOGUE NO. 114, The Chapin-Stephens Co., 1914 (Reprinted by the Ken Roberts Publishing Co.)

TABLE 2 **TABLE 4** **TABLE 5**

TABLE 3

Carpenters' And Engineers' Slide Rules
Errors in Data Tables

By Philip E. Stanley

■ *An electrical engineer by profession, Phil Stanley is one of the mavens of the facinating avocation of rule collecting.*

Articles by Roberts (Ref. 1) and the author (Ref. 2) have described and illustrated the various types of engineers' rules which were invented during the 19th century, and analyzed the data tables on the most common of these, Routledges' engineers' rule. Even more interesting than the history and use of these rules, however, is the large number of errors which can be found in their data tables.

The author has recently completed the examination of 18 engineers rules from a number of manufacturers, and has found every one of them to have between 3 and 10 errors in the values given in their tables. Some of these are random errors, introduced during the hand stamping of the values on a particular rule, and occur on only one or two examples (see Figure 1). Others, on the other hand, are common errors, occurring in several or even all of the 18 rules which have these particular parameters (see Figure 2).

Of the 18 rules examined, ten were made by Stanley; seven Stanley Rule & Level Co. rules (examples nos. 1 through 7), one A. Stanley & Co. rule (example no. 8), and two James Hogg "Improved Slide Rules" manufactured by the Stanley Rule & Level Co. (examples 9 & 10). Six of the 18 were from other American makers; one Standard Rule Co. rule (example no. 11), two E. A. Stearns & Co. rules (examples nos. 12 & 13), two H. Chapin rules (examples nos. 14 & 15), and one Belcher Brothers & Co. rule (example no. 16). Two of the 18 were engineers rules from English makers, one (example no. 17) from Sampson Astin, the other (example no. 18) having no makers name, but being clearly of English origin.

The errors found on these 18 rules can be classified into four different types:

TYPE 1: Errors where one digit in a value is wrong, and the wrong digit physically resembles the correct one. This is the most common error, and was undoubtedly caused by the workman mistaking one number for another when selecting the next stamp. Typical of this would be to find a value of 386 in place of 286 (error no. R-7, observed on example no. 11), or 6132 in place of 6182 (error no. R-12, observed on example no. 13).

TYPE 2: Errors where one digit in a value is omitted. This type of error is less common, and was probably caused by hurry or carelessness. Typical of this would be to find 14 in place of 149 (error no. R-6, observed on example no. 5), or 35 in place of 235 (error no. R-2, observed in example no. 8).

TYPE 3: Errors where a value is stamped in the wrong location in a table, most likely one of the four locations adjacent to the correct one. Again, this is probably the result of carelessness or lack of concentration. Typical of this would be to find 163 in place of 16 in the IMP GALS/FFF location in Table 1, directly below the correct value of 163 in the ALE GALS/FFF location (error no. R-5, observed on examples nos. 10, 17, and 18).

TYPE 4: Errors where the entire value is totally wrong. This type of error is relatively rare, and the cause is not clear. The workman undoubtedly had a sample table of gauge points to work from; perhaps due to hurry or overconfidence he attempted to stamp the values from memory, and did not know the tables as well as he thought. Typical of this would be to find 532 in place of 372 (error no. R-9, observed on example no. 18), or 179 in place of 183 (error no. R-13, observed on example no. 17).

Random Errors:

18 random errors were found in the 18 rules examined. 13 of these were unique, occurring only on a single rule; one (error no. R-10) occurring on two of the rules, and one (error no. R-5) occurring on three of the rules. Each occurrence of these last two is considered a separate error, since each appeared on a rule from a different maker (if all occurrences of either of these errors had been on the rules from a single maker, they would have been considered common errors, and shown instead in the table of Figure 2).

Since the 168 values in these tables are made up of 501 separate digits, these 18 errors represent an error rate of approximately 0.2%, a not unreasonable figure for rules produced in quantity almost completely by hand.

It is interesting to note that there seems to be a wide variation from maker to maker in random error rate. The two H. Chapin rules had no random errors, and the ten Stanley rules only 5, while on the other hand the one Standard rule had 2, and the two English rules had 7. This cannot be taken as conclusive evidence of good or bad quality control; except in the case of the Stanley rules the sample sizes were too small to allow valid statistical inferences to be drawn. Additional research, including the examination of many more non-Stanley rules, will be required to support or refute this indication.

Common Errors:

More interesting than the random errors found in these engineers' rules, however, are the common errors, errors which are present in some or most of the examples studied. Nine such errors were found on the 18 rules (see Table, Figure 2).

ERROR #	TABLE	COLUMN/ROW	IS	SHOULD BE	STANLEY 1	2	3	4	5	6	7	8	9	10	STD 11	STEARNS 12	13	CHAPIN 14	15	BEL 15	ENGLISH 17	18
R-1	1	II/CU IN	123	1273							X											
R-2	I	FII/ALE GAL	35	235								X										−
R-3	I	FII/CU IN	85	83																		X
R-4	I	III/WINE GAL	131	231																		X
R-5	V	FFF/IMP GAL	163	16										X							X	X
R-6	2	FI/GOLD	14	149					X													
R-7	I	F/SILVER	386	286											X							
R-8	I	FI/WRT IRN	478	378																X		
R-9	I	FI/STEEL	532	372																		X
R-10	V	FFF/SILVER	14	15										X						X		
R-11	3	6	298	2598											X					−		
R-12	V	9	6132	6182													X			−		
R-13	5	10	179	183								−								−	X	
R-14	I	18	595	591								−				X				−		
R-15	V	19	691	661								−				X				−		

(X): Error
(−): This Parameter Not On Rule

Figure 1: Random Errors in the Engineering Tables

From *The Chronicle of the Early American Industries Association*, Vol. 40, No. 1 (March, 1987), by permission

ERROR #	TABLE	COLUMN/ROW	IS	SHOULD BE	\multicolumn{10}{c}{STANLEY}	STD	\multicolumn{2}{c}{STEARNS}	\multicolumn{2}{c}{CHAPIN}	BEL	\multicolumn{2}{c}{ENGLISH}
					1	2	3	4	5	6	7	8	9	10	11	12	13	14	15	15	17	18
C-1	1	F/CU IN	105	1105	X	X	X	X	X	X	X	X	X	X	X			X		X	X	X
C-2	V	F/WINE GAL	235	255	X	X	X	X	X	X	X	X			X	X	X	X	X	X		X
C-3	2	FII/BRASS	218	278	X	X	X	X	X	X	X	X	X	X	X	X	X	X	X		X	X
C-4	I	II/MERCURY	25	258	X	X	X	X	X	X	X	X			X			X	X			
C-5	V	II/COAL	22	280	X	X	X	X	X	X	X	X	-	X	X				X			
C-6	4	C & D	5.141	3.141	X	X	X	X	X											-		
C-7	5	10	176	183	X	X	X	X	X	X	X	X	-		X			X	X	-		
C-8	I	19	695	661	X	X	X	X	X	X	X	X	-		X		X	X	X	-	X	X
C-9	V	24	406	106	X	X	X	X	X	X	X	X	-		X	X	X	X	X	-	X	X

(X): Error
(-): This Parameter Not On Rule

Figure 2. Common Errors in the Engineering Tables

It is not possible that this many occurrences of these errors could occur from coincidence. Of the nine errors, one (no. C-6) occurred 5 times, one (no. C-5) occurred 11 times, two (nos. C-4 and C-7) occurred 12 times, two (nos. C-1 and C-8) occurred 15 times, two (nos. C-2 and C-9) occurred 16 times, and one (no. C-3) occurred 17 times!

The only reasonable explanation for this common error rate must be plagiarism, each maker copying the tables on his engineers' rule from those on some other makers rule already in production. If, as seems probable, these tables were not checked for errors, then a random error on the sample rule would thus become a common error on the rules which used it as a prototype. Even worse, if the sample rule contained its own common errors, due to having been copied from an even earlier rule, then they would also be reproduced on the new rules.

Some of these errors date back to the near-origin of the engineers' rule. Routledge's book of instructions for the use of his engineers' rule (1823, Ref. 3) reproduces a copy of Table 5, and in that table, published by the inventor himself, are two of the common errors (nos. C-8 and C-9) found on the rules themselves! The Chapin (1858, Ref. 4) and the Rabone (1867, Ref. 5) instructions not only reproduce both of these errors, but also errors nos. C-1, C-2, and C-3 as well! No wonder these errors were close to universal.

It would have been relatively easy for a maker to check the figures in the tables, even given the significantly lower level of mathematical expertise prevalant during the 19th century. In Tables 1 and 2 the values in any row have to be consistent with the laws of geometry and the 12:1 relationship of Feet to Inches (e.g.: if, as in the BRASS row, the III value is 333, then the FII value must be 278 (333/12 = 277.5), and cannot be 218. Tables 3 and 4 would have similarly yielded to geometrical analysis. In Table 5, it would have been easy to establish that the Gauge Points were related to the Diameters by the square law relationship:

$$G.P. = \frac{Pi \times (DIA)^2}{4} \times 23.289... = (DIA)^2 \times 18.291...$$

...and the DIA value of 24 would have to have a G.P. of 106,, and not 406 (even a cursory examination of the sequence of G.P.'s between 21 and 27 (..., 81, 885, 97, 406, 114, 124, 134, ...) would have detected this erroneous value!)

The magnitudes of the various errors are sometimes quite large. The worst, of course, is the value of 406 in place of 106 (383%), but others were almost as bad: 218 in place of 278 (79%), or 22 in place of 280 (79%). It is to be hoped that some were so obvious as to be recognized by the user (such as the value of 5.141 for Pi) and corrected for automatically during use, but others are less obvious, and must have more than once caused designs to fail when used in computations. Perhaps these tables were not used very often (like some of the exotic functions on the modern electronic calculator), buyers simply choosing the engineers' rule over the carpenters' rule because it looked impressive.

As we have seen, the typical random error rate for stamping the tables on the engineers' rule seems to be about 1 per rule. It is tempting to try to use the number of common errors on a rule as a measure of the number of generations of copying which its' tables represent, but this is risky; as was observed earlier, some rules have been found with no random errors, while others had as many as 4. It is worthwhile observing, however, that the two makers whose rules exhibit the largest number of these common errors, The Stanley Rule & Level Co. (all 9 errors) and the Standard Rule Co. (8 out of 9), are, relatively speaking, the newcomers to the manufacture of Routledges' rule. Stanley began making these rules in 1855, and Standard did not begin until at least 1872, while Chapin began rulemaking in 1835, Stearns in 1838, and Belcher Brothers in the 1820's. Stanley and Standard must have inherited all of the mistakes of their predecessors.

Even catalogue illustrations were not immune to this tendency toward accumulated common errors. In the illustrations depicting Routledges' rule in the 1892 John Rabone & Sons catalogue (Ref. 6) and 1901 Edward Preston & Sons catalog (Ref. 7), it is possible to detect no fewer than 5 of the common errors listed in Figure 6. The Preston illustration shows errors nos. C-1, C-2, C-3, C-8, and C-9. The Rabone illustration, which shows only about 75% of the rule, contains errors nos. C-1, C-2, C-8, and C-9; the section of the table where error no. C-3 might exist is not present. The artist made errors of his own, as well (such as substituting ITS for 115 in the table of gauge points of a circle, in the Preston illustration) but these 5 must be the result of making the drawing using an actual rule as a model, and thus perpetuating them even further!

ACKNOWLEDGEMENTS

The Author wishes to acknowledge the assistance of Gene Frankio, Paul Kebabian, Dave Knight, Rubin Morrison, Ken Roberts, and Bud Steere, all of whom allowed him to examine rules in their possession, and of Ken Roberts and Jim Hill for helpful criticism.

REFERENCES

1. CARPENTER'S AND ENGINEER'S SLIDE RULES (PART I, HISTORY); Kenneth D. Roberts, P.E.; The Chronicle of the Early American Industries Association, Vol. 36, No. 1 (March 1983), pp. 1-5
2. CARPENTER'S AND ENGINEER'S SLIDE RULES (ROUTLEDGES' RULE); Philip E. Stanley; The Chronicle of the Early American Industries Association, Vol. 37 No. 2, pp 25-27
3. INSTRUCTIONS FOR THE ENGINEERS IMPROVED SLIDING RULE, WITH A DESCRIPTION OF THE SEVERAL LINES UPON IT, AND DIRECTIONS HOW TO FIND ANY NUMBER THEREON (6th Edition); J. Routledge, Engineer, Bolton; 1823
4. INSTRUCTIONS FOR THE ENGINEER'S IMPROVED SLIDING RULE, WITH EXAMPLES OF ITS APPLICATION; Published by Hermon Chapin, Pine Meadow, Conn.; 1858
5. INSTRUCTIONS FOR THE USE OF THE PRACTICAL ENGINEERS' & MECHANICS' IMPROVED SLIDE RULE, AS ARRANGED BY J. ROUTLEDGE, ENGINEER; Published by John Rabone and Son, Birmingham; Ca. 1867 (Reprinted by the Ken Roberts Publishing Co.)
6. CATALOGUE OF MEASURING RULES, TAPES, STRAIGHT EDGES, AND STEEL BAND CHAINS; SPIRIT LEVELS, & C., John Rabone & Sons, July 1892 (Reprinted by the Ken Roberts Publishing Co.)
7. ILLUSTRATED CATALOGUE OF RULES, LEVELS, PLUMBS & LEVELS, THERMOMETERS, MEASURING TAPES, PLANES, IMPROVED WOODWORKERS' AND MECHANICS' TOOLS, & C. (6TH EDITION), Edward Preston & Sons Ltd., July 1901 (Reprinted by the Ken Roberts Publishing Co.)

Joshua Routledge   1775-1829

John V. Knott

The Routledge two-foot, folding Engineers Slide Rule, made of boxwood, was invented by J. Routledge in 1809. A report of 1888 indicates that "although there is information on the Routledge Slide Rule, none can be found of the inventor". It Appears that a large quantity of letters, etc., were burned after his death, and only his journal, of his business transactions in Bolton, Lancashire, England, and his last diary (for 1828 until his death in 1829) remain.

His invention consisted of changing the carpenter's slide D scale, which has a single radius running from 4 to 40, to a single radius scale starting at 1 at the left hand index. He also introduced gauge point tables for squares, cylinders, globes, polygons having 6 to 12 sides, circles, and pumping engines.

No patent can be found, although a record does exist for four different editions of his *Instructions for the Engineers Improved Slide Rule*. The fourth edition dated 1813 was printed in London.

The following is a chronology of his life:

- 1773   Routledge was born in Ricall in Yorkshire on April 27th

- 1800   At the age of 27 he became manager for Murray & Wood in Leeds

- 1808   He invented the Engineer's Slide Rule

- 1810   He came to Bolton and worked for the Thomas Swift & Co. Foundry.

- 1814   He started and ran his own business as an ironmonger until 1818

- 1818   He invented the steam rotary engine, Patent No. 4232

- 1824   He left England for Warsaw under an engagement with Thomas Evans & Co., Engineers of Warsaw, Poland, at a salary of £300. He was engineer in charge of erecting a large steam corn mill. The machinery for this mill was made by Hey Foundry of Wigan, Lancashire.

- 1828   Routledge contracted a tumor. His diary indicates that Thomas Evans owed him money, and the tumor was very bad. Apart from the pain of his illness, both he and his wife were suffering from lack of food and from having to beg for money he had justly earned (and then only receiving small installments, barely sufficient to keep him and his family alive).

- 1829   He died in pain and poverty on the 8th of February, at the age of 55. He was buried in the Lutheran burial ground in Warsaw.

From the *Journal of the Oughtred Society*, Vol. 4, No. 2 (October, 1995), by permission

Some Notes on the History and Use of Gunter's Scale

Bruce E. Babcock

If you are interested in the history of slide rules, chances are that, at some time in your life, you had some instruction, either formal or informal, on their use. But what about your knowledge of the use of the Gunter scale? Have you ever tried to solve a mathematical problem with a pair of dividers and just one logarithmic scale?

The Gunter scale was invented by Edmund Gunter in 1620 and was the link between Napier's logarithms, which were invented (or should I say discovered?) in 1614, and Oughtred's slide rule which was invented around 1630. Gunter created the scale by laying out a table of logarithms on a rule using marks to designate the values of the logarithms. On this rule, the locations of the marks were proportional to the values of the logarithms. The end result was almost identical to the A or B scales on modern slide rules and was referred to as Gunter's line of numbers, and was usually designated on the rule as ``NUM".

In addition to the line of numbers, Gunter scales eventually evolved to include at least seventeen other scales for use in trigonometry and navigation. Cajori (1920) quotes a description of a rather simple Gunter scale from 1624 as having, in addition to the line of numbers, a line of tangents, a line of sines, a line one foot in length, divided into 12 inches and tenths of inches, and a line one foot in length divided into tenths and hundredths. Cajori (1909) includes a copy of a drawing of a slightly more elaborate Gunter scale from M. Bion's book of 1723, showing the same lines, arranged in the same order, as is shown in the lower figure below. In a mathematical text printed in 1799, Alexander Ewing described a nearly identical Gunter scale consisting of the same eight lines as shown in Bion's illustration. Both Bion's illustration and Ewing's text includes four additional scales not found on the 1624 rule. These are: S.R. (sines of the rhumbs), T.R. (tangent of the rhumbs), MER. (meridian), and E.P. (equal parts) which, according to Ewing (1799), are used only for navigation.

The above authors make no reference to any lines on the reverse side of these rules. The Gunter scale owned by the author, which was made by J. D. Potter sometime between 1830 and 1861 (Taylor, 1966), (drawings of which are shown below) contains 18 scales of various types with scales appearing on both sides of the rule. This arrangement appears to be nearly identical to a late, 17th-century Gunter scale in the Science Museum that is

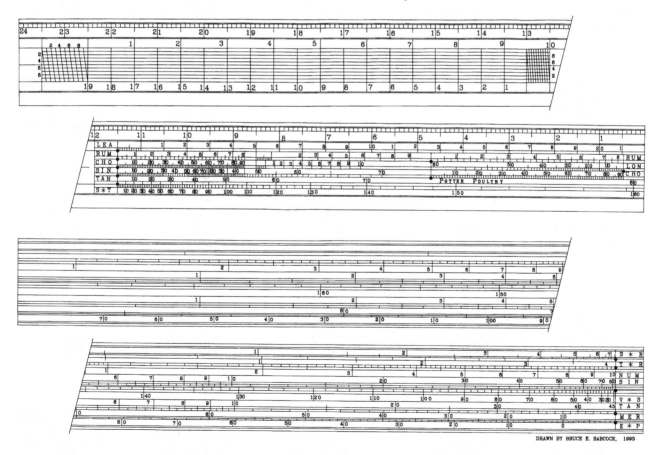

DRAWN BY BRUCE E. BABCOCK, 1993

From the *Journal of the Oughtred Society*, Vol. 3, No. 2 (September, 1994), by permission

described by Baxandall and Pugh. Jeremiah Day, the President of Yale College, in the third edition of his book *The Principles of Plane Trigonometry, Mensuration, Navigation and Surveying* written in 1831, indicates that two different names were used for the two sides of the Gunter scale. He states:

> "To facilitate the construction of geometrical figures, a number of graduated lines are put upon the common two feet scale; one side of which is called the Plane Scale, and the other side, Gunter's Scale."

Gunter scales were usually made on strips of boxwood 24 inches long, approximately 1¾ inches wide and roughly ¼ inch thick. Many of them had small brass pins inserted into the wood at points of heavy use to protect the wood from damage by the points of the dividers that were used to measure off values on the scales. These pins are shown as solid black dots in the figures. There were also variations in size and materials. Baxandall and Pugh list a brass Gunter scale by Andrew Yeats as being in the Science Museum in London. Belcher Brothers and Company of New York offered both one- and two-foot Gunter scales of both boxwood and satinwood in their 1860 catalog. The two-foot scales were available unbound, half-bound and bound. According to Turner's *Antique Scientific Instruments*, Gunter's line of numbers (or NUM scale) also appeared on some sectors, and Kebabian (1991) describes a two-foot, two-fold carpenter's rule with Gunter's line of numbers inscribed on one leg, obviously a precursor to the Gunter slide that later became so common on these rules.

After Gunter created his rule, all that remained for Oughtred to do to invent the slide rule was to lay two of Gunter's lines of numbers side by side, as A and B scales. Having done this, he created a device that would be indispensable to businessmen, scientists, astronomers, navigators, engineers and others for nearly 350 years. In fact, the two-foot, two-fold wooden, brass or ivory carpenter's and engineer's rules with slide rules built into them, were known throughout the 19th century as Gunter rules, even though they were truly slide rules.

Instructions on the use of the Gunter scale were included in early books on arithmetic and navigation. The *Encyclopaedia Britannica*, in its 1771 edition, dedicated nearly a full page to instruction and descriptions relating to Gunter's line and Gunter's scale. Two examples of these texts are mentioned by Cajori (1909) in his book on the history of the logarithmic slide rule. One of these texts is Bowdich's *Navigator*, and the other is Nicholas Pike's *A New and Complete System of Arithmetic*. Pike's book, which was published by the colonial printer Isaiah Thomas in 1788, is the earliest book printed in America that is mentioned by Cajori as including instructions on the use of the slide rule. Not having seen the first edition of this book, the author recently acquired a copy of the second edition which was published in 1797, and was surprised to find that this edition contained no description of the slide rule and no detailed instructions on the principles and use of either the slide rule or Gunter's scale. Instead, Pike seems to have assumed that the reader had a working knowledge of both of these devices, and proceeded merely to give the settings required to solve a variety of problems using both Gunter's scale and the sliding rule. It seems very interesting that this book, published before 1800, starts out with the basic rules of addition and subtraction, and covers trigonometry, geometry, currency conversions, compound interest, present value, etc., and yet assumes that the reader has a working knowledge of both the slide rule and the Gunter scale.

What is also surprising about this book is simply the fact that, along with the settings for the solutions of various problems by the use of the ``sliding rule", are the descriptions of the manipulations of the dividers required to solve the same problems using Gunter's scale. The references to the sliding rule and to the Gunter are to be found only in the section of the book on the measurement of surfaces and solids. This is the section of the book that also includes references to the practice of gauging.

Before proceeding to Pike's use of the Gunter scale in the solutions to some common problems, it may be worthwhile to cover some of the basic operations using the line of numbers on the Gunter scale. Multiplication is accomplished by measuring the distance from the left index to the first number, and then adding this to the distance from the left index to the second number. Thus to multiply 12 times 5, you would set the dividers to span the distance from the left index to 12 and them move them to where the left leg is on 5 and the result, 60, will be found under the right leg. Or, in other words, the sum of the distances from the left index to each of the numbers to be multiplied is equal to the distance from the left index to the product. (This is true because the sum of the logarithms is equal to the log of the product.) Division, likewise, is accomplished by merely measuring the distance of the denominator from the left index, and subtracting it from the distance from the left index to the numerator. Thus, to divide 18 by 3, you would merely set the dividers to the distance from the left index to 3 and then move the dividers to where the right leg is on 18 and the answer, 6, will be found under the left leg. Squaring is very simple; it is accomplished by merely flipping the dividers once over. The square root is slightly more complicated in that it is necessary to set the dividers to one half of the distance from the left index to the number of which the root is to be taken. Cube roots would be one third, fifth roots one fifth, etc. It is the author's guess that the E.P. (equal parts) scale could have been used in the calculation of powers and roots. For example, to calculate the cube root of 27, first set the dividers to span from the left index to 27. Then measure the span of the dividers on the equal parts scale. In this case it would be 132 units. Divide this by three (to get the cube) and get 44 units. Now set the dividers to 44 on the E. P. scale and transfer this to the line of numbers (NUM scale) and with the left leg of the dividers on 1, the right leg will be on 3, the cube root of 27. A set of proportional dividers was sometimes used to assist in calculating both roots and powers.

Nicholas Pike's book of 1797 includes many examples of solutions of problems through the use of first the sliding rule and then the Gunter scale. Two of these, which were selected because they are rather simple, have been included here. These problems illustrate not only the use of the sliding rule and the Gunter, but they also include some simplistic insights into 18th century arithmetical methods. For an example, Pike did not use the units of "square inches" or "square feet" in these problems nor did he use π in the problem involving the area of a circle. Also, his harsh economy of words in describing the solutions seems to be typical of that era.

The first example, from page 438, involves calculating the area of a circle when the diameter is known.

"The Diameter being given, find the Area of a Circle without finding the Circumference. Rule - Multiply the square of the diameter by .7854, and the product will be the area of the circle, whose diameter was given. Example. The diameter of a circle being 12, to find the area?

$$12 \times 12 = \begin{array}{r} .7854 \\ 144 \\ \hline 3\ 1416 \\ 31\ 416 \\ 78\ 54 \\ \hline 113.0976 \end{array} = \text{area}$$

BY THE SLIDING RULE
Set 1 on A to the diameter on B, then find .7854 (which expresses the area of the circle whose diameter is 1) on A, against which on B is a 4th number, then find this 4th number on A, against which on B is the area.

BY GUNTER'S SCALE
The extent from 1 to the length of the diameter reaches from .7854 to a 4th number, and from that 4th number to the area."

The second example is taken from page 427 and simply involves the calculation of the area of a square.

"Rule. - Multiply the side of the square into itself, and the product will be the area of superficial content, of the fame name with the denomination taken, either in inches, feet, or yards, respectively. Let ABCD represent a square, whose side is 12 feet. Multiply the side 12 by itself, thus,

$$\text{Area} = \begin{array}{cc} 12 \text{ inches} & 12 \text{ feet.} \\ 12 \text{ inches.} & 12 \text{ feet.} \\ \hline 144 \text{ inches.} & 144 \text{ feet.} \end{array}$$

BY THE SLIDING RULE
Set 1 to the length on B, then, find the breath on A, and opposite to this on B, you will have the content.

BY GUNTER'S SCALE
Extend the dividers from 1, on the line of numbers, to the length; that distance, laid the same way from the breadth, will point out the answer."

An excellent, and somewhat more recent, description of the use of the Gunter scale can be found in Ken Roberts' recent reprint of Rabone's *The Carpenter's Slide Rule, Its History and Use*, which was originally printed in 1880.

By using a pair of dividers along with the B scale on the slide out of almost any slide rule, the reader can quickly confirm two things. First, that yes, calculations can be made in this manner, and secondly, that Oughtred had good reason to try to find a more convenient way to use Gunter's scale.

The question, however, that seems to beg for an answer is: why was there still a market for Gunter scales more than 250 years after the slide rule was invented? It would seem that the slide rule, with it's greater ease of use, would have replaced Gunter's scales some time prior to 1700. Oddly enough, some were still being sold near the end of the 19th century, and some arithmetic, surveying and gauging texts still contained descriptions of their construction and instructions on their use well into the 19th century. For example, Davies' *Elements of Surveying and Navigation*, in both the 1853 and 1868 editions, contains two paragraphs on the use of Gunter's scale. One paragraph mentions only the scale of equal parts, the diagonal scale of equal parts, and the scale of chords. The other paragraph is dedicated to the use of the line of meridional parts. Oren Root's *New Treatise on Surveying and Navigaton* of 1867 allows a page and a half for discussion of the use of Gunter's scale, but prefaces it with the following comment: "Gunter's scale is commonly two feet in length, containing the plane scale, and the scale of sines, chords, and tangents on one side of it, and the scale for logarithms of numbers, sines and tangents on the other. The logarithmic scale is not much used". John Rabone and Sons, rule makers in Birmingham, England, listed a 24" Gunter's navigation scale in their catalog as late as 1892. However, not all rule makers offered Gunter scales at this time. Even though the Stanley Rule and Level Company was a prolific manufacturer of rules in the second half of the 19th century, it does not appear that they offered any Gunter scales. They did, however, offer folding engineer's and carpenter's rules with Gunter slides (See *The Stanley Catalog Collection* 1989).

It would seem that for Gunter scales to be offered for

sale as late as 1892, they must have had some advantage over slide rules. One advantage may have been the result of a combination of price and accuracy. In the 1860 Belcher Brothers & Co. catalog, two-foot, two-fold carpenter's rules with Gunter slides were $10.50 to $12.50 per dozen (88¢ to $1.04 each), depending on the type of joint chosen. In the same catalog, two-foot Gunter scales were $10.00 per dozen (83¢ each). In Rabone's 1892 catalog, both of these rules were $20.00 per dozen ($1.66 each). Keuffel and Esser, in their catalog dated the following year, priced their 10-inch Gunter slide rule at $3.50 each and their 10-inch and 20-inch Mannheim rules at $4.50 and $10.50 each, respectively.

A second advantage may have been greater accuracy. It would seem that the Gunter scale, being twice as long as the slide on the carpenter's rule, could have been at least twice as accurate.*

A third advantage of the Gunter scale may have been the simple fact that the use of the Gunter scale was so ingrained in the habits of those who used them that they merely resisted the change to the slide rule. Williams (1985) in his book *A History of Computing Technology* makes a comment that supports this idea quite clearly:

> "Indeed the early examples of slide rules which still survive usually show unmistakable signs of having been used, not in the intended way, but by having had a pair of dividers pick off lengths along the logarithmic scales."

It may be interesting to note that a book published by Dietzgen as late as 1905 (Rosenthal, 1905) includes instructions on how to use a pair of proportional dividers to determine powers on a slide rule.

A fourth possible advantage of the Gunter scale over the slide rule may have been that, because it had no moving parts, it was not subject to sticking and binding in humid nautical settings. Its use in navigation is confirmed by Joseph Bateman, an expert on the slide rule (Taylor, 1966) who observed about 1840 that navigators still used a Gunter scale with compasses. As mentioned above, the Gunter scale listed in the 1892 catalog was described as a "Gunter's Navigation Scale."

And one last advantage may have been simply that with the large number of scales, the Gunter scale may simply have filled a need that was not filled with the relatively few scales found on slide rules. Attempting to compare a slide rule with a Gunter scale may be analogous to comparing an adding machine to a book of mathematical tables.

It appears that through the 18th and 19th centuries the Gunter scale may have been used more for general trigonometric calculations and navigational calculations and that the slide rule was used more for specific technical applications. Many slide rules made prior to 1880s have special scales and gauge points for timber, tax assessment, pumping engine design, pipe sizing, etc.

The author does not know how common Gunter scales are today, and would be interested in any information that might be available from other scholars or collectors.

---

* The carpenter's rule also has two cycles on the scales commonly used for multiplication. *Ed.*

Bibliography

1. Anonymous. *The Stanley Catalog Collection.* Morristown NJ: The Astragal Press, 1989.
2. Baxandall, D. and Pugh, Jane. *Catalogue of the Collections in the Science Museum; Calculating Machines and Instruments.* London: The Science Museum, 1975.
3. Belcher Brothers and Co. *Price List of Boxwood and Ivory Rules*, 1860. (reprinted by Ken Roberts) Fitzwilliam NH: The Ken Roberts Publishing Co., 1982.
4. Bowditch. *Navigator.* 1802.
5. Cajori, Florian. "On the History of Gunter's Scale and the Slide Rule During the Seventeenth Century." *University of California Publications in Mathematics*, I:9 (February 20, 1920). (Reprinted by the Astragal Press) Morristown NJ: The Astragal Press, 1994.
6. Cajori, Florian. *A History of the Logarithmic Slide Rule.* New York: The Engineering News Publishing Co., 1909.. (Reprinted by the Astragal Press) Morristown NJ: The Astragal Press, 1994.
7. Davies, Charles. *Elements of Surveying and Navigation, with Descriptions of the Instruments and the Necessary Tables.* A.S. Barnes & Co., 1853.
8. Davies, Charles. *Elements of Surveying and Navigation, with Descriptions of the Instruments and the Necessary Tables.* A.S. Barnes & Co. 1868.
9. Day, Jeremiah. *The Principles of Plane Trigonometry, Mensuration, Navigation, and Surveying. Adapted to the method of instruction in the American colleges.* Third edition. New Haven: Hezekiah Howe, 1831.
10. *Encyclopaedia Britannica*, Volume II. Edinborough: A.Bell and C.McFarquahar, 1771.
11. Ewing, Alexander. *A Synopsis of Practical Mathematics.* 4th Edition. London: 1799.
12. Kebabian, Paul B. ``The English Carpenter's Rule; Notes on its Origin," in *Selections From The Chronicle: The Fascinating World of Early Tools and Trades.* Morristown NJ: The Astragal Press, 1991.
13. Keuffel & Esser Co. *Catalog and Price List.* 24th Edition. 1893.
14. Pike, Nicholas. A New and Complete System of Arithmetic. 2nd Edition. Worcester: Isaiah Thomas, 1797.

15. Rabone, John, and Sons. *Catalog of Rules and Tapes*, 1892. (reprinted by Ken Roberts) Fitzwilliam NH: The Ken Roberts Publishing Co., 1982.
16. Rabone, John, and Sons. *The Carpenter's Slide Rule, Its History and Use*. Third Edition, 1880. (reprinted by Ken Roberts) Fitzwilliam NH: The Ken Roberts Publishing Co., 1982.
17. Root, Oren. *A New Treatise on Surveying and Navigation, Theoretical & Practical*. Ivison, Phinney, Blackman & Co., 1867.
18. Rosenthal, L.W. *Mannheim and Multiplex Slide Rules*. Chicago: The Eugene Dietzgen Co., 1905.
19. Taylor, E.G.R. *The Mathematical Practitioners of Hanoverian England 1714-1840*. Cambridge: The University Press, 1966. (reprinted by the Gemmary). Redondo Beach CA: The Gemmary, 1989.
20. Turner, Gerard L'Estrange. *Antique Scientific Instruments*. Poole: Blandford Press, 1980.
21. Williams, Michael R. *A History of Computing Technology*. New York: Prentice-Hall, 1985.

Further Notes on the Operation of Gunter's Rule

Dieter von Jezierski

In addition to Bruce E. Babcock's article (*Journal*, Vol.3 No.2) the "German viewpoint" can be found in the well known book of Captain Ludwig Jerdmann *Die Gunterscale*, 1888, Hamburg, Eckardt & Wesstorff. The subtitle in translation is *Complete Explanation of the Gunterscales and Proof of Their Origin with Numerous Examples for Practical Use*. Unfortunately there is no English version, but I will give some interesting examples from his book, which should be a rich source for all Gunter fans.

Also, reading this book and following the intentions of L. Jerrmann, I am convinced that the survival of Gunter's rule (besides the later "sliding Gunter" and then "slide rules") over two centuries is based on its use for nautical problems. For instance, the *Oxford English Dictionary* of 1933 states "... the Gunter's rule commonly (was) called The Gunter by seamen (Phillips 1706, de. Kersey)." From that time on, it was the admired and often used "Gunter" in seafaring! For another example, A. Rohrberg emphasizes in his book *Der Rechenstab im Unterricht aller Schularten* the Gunter's employment for navigation, and mentions that Lord Nelson used it. I suggest that because navigators were very familiar with the use of dividers, such facility made it easy for them to use the "Gunter."

We do not have biographical information for Ludwig Jerrmann, but probably when he wrote his book he was a teacher at the navigation school located in Hamburg. In the introduction of his book he points out that it was his intention "to give practical instructions to the students of navigation science in using the Gunterscale". He also wrote about the exclusive historical precedent of Edmund Gunter in the design of the Gunterscale in 1620.

In Jerrmann's time, seamen often used Donn's scale. In the introduction to his book Jerrmann stated that he could not justify the imprint *Navigation Scale improved by Benjamin Donn*[1] without any reference to E. Gunter.

Jerrmann's description and explanation of all scales is excellent and detailed: it takes nearly 20 pages of his book, and is followed by a chapter "Development of the Gunter-lines."

There are also a number of practical examples. The first of three of these is as follows:

Example Ia. From a ship a bearing of SW 3/4 W is taken of Lighthouse Arcona. After having sailed S 3/4 W 21.6 miles, a second bearing of WNW 1/4 W is taken. What was the distance from the starting position to Arcona?

Example Ib. What is the direction by compass and the distance to Arcona after having sailed SSE 12.6 miles after the second bearing is taken?

In order to understand what is happening, it is necessary to discuss both the compass and *rhumbs*. As shown in the above illustration, the compass has 32 major points, the old name for these being rhumbs. The term rhumb here means a compass point equivalent to 11.25° or 1/32 of a circle. Thus, eight rhumbs is equivalent to 90°. The term rhumb will be abbreviated by a simple "r" in the examples. Note that the compass has three divisions between each rhumb, making a total of 128 equally spaced quarter rhumbs for the full circle in the rhumb system of angular measurement.

Each sighting angle, in terms of the standard compass readings, would have been converted to rhumbs and quarter rhumbs for making calculations.

Jerrmann has three pages for the exact construction and calculation by means of the Gunterscale. It is not possible to add the whole translation here, nor to explain all three problems. But, after having explained the graphical solution, Jerrmann also describes how to find the two distances by mathematical procedures.

The sketch that follows illustrates problem Ia.

---

[1] Donn, a British mathematician, only added one scale — "Guns Diamr" — to find out the caliber (pounds) of a gun with a determined internal diameter, a scale that was no longer up to date at that time!

Sketch of Problem Ia:

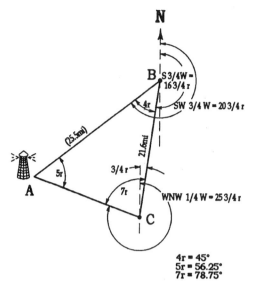

Employing trigonometry, note that the distances are to be found as a proportion:

$$\sin(5\ r) : \sin(7\ r) = BC : AB$$

or,

$$\sin(56°\ 15') : \sin(78°\ 45') = BC : AB$$

Solving:

$$
\begin{aligned}
\log \sin(78°\ 45') &= 9.991574 - 10 \\
\log BC &= \log(21.6) = 1.3345345 \\
\log \sin(56°\ 15') &= 9.919846 - 10 \\
\log AB &= \log BC + \log \sin(78°\ 45') \\
& \quad - \log \sin(56°\ 15') \\
&= 1.4062 \\
AB &= 25.5 \text{ miles}
\end{aligned}
$$

For the second part of the problem:

$$\sin(5\ r) : \sin(4\ r) = BC : AC$$

Or,

$$\sin(56°\ 15') : \sin(45°) = BC : AC$$

Solving:

$$
\begin{aligned}
\log \sin(45°) &= 9.849485 - 10 \\
\log BC &= \log 21.6 = 1.33445 \\
\log \sin(56°\ 15') &= 9.919846 - 10 \\
\log AC &= \log BC + \log \sin(45°) \\
& \quad - \log \sin(56°\ 15') \\
&= 1.26482 \\
AC &= 18.4 \text{ miles}
\end{aligned}
$$

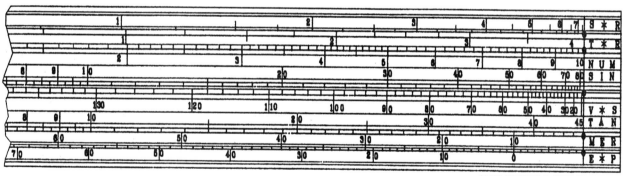

DRAWN BY BRUCE E. BABCOCK, 1993

These turn out to be very simple to work out on the Gunter scale.

In order to solve the first part of the problem on the portion of the Gunter rule shown above, go to the scale *S*R*. It is the sine in rhumbs: you take by divider the distance from 5 to 7 rhumbs on this scale and then with one divider point placed on 21.6 on the *NUM* scale, place the other point of the divider to the right and find the result 25.5 = AB. For the second part of the problem, take the distance from 5 to 4 rhumbs on *S*R* and then from 21.6 on NUM to the left yields AS = 18.4.

The following examples are not solved, but are included to give you an idea of the types of problems that were of interest.

Example II: There is a distance from a traffic beacon to a lighthouse NE 11.4 miles. You can take a bearing on the lighthouse WSW and find the angle between lighthouse and traffic beacon to be 130°50'. What was the ship's distance from both objects? This took one page of instructions.

Example III: Draw the orthographic projection of the globe for the meridian of Greenwich with the radius of 1.3'. This required two pages of instructions.

I hope that this will help the reader understand the use of the rule.

William Cox on the Sector

Philip Stanley

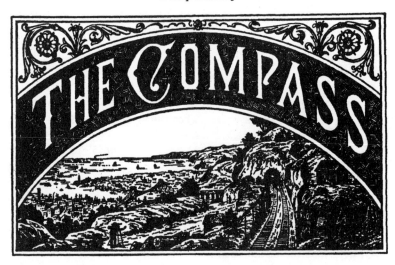

The exact relationship between William Cox and Keuffel & Esser is not clear. He was not, apparently, an employee, but instead an independent consultant whose services were used extensively but not exclusively by the company.

Cox was originally English, but why and when he emigrated to the United States is not known. It has been speculated that he originally worked for W.F. Stanley, the English instrument makers who made the paper scales for the Thacher slide rule, and that he came over to this country some time after 1881 to show K&E how properly to fasten these scales to that instrument. This has not been proven, however, and must remain a subject for future enquiry.

His expertise seems to have been in the area of slide rules and computation, and he wrote about these subjects extensively. Beginning in 1890 he published a series of articles in *Engineering News* popularizing slide rules. He was the inventor of the duplex slide rule, for which he received a U.S. patent on October 6, 1891, and which he licensed to K&E for manufacture. He was the author of the K&E instruction book for many slide rules. He was the editor (and probably primary writer) of the short-lived K&E newsletter, *The Compass*.

The Compass was a small journal (16 pp. per issue) published monthly by K&E from 1891 to 1894. It was published on the company premises in New York City. Its contents were a mixture of product descriptions and announcements, articles on subjects of interest to the engineering, surveying, and drafting professions, extracts from foreign journals, letters from readers and replies thereto, and technical book reviews.

The Compass was discontinued after July, 1894, the company stating that in the future the catalog would be used to describe instruments and products and to provide useful information.

Articles were not necessarily limited to a single issue, but if lengthy would be serialized in successive issues. This article on the sector is one of those; it appeared in four parts between February and June in 1893. Other long, multipart articles dealt with "Light and Its Reflection and Refraction", "Verniers, Adjustments of Surveying Instruments", "The Solar Transit", etc. The Cox article on the sector has been assembled from such a set of parts. Some smoothing has been done to the text at the junctures.

The sector was one of the most important calculating devices of the seventeenth and eighteenth centuries. Invented almost simultaneously by Hood (1) and Galileo (2), its operation was based on the similarity of isoceles triangles whose vertex angles are equal. While not particularly accurate, its operation was so simple that it rapidly became widely used and remained so until well into the nineteenth century.

In the form developed by Edmund Gunter (3) (called the "English" sector by Bion (4); see the accompanying figure) it was not only sold separately, but was a standard inclusion in the pocket case of drawing instruments remaining popular until almost 1900.

The sector sold by K&E – described in this article – is a simplified version with some scales omitted, and apparently intended as a teaching tool.

References

(1) Hood, Thomas. *The making and use of the Geometrical Instrument, called a Sector. Whereby many necessarie Geometricall Conclusions concerning the proportionall description, and division of lines, and figures, the drawing of a plot of ground ... may be mechanically performed with great expedition, ease, and delight ...*, London, John Windet, 1598. Reprinted, 1973, by Da Capo Press.

(2) Galilei, Galileo. *Operations of the Geometric and Military Compass*, 1606. Translated by Stillman Drake. Washington DC:

Smithsonian Institute Press, 1978.

(3) Gunter, Edmund. *Use of the Sector, Crosse-Staffe, and Other Instruments,* London: William Jones, 1624. (Reprinted, 1971, by Da Capo Press.)

(4) Bion, Nicholas. *The Construction and Principal Uses of Mathematical Instruments. To which are added The Construction and Uses of Such Instruments as are omitted by M. Bion, particularly of those Invented or Improved by the English.* Translated by Edmund Stone. The second edition. *To which he has added a Supplement Containing a Further Account of Some of the Most Useful Mathematical Instruments as Now Improved.* London: J. Richardson, 1758. (Reprinted, 1995, by the Astragal Press.)

The Sector

William Cox

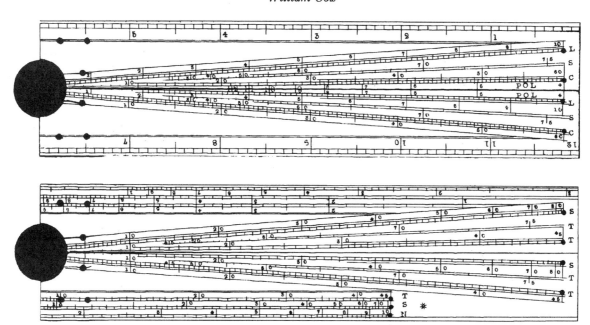

BEFORE PROCEEDING to describe the Sector it will not be inappropriate to say a few words about its inventor.

Edmund Gunter, whose name is so well known to the Surveyor, was born in Hertfordshire in 1581 of Welsh parents. His early education was acquired at the ancient Westminster school, as a Queen's foundation scholar, from which he was elected to an exhibition at Oxford, where he went in 1599 as a student of Christ Church. He graduated in due course bachelor and master of arts, then took holy orders and began to preach in 1614, and at the close of 1615 took the degree of bachelor of divinity. In his youth he had shown a great taste for mathematics, and at this time he allowed his early inclinations, which he had continued to foster, to influence his future course of life. He was appointed in 1619 to be Professor of Astronomy in Gresham College, which post he held to his death, which occurred December 10, 1626.

To Gunter's genius are due many inventions which have had an important influence upon the progress of mathematical science, amongst them being the well known Gunter's Chain, consisting of 100 links, the chief merit of which lies in the fact that 10 square chains make an acre; Gunter's Scale, which was the first attempt to make use of graphic logarithmic and other scales for the purpose of computation, and which was the origin of the now well known Slide Rule; the Sector, invented about the year 1606, which is a combination of nearly every scale required for geometrical and trigonometrical computations. It is generally believed that Gunter was also the discoverer of the variations of the declination of the magnetic needle. He it also was who introduced the terms cosine, cotangent, etc., to designate the sine, the tangent, etc., of the complement of an angle.

The Sector represented above, by means of which a great number of problems in arithmetic, geometry, trigonometry and navigation may be solved, consists generally of a folding boxwood or ivory foot rule, both sides of which are covered with various scales or lines, some being parallel with the edges of the rule, while the others radiate from the centre of the joint.

The *parallel* scales are single lines, that is, a line or scale on one leg is complete by itself, while the *Sectoral* or *radiating* scales are *double lines,* being in pairs, one line being on one leg and a similarly graduated line on the other leg, the two being used together for the purposes of computation.

The following are the different scales usually found on the Sector:

SECTORAL LINES, (RADIAL.)

On one side,

1. Lines of lines, (equal parts) marked .... L.
2. Lines of natural chords, marked .... C.
3. Lines of natural secants, marked .... S.
4. Lines of Polygons, marked .... Pol.

On the other side.

5. Lines of natural sines, marked .... S.
6. Lines of natural tangents, 0° to 45° marked ... T.
7. Lines of natural tangents, 45° to 75° marked . T.

PARALLEL LINES. (SINGLE SCALES.)

On one side.

8. Scale of 12 inches, subdivided to tenths of an inch.
9. Scale of 1 foot, divided in 10 × 10 = 100 equal parts, called the decimal scale, on the outer edges of the rule.

On the other side.

10. Line of logarithmic numbers, marked .... N.
11. Line of logarithmic sines, marked .... S.
12. Line of logarithmic tangents, marked .... T.

The principles of the Sector depend upon the proposition which we have already had occasion to state, namely,

When two sides of a triangle are proportional to the two sides of another triangle, each to each, and their included angles are equal, then the third side of the one bears the same proportion to the third side of the other.

To understand fully, however, the application of this principle, it is necessary that we first examine the different scales or lines which are found upon the Sector, and ascertain how they are set out.

THE TRIGONOMETRICAL scales found on the Sector are, as stated above, the scales of Sines, Tangents, Secants and Chords. The method of constructing them will be readily understood from the diagram, Fig. 2.

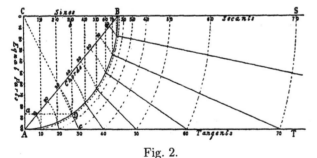

Fig. 2.

1. Let A C be the radius of any circle, and with C as centre and radius A C, describe the quadrant arc A D B, the angle at C being consequently a right angle. Divide the arc A D B into 9 equal parts, each part being then subdivided into tenths (these as well as other subdivisions are omitted in the diagram on account of the limited space). The arc A D B represents a scale of 90 *degrees.*

2. Join A B and with centre A describe a series of short arcs between the chord A B and the arc A D B, thus transferring to A B the chords of 10°, 20°, 30°, etc. These divisions on A B will now represent a scale of chords to radius A C = chord of 60°.

3. From the different divisions of the arc A D B draw perpendicular lines to C B, and number the points of intersection from 0° at C to 90° at B, when we have a scale of sines to radius A C.

4. Prolong the line C B indefinitely to S, and draw the line A T parallel to C S. From C as centre draw radial lines through the several divisions of the arc A D B until they meet the line A T, and number the points of intersection from 0° at A to 70° near T (the limits of our diagram not allowing of their being carried further); we then have on A T a scale of *tangents,* also to radius A C.

5. With centre C describe a series of short arcs between A T and B S, commencing with the division 10° on A T; and proceeding to 70°. The points of intersection of these arcs with B S should be numbered from 0° at B to 70° near S, when we have a scale of *secants* to radius A C.

6. To complete the scales, divide the radius A C into 10 equal parts, each one of these being then further subdivided into 10 other *equal parts,* making 100 in all.

Such are the various trigonometrical scales found on the Sector. A glance at the diagram will make them clear. In the triangle C a D, let C D = radius = 100 by the scale of Equal Parts C A, and the angle D C a = 30°; then D a = C b = sine 30°. Now take a pair of Compasses and with one leg on centre C, take in C b, then with the same centre, transfer this distance to the scale of Equal Parts, when the other leg will fall upon the division answering to 50, which is the Sine of 30° to radius 100. The Cosine of 30° to the same radius is D b = C a = 86.6, while the Tangent is A c, and the Secant C c, the lengths of which may also be taken off the scale of Equal Parts.

These scales, as shown in the diagram, may be applied to several uses, but as on the Sector they are all double ones, their usefulness is considerably increased, the measurements not being confined to the fixed scale of equal parts.

The solution of the various computations to which the Sector is applicable, is effected by means of a pair of ordinary dividers, although the most suitable for the purpose are those known as Hairspring Dividers, as the points may be set to take in a given measurement with greater ease and precision.

Fig. 3 will make the principles of the use of the Sector clear. We will suppose that O is the centre of the joint, and that O A and O B are the sectoral lines of lines, divided into 100 equal parts. We will also suppose that the legs of the instrument have been opened out until the distance from A to B is equal to 50 parts of the line of lines. Now from the properties of similar triangles

$$OA : AB :: Oc : cd.$$

Let d be at the 60th division of the line of lines. then we have

$$100 : 50 :: 60 : 30 = cd$$

also

$$50 : 100 : 30 : 60 = Od$$

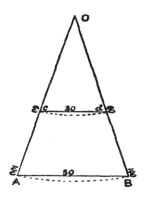

Fig. 3.

In the above figure, the distances or measurements O d and O B are called *lateral* (or side) distances, while the distances A B and c d are called *transverse* (or cross) distances. Measurements are taken in the former case from the centre of the joint, and in the case of transverse distances from that line of the scales which is nearest to the inner edge of each leg, as this is in the case of every scale, the *radial* line proceeding from the centre of the instrument.

We will now give a few examples of the methods of using the Sector.

*1. To find the fourth term of a proportional,
as $8 : 14 :: 36 : x$.*

With the dividers take in 8 parts on one of the lateral scales or lines of lines; then set one of the points of the dividers upon 4 of the same lateral scale, and open out the two legs of the Sector until the other point of the dividers reaches to 14 of the other lateral scale. Now open the dividers to take in 36 parts from one of the lateral scales, and make this a transverse distance from one lateral scale to the other, when the points of the dividers will be found to reach from 63 of one scale of lines to 63 of the other scale of lines; this is therefore the fourth term sought.

If the first term of the proportion is greater than the second, it will be better to make the first term a lateral distance, that is to measure it from the centre upon one of the lateral scales, and make the second or lesser term a transverse distance, that is, measure it across from one lateral scale to the other. The third term will also be a lateral distance, while the fourth term will be a transverse distance.

*2. To divide a given straight line into
a number of equal parts, as 7.*

Take between the points of the dividers the length of the given line, and open out the Sector legs until this distance becomes a transverse distance from 70 to 70. The transverse distances 10 to 10, 20 to 20, etc., from one lateral scale to the other, will then be the divisions required.

3. To measure an angle B A C, Fig. 4.

This may be done in several different ways, each of which we shall describe.

First: With the lines of chords. Take the dividers, and with A as convenient distance A D and A E, then open out the Sector until this same distance becomes a transverse distance from 60 to 60 on the lines of chords. Now take in the dividers the chord of the angle from point to point just laid off on the sides, *i.e.* D E, and apply this distance transversally to the lines of chords on the Sector, when the reading on each leg will be the number of degrees in the angle sought for.

Second: With the lines of sines. Lay off a convenient distance on A B from A, say A D, and from D draw D F perpendicular to and intersecting A C in F. Now with the dividers make A D a transverse distance from 90 of one line of sines to 90 of the other line of sines, then take in the dividers the perpendicular DF and apply it transversally to the lines of sines on the Sector, when the coinciding points on each leg will be the number of degrees in the angle B A C.

Third: With the lines of tangents. Lay off a convenient distance on A C from A, say A F, and erect upon A C the perpendicular D F. Now with the dividers make A F a transverse distance between 45 of each line of tangents on the sector, then take in the dividers the length and apply it transversally to the lines of tangents until similar points on each line are found which take in between them exactly this distance D F; the readings of these points will be the number of degrees in the angle BA C.

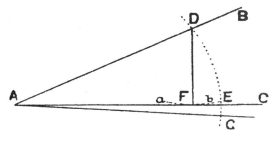

Fig. 4.

In these examples it must be distinctly borne in mind that a transverse distance means a distance from a given number on one leg to the same number on the other leg.

All transverse distances will therefore be lines parallel to D E.

From a slight examination of the sector it will be seen that for angles greater than about 70°, some indirect method of measuring an angle must be followed. In this, as in other cases, where the simple measurement of the angle is required, irrespective of any of its trigonometrical functions, the lines of chords will be found to be the most suitable to use, as the differences between successive degrees are much more uniform than is the case with either the sines or the tangents of the angles. If, therefore, we wish to measure an angle of what we should suppose to be about 150°, we should consider the simplest method would be with the lines of chords, to continue one side of the angle beyond the point of intersection, and then to measure the small angle thus laid down and deduct it from 180°, which would then give the value of the larger angle. If we wished to measure an angle of say about 100°, we should first strike a convenient arc intersecting the sides of the angle and measure off 60° with the same diameter upon this arc; then measure in the manner described the remaining angle, which, added to 60° will give the total angle.

To measure very small angles, say less than 5° or 10°, proceed as above, and deduct the value of the larger from 60°, when the remainder will be the number of degrees in the smaller angle.

Fourth: To protract an Angle. Let it be required to protract an angle of 25° upon A C. Take in the dividers any convenient length, say A E, and describe with A as centre the arc D G, then make this same length a transverse distance between 60 and 60 of the lines of chords. Now take in the dividers the transverse distance between 25 and 25, and with E as centre describe a small arc cutting D G in G, and join A D, then the angle D A E will be the required angle of 25°. By similar methods, and noting the directions already given for measuring angles, any required angle may also be protracted by means of the scales of sines and tangents. In those cases where it is required to know the sine or the tangent of an angle, the lines of these functions will be used instead of the lines of chords; thus, let A D to a convenient scale = 150, and let it be required to obtain the sine of an angle at A of 25° to radius 150. Make A D a transverse distance between 90 and 90 of the lines of sines, then take in the dividers the transverse distance 25, and with this as radius and D as centre, describe the small arc a b, and draw A F, and D F perpendicular to A F; then will the angle D A F = 25°, and D F will be the sine of the same to radius A D, while A F will be the sine of A D F = 65° to the same radius. On measuring D F, it will be found = 63 +, while sine(25°) × 150 = 63.4.

Fifth: To construct a regular polygon. This is of course the same as dividing the circumference of a circle into a given number of equal parts. As the radius of a circle and the side of an inscribed regular hexagon are equal, the legs of the Sector must be opened out until the radius of the circle becomes a transverse distance between 6 and 6 of the lines of polygons; then take in the dividers the transverse distance corresponding to the number of sides the required polygon is to have, and with this opening prick off on the circumference of the circle the given number of equal parts; join these points by straight lines, and the polygon is complete.

If it should be required to construct, say a heptagon, upon a given straight line, first take in the dividers the length of the line and make it a transverse distance between 7 and 7 of the lines of polygons, then take in the dividers the transverse distance 6 to 6, and with this as radius describe a circle whose circumference shall touch each end of the given straight line, then with the dividers transfer the length of the straight line round the circumference when it will be found to have been divided into 7 equal parts.

Regular polygons may also be constructed by means of the lines of chords, by dividing 360° by the required number of sides, then finding the chord of the centre angle to a given radius, and pricking off this chord round the circumference of a circle, described with the given radius.

As MAY HAVE been gathered, a variety of problems can be solved by means of the Sector. With it, a pair of dividers and a finely divided or a diagonal scale, the surveyor or engineer, who does not wish to be burdened in the field or on a journey with a number of instruments, may make his sketches, protract his angles and make his geometrical or trigonometrical computations with ease and a fair degree of accuracy. For purposes of calculation, however, the Sector must undoubtedly give place to the Slide Rule. We shall now close the subject with a few practical applications.

To make a vernier. On a given straight line lay off say an inch and divide it into 10 equal parts, as described page above, then take in the dividers 9 of such parts and open out the Sector legs until this distance becomes a transverse distance from 10 to 10 on the lines of lines. The transverse distances 1 to 1, 2 to 2, etc., from one lateral scale to the other will then be divisions of the vernier, the lowest count being $\frac{1}{100}$th of an inch.

To use the Sector as a scale of inches and chains or inches and feet, etc. Let it be required, for instance, to take off measurements from a drawing made to a scale of 1 inch = 3 chains. Take in between the points of the dividers one inch, and open out the Sector legs until it becomes a transverse distance between 3 and 3 on the lines of lines. The Sector is now "set" to serve as a scale of 3 chains to the inch, and all transverse distances, read on the lateral scales of lines, from a main division on one leg to the corresponding main division on the other leg represent chains, while the transverse distances from and to subdivisions represent chains and links.

In the same way scales of inches and feet, inches and miles, inches and metres, etc., may be set, and corresponding scales laid down on the drawing if desired.

Required the height of A B, Fig. 5, its base being inaccessible. Let CD = 60 feet, A C B = 60° and A D B = 35°, then we have C A D = A C B - A D B = 60° - 35° = 25°.

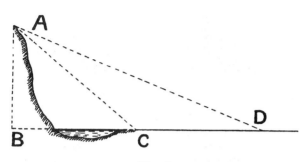

Fig. 5

Now we have, according to the rules of Trigonometry,

(1). Sine C A D : C D :: Sine A D C : A C and
(2). Sine A B C : A C :: Sine A C B : A B.

With the Sector the operations are:

First. Take in the dividers from the line of lines or any other convenient scale 60 parts = C D, and open out the legs of the Sector so as to make them a transverse distance between 25 on one line of sines to 25 on the other line of sines = Angle C A D; then without disturbing the Sector, take in the dividers the transverse distance 35 to 35 of the lines of sines = angle A D C, and apply this to the same scale of equal parts, when $81\frac{1}{2}$ will be read off = A C, now

Second, with the dividers as just set = A C, make their opening coincide with the points 90 and 90 of the lines of sines = angle A B C, by opening out the legs of the Sector the necessary quantity; now take in the dividers the transverse distance 60 to 60 of the lines of sines = A C B. and apply it to the scale of equal parts, when $70\frac{1}{2}$ will be read off, which is the height of A B.

To determine the distance A B, the point A being inaccessible.

Measure off from B, at right angles to A B, a distance of say 100 feet = B C, then having measured the angle A C B = say 40°, we have C A B = 50°, whence

Sine C A B : C B :: Sine A C B : A B.

Take in the dividers as before, from a convenient scale, 100 parts and make them a transverse distance between 50 and 50 of the lines of sines. Now without disturbing the Sector, take in the dividers the transverse distance between 40 and 40 of the lines of sines, when upon applying this to the same scale of equal parts, we shall read off 84 feet as the distance A B.

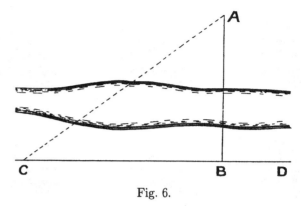

Fig. 6.

The reason of these methods of working, with their results, will be clear when we remember that a given lateral distance is to its corresponding transverse distance, as any other lateral distance is to its corresponding transverse distance; and also that in any triangle the sine of the angle opposite a given side : the given side :: the sine of the angle opposite a second side : the second side :: the sine of the angle opposite the third side : the third side; so that all the sides of the triangle become transverse distances and the sines of the angles become lateral distances. It is immaterial whether we take the transverse distance from the lines of lines, or from any other and more convenient scale: the result will always be the same.

Before closing it is necessary that we say a few words about the lines of logarithmic numbers, sines and tangents (scales 10, 11 and 12). If the Sector be opened out to its full extent, it will be seen that these scales correspond identically with those of the slide rule: they may be used in the same way by taking off the distances with the dividers or on a slip of paper. The chief points to be borne in mind are that multiplication is effected by adding the distance represented by one factor to the distance represented by the other factor, when the total resulting distance represents the product, and that in division the distance corresponding to the divisor is subtracted from the distance corresponding to the dividend, the quotient being represented by the remaining distance. Multiplication is, therefore, always performed forwards to the right, and division backwards to the left.

DIALLING (Figuring Time)
William J. Baader

Dialling, the art of drawing dials on the surface of any given body or plane. The Greeks and the Latins call this art gnomonica by reason it distinguishes the hours by the shadow of the gnomon. Some call it photo-sciatherion because the hours are sometimes shown by the light of the sun. Lastly others call it horologiography.

Dialling is a most necessary art; for notwithstanding we are provided with moving machines, such as clocks and watches to show time; yet these are apt to be out of order, go wrong and stop. Consequently they stand frequently in need of regulation by some invariable instrument as a dial; which being rightly constructed and duly placed will always, by means of the sun, inform us of the true solar time; which time being corrected by the equatron table published annually in the ephemerides, almanacks, and other books, will be the mean time to which the clocks and watches are to be set.

The antiquity of dials is beyond doubt. Some attribute their invention to Anaximenes Milesius and others to Thales. The discus of Aristarchus was a horizontal dial with its limb raised up all around to prevent the shadow stretching too far. The first sun dial at Rome was set up by Papirius Cursor about the year of the city 460 before which time says Pliny there is no mention of any account of time but by the suns rising and setting.

The Dialling Sector contrived by Mr. Benjamin Martin is an instrument by which dials are drawn in a more easy, expeditious, and accurate manner. The principal lines on it are the line of latitude and the line of hours. They are found on most of the common scales and sectors; but in a manner that greatly confines and diminishes their use. First they are a fixed length, and secondly too small for any degree of accuracy. But in this new Sector, the line of latitudes is laid down as it is called, sector-wise viz. one line of latitudes upon each leg of the sector, beginning in the center of the joint and diverging to the end (as upon other sectors) where the extremes of the two lines at 90° and 90° are nearly one inch apart, and their length 11½ inches: which length admits of great exactness; for the 70th degree of latitude, the divisions are to quarters of a degree or 15 minutes. This accuracy of the divisions admits of a peculiar advantage, namely that it may be equally communicated to any length from 1 to 23 inches, by taking the parallel distance (*see figure 2*), viz. from 10 to 10, 20 to 20, 30 to 30 and so on as is done in like cases on the lines of sines, tangents, etc. Hence its universal use for drawing dials of any proposed size. The line of hours for this end is adapted and placed contiguous to it on the sector, and of a size large enough for the very

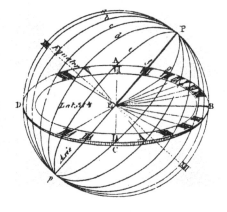

minutes to be distinct on the part where they are smallest, which is on each side of the hour of III.

From the construction of the line of hours before shown the divisions on each side of the hour III are the same to each end, so that the hour line properly is only a double line of three hours. Hence a line of 3 hours answers all the purposes of a line of 6, by taking the double extent of 3, which is the reason why upon the sector the line of hours extends only to 4½.

To make use of the line of latitude and line of hours on the sector: as single scales only, they will be found more accurate than those placed on the common scales and sectors, in which the hours are usually subdivided, but into 5 minutes, and the line of latitudes into whole degrees. But it is shown above how much more accurately these lines are divided on the dialling sector. As an example of great exactness with which horizontal and other dials may be drawn by it on account of the new sectoral disposition of these scales, and how all the advantages of their great length are preserved in any lesser length of the VI o'clock line CE and AF, *figure 30*: Apply either of the distances of CE or AF

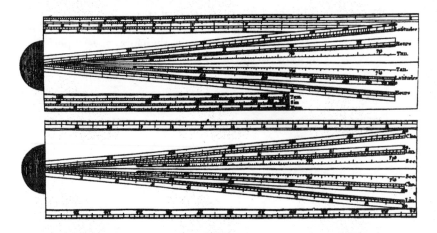

From the *Fine Tool Journal*, Vol. 37, No. 2 (November/December, 1988), by permission

to the line of latitude at the given latitude of London, suppose 51 degrees 32 minutes on one line to 51 degrees 32 minutes on the other, in the manner shown in figure 5, and then taking all the hours, quarters, etc. from the scale by similar parallel extents, you apply them upon the lines ED and FB as before described.

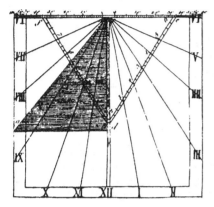

As the hour lines on the sector extend to but 4½, the double distance of the hour 3, when used either singly or sectorally must be taken, to be first applied from 51 degrees 32 minutes on the latitudes, to its contact on the XII o'clock line, before the several hours are laid off. The method of drawing a vertical north or south dial is perfectly the same as for the above horizontal one; only reversing the hours as in *Figure 1* and making the angle of the stile's height equal to the complement of the latitude 38 degrees 28 minutes.

The method of drawing a verticle declining dial by the sector, is almost evident from what has been already said in dialling. But more fully to comprehend the matter, it must be considered there will be a variation of particulars as follows: 1. of the substile or line over which the stile is to be placed; 2. the height of the stile above the plane; 3. the difference between the meridian of the place and that of the plane, or their difference of longitude. From the given latitude of the place, and declination of the plane, you calculate the three requisites just mentioned, as in the following example. Let it be required to make an erect south dial declining from the meridians westward 28 degrees 43 minutes in the latitude of London 51 degrees 32 minutes. The first thing to be found is the distance of the substilar line GB figure 31 from the meridian of the plane G XII. The analogy from this is: As radius is to the sine of the declination so is the co-tangent of the latitude to the tangent of the distance sought, viz. As radius: 28 degrees 43 minutes: : tangent 38 degrees 28 minutes: tangent 20 degrees 55 minutes. This and the following analogy may be as accurately worked on the Gunter's line of sines, tangents et. properly placed on the sector, as by the common way from logarithms. Next, to find the planes difference of longitude. As the sine of the latitude is to radius, so is the tangent of the declination to the tangent of the difference of longitude viz. As s 51 degrees 32 minutes; radius: : tang. 28 degrees 43 minutes : 35 degrees 0 minutes. Lastly, to find the height of the stile: as radius is to the co-sine of the latitude, so is the co-sine of the declination to the sine of the stile's height, viz. Radius: s 38 degrees 27 minutes ; ; s 61 degrees 17 minutes : 53 degrees 5 minutes.

The three requisites thus obtained, the dial is drawn in the following manner: Upon the meridian line G XII, with any radius G C described the arch of a circle, upon which set off 20 degrees 55 minutes from C to B, and draw G B, which will be the sub stilar line, over which the stile of the dial must be placed.

At right angles to this line G B, draw A Q indefinitely through the point G; then from the scale of latitudes take the height of the stile 35 degrees 5 minutes, and set it each way from G to A and Q. Lastly, take the double length of 3 on the hour line in your compasses, and setting one foot in A or Q, with the other foot mark the line G B in D, and joint ADQD, and then the triangle ADQ is completed upon the substile G B.

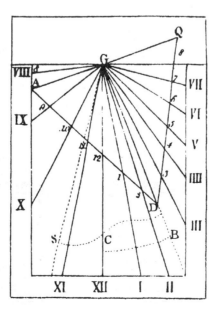

To lay off the hours, the plane's difference of longitude being 35 degrees, equal to 2 hours 20 minutes in time, allowing 15 degrees to an hour, so that there will be 2 hours 20 minutes between the point D and the meridian G XII, in the line A.D. Therefore, take the first 20 minutes of the hourscale in your compasses, and set off from D to 2; then take 1 hour 20 minutes, and set off from D to 1; 2 hour 20 minutes, and set off from D to 12; 3 hour 20, from D to 11; 4 hour 20 minutes from D to 10; and 5 hour 20 minutes from D to 9, which will be 40 minutes from A.

Then, on the other side of the substilar line G B, you take 40 minutes from the beginning of the scale, and set off from D to 3; then take 1 hour 40 minutes, and set off from D to 4; also 2 hour 40 minutes, and set off from D to 5; and so on to 8, which will be 20 minutes from Q. Then from G the centre, through the several points 2, 1, 12, 11, 10, 9, on one side, and 3, 4, 5, 6, 7, 8, on the other, you draw the hour lines, as in the figure they appear. The hour of VIII need only be drawn for the morning; for the sun goes off from this west decliner 20 minutes before VIII in the evening. — The quarters, etc., are all set off in the same manner from the hour scale as the above hours were.

The next thing is fixing the stile or gnomon, which is always placed in the substilar line G B, and which is already drawn, The stile above the plane has been found to be 33 degrees 5 minutes; therefore with any radius G B describe an obscure arch, upon which set off 33 degrees 5 minutes from B to S, and draw G S, and the angle S G B will be the true height of the gnomon above the substile G B. ■

Encyclopaedia Britannica, 4th Edition. Edinburgh: 1810.

Ashton Blaine French
Scaler
Log Rule Maker

Dale Butterworth
Clarence Blanchard

Ashton Blaine French and his new bride, Violet Lovejoy, of Milo, Maine, spent the winter of 1920 in a small cabin in the deep woods just outside the small Northern Maine town of Monticello. French had built the small cabin the previous fall to give him and his bride some privacy from the main Great Northern Paper Company (GNP) camp where he and his brother were stationed. That was no doubt a long, hard winter as both French and his wife were far from their hometowns and deep in the Maine wilderness.

French had joined GNP in 1918 just before being drafted into the U.S. Army. His army career was spent at Fort Devens, Massachusetts, where he was quarantined by the influenza epidemic. The war ended and French was mustered out without him ever getting beyond the fort.

In 1919, French returned to GNP, where he trained as a timber cruiser and scaler. He worked in numerous camps throughout the Maine woods while steadily advancing his career. In 1930, French was promoted to an office position in GNP's Bangor office.

In the early 1930s, French purchased and moved into a home on Summer Street in the Hampden Highlands. This location allowed for a short commute to his Bangor office. At about that same time, French began to make log rules for the lumber companies around Maine.

From the *Fine Tool Journal*, Vol. 51, No. 1 (Summer, 2001), by permission

French's most popular product was the 4-foot square rule. These relatively common rules can be found in several styles with different scales. He also made special-order tree calipers and at least one production run of log calipers with wheels. French rules are found with several scales, but most are stamped with the Holland or Maine Log Rule scale. All of his work is of top quality and his products were made with the finest craftsmanship. The French 4-foot scales are well known and were clearly widely accepted throughout the Northern Maine woods.

Ashton Blaine French was born in Levant, Maine, on May 16, 1896. He died at the age of 63 on June 15, 1959.

French Rules and Sticks

French rules were made for both the lumber and the pulpwood industries. The lumber rules measured the diameter of the log at the small end and were read in board feet. Pulpwood scales were also used on the small end of the log and were read in units of a cord—100 units made a full cord equal to 128 cubic feet in volume. Most rules observed to date are square in section and measure from about 7/8 to 1-1/4 inches per side. Flat rules do exist but are much less common. Lengths range from 2 to 6 feet, with the 4-foot scale being the most common.

Numerous lumber scales existed but by the 1930s, when French was making rules, the most common scales in Maine were the Holland, Maine, and Bangor scales. In fact, by the 1930s, these three scales were just different names for the same set of tables. Both the loggers and the yards had accepted these tables as the standard for measuring the board-foot yield in a given log.

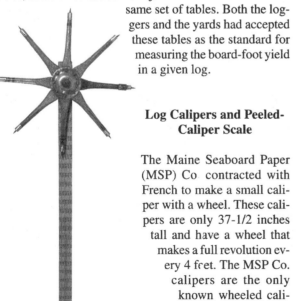

Log Calipers and Peeled-Caliper Scale

The Maine Seaboard Paper (MSP) Co contracted with French to make a small caliper with a wheel. These calipers are only 37-1/2 inches tall and have a wheel that makes a full revolution every 4 feet. The MSP Co. calipers are the only known wheeled calipers made by French.

The wheeled caliper is unique and it

THE SCALER

The scaler is a most important factor in the pulpwood industry. His scale on the landing, is the final word.

Years ago, ninety-seven per cent of all scalers hailed from Old Town; in deed it was as natural for Old Town to turn out scalers as for game wardens to come from Dover. The other three came from Milford. Of late, however, every section of Maine has been represented in the profession and even Bangor has come to the fore.

The scaler may be a company man or he may be a stumpage man, looking out for the interests of the land owner. In either case he is sure to be an expert in the art of gathering spruce gum and a cribbage player of the deepest dye. The tonnage of the spruce gum that has been sent down river to the wives and sweethearts of scalers reveals a staggering total; the number of times that the mystic words, " Fifteen-two, fifteen-four," have been uttered will never be ascertained.

There has long been a rumor about that scalers have a marked aversion to work of any kind. This is not always true. The scaler usually shares the office with the timekeeper. Here, it must be admitted, he has a curious habit of regarding himself as the star boarder and will cheerfully watch the mercury dropping to forty below before he will manipulate the other end of the crosscut saw. There are exceptions. Then there are the fellows who require an unearthly amount of sleep. It was of them that the old chap remarked that he would like to found a Home for Tired Scalers, preferably on the Island of Guam.

But these are matters of minor importance. The scaler's greater usefulness is naturally out of doors. There, armed with his staff of office, the scale-rule or guessing-stick, and holding a lumber crayon firmly in one hand, he traverses his allotted area. The jobber beams at his approach, fondly hoping for favors. The scaler throws his rule dispassionately along the pile, records the results in his little book and calmly then goes his way. The cordcutter hopes he will not discover that two of his piles are built over stumps. Foolish man! The scaler not only discovers it at once but finds that much of his wood is poorly knotted and as a parting favor discounts him five per cent for rot. This is the reason why an observant eye and an alert mind are most essential parts of a scaler's make-up. The variety of ocular illustrations that woodsmen can invent in the hope of increasing their scale illicitly is very nearly endless; it is a chapter of itself.

The scaler, therefore, protecting the interest of his company, has an important role. The fact that now and again a man may kick on his scale is point in his favor. For it is an old adage that if the crew kicks on a scaler, he is a good scaler.

From *The Northern.* April 1925. Company news letter for Great Northern Paper Co.

is believed that only one production run was ever produced. In fact, French once told his son that these calipers were extremely time-consuming to make and impossible to make a profit on. Therefore, it is reasonable to assume that once the contract was filled, French did not continue to make the wheeled caliper. The calipers are extremely rare, which further supports this conclusion.

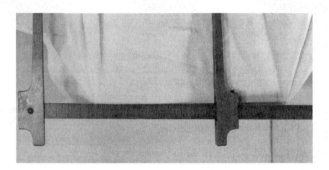

All of the known log calipers with wheels are marked with the French handwritten ink stamp and MSP Co. The aluminum plates that reinforce the fixed jaw of the caliper are stamped with a number. Four calipers are known, with numbers 1, 6, 7, and 10. Therefore, the batch of calipers made for the MSP Co. could have been either 10 or 12\; certainly, not many more exist.

The MSP Co. was a regular customer and also ordered a special peeled-caliper scale rule. These rules are often equipped with a 10-inch steel hook, which was used for scaling logs in the water. The length of the hook aided the scaler when measuring partly submerged logs.

The MSP Co. was a small paper mill located in Bucksport, Maine. Its demand for scale rules was limited to its own mill use and forest operations. Therefore, all marked MSP Co. scales and calipers are considered rare.

Tree Calipers

The French tree calipers were made only on special order and are quite rare. As with the wheeled caliper, each tree caliper had a serial number. To date, the highest number known is 18.

The tree caliper was used by the cruiser to estimate the yield of a plot of land. First, a section of a quarter-acre was laid out. The caliper rule was then used to measure the diameter of standing trees at about 4 feet above the ground. This measurement was known as the diameter at breast height (DBH). Using a hypsometer, the height of each tree was also measured. The two measurements were tallied by species for all the marketable trees in the quarter-acre. Back in camp, various tables were used to determine the amount of lumber that could be harvested from the plot. Once the yield per quarter-acre was known, it was a simple matter to estimate the yield for any given size plot.

The tree-caliper jaws were fastened to the beam with wing nuts, and the entire unit could be disassembled for backpacking or shipping. All parts of the caliper were numbered to avoid mixing them up during reassembly.

Making Scales

Several years ago, an interview with Gerald French, A.B. French's son, gave us a firsthand account of the steps involved in making rules. Gerald lived at home during some of his father's most productive years and had clear memories of the business.

A.B. French did not apprentice with a rule maker but rather learned the trade on his own by doing it. He started business about the time another Maine rule maker, V. Fabian, stopped making rules. Certainly, French knew of Fabian and may have even visited his shop. Fabian's shop was in Milo, Maine, and French's wife was from the same town.

Although no documented connection between French and Fabian has been established, it is clear that when French was contracted to make the MSP Co. caliper, he used a Fabian caliper as a model. His calipers are nearly exact design copies of the Fabian. The interesting difference is that the French caliper is a much stronger and better-built caliper. Did French know only too well from firsthand experience that the Fabians did not hold up in the Maine woods?

French's earliest rules are all hand-stamped, one number at a time. The work was slow but clearly done with a steady hand; very seldom do you see a number that is even slightly out of line. However, stamping numbers was a time-consuming job and soon a holder for several numbers in a group was procured. The holder was loaded with the desired digits and hit with a ball-peen hammer,

imprinting the entire number with a single blow.

Finding good maple wood for making rules was a constant concern. Only rock maple with a good straight grain, uniform color, and no stain was acceptable. French and his son spent many trips searching far and wide for suitable maple for his rules.

The lines on French rules are all hand-cut. The lines going around the rule were scribed with a knife and combination square. The lines running the length of the rule were marked with gauges set to the proper widths.

Darkening the numbers was always a problem for French. He experimented with several ideas but never found a suitable method. The numbers on French rules are all colored in with a soft lead pencil, one digit at a time. This was obviously slow going, but it was the only method ever found to achieve the desired result. Between hand-stamping and hand-coloring, it is clear that making the rules was very labor-intensive. Once the numbers had been darkened, the rule was signed in ink with French's name and address. The rules were then finished with a high-quality spar varnish and hung to dry.

Nearly all French rules observed to date have one or more small holes drilled through the brass end plate. These holes are unique to French rules and identify the maker as clearly as his longhand signature or stamp. French would drive a small nail in the hole, twist a loop in the nail, and hang the rule up for drying—a nice solution that left a telltale hole.

The French calipers are of maple, brass, and aluminum construction. The fixed jaw is attached to the beam with two aluminum ell plates. These plates are nearly 1/8-inch thick and are let into the beam, resulting in a very strong joint. The movable jaw has brass plates with wooden filler at right angles to the jaw and parallel to the bottom of the beam. A hard rubber roller on the top travels along the top of the beam. Tension between the movable jaw and beam is maintained by a brass friction plate with an adjustment screw. The tips of the jaws have brass wear plates.

The caliper is laid out with Holland Rule tables for logs from 10 to 32 feet. The movable jaw has tally holes from 100 to 900 board feet by hundreds and from 1,000 to 10,000 board feet by thousands. Tally pins are stored in two holes at the top end of the movable jaw.

The wheels have eight spokes and travel 4 feet per revolution. The hubs are pressed brass and the spokes are held in position with round-head iron rivets. The spokes are made of ash and appear to have been hand-carved. Each spoke has a steel tip and a brass ferrule cut from tubing. Extra-heavy tubing was used for the weighted spoke. Every other spoke has one, two, or three painted bands to mark the feet of travel. The weighted spokes marked 4 feet of travel and have been observed in both red and black.

Overall, the wheeled calipers are very well made and appear to be sturdy. It is interesting to note that French made these calipers at a time when the practice of scaling the center of the log had been abandoned. The wheels would have been used to find the overall length of the log, not the center—as would have been the practice a few years earlier.

Just how extensive a sales program A.B. French had is not known. No catalogs have surfaced that offered his rules; however, the 4-foot sticks are quite common and appear to have been widely used in Maine. Certainly, his employment at GNP would have been a heads-up for getting his rules in there. Also, he was making rules when the demand for paper and other wood products was on the rise, but when few others were making rules. If a scaler broke a rule, he would have had to make a replacement himself or seek out Mr. French.

Snow & Neally Co. of Bangor, Maine, a long-time supplier to lumbermen, did carry 4-foot French rules. However, to date, a catalog listing by Snow & Neally for the rules has not been found. It is possible that the rules found without the French mark were sold by Snow & Neally.

Ashton Blaine French made log rules for only about 25 years. His products were first-rate and over top quality. He certainly earned his place with the top log rule makers of the 20th century.

Flat Log Rule with French Ink Stamp with Holland stamped with Individual Letters Above.

Wm. Greenleaf: Talented, Eccentric

by Bill Holden

The beautiful wood and brass calipers pictured in Figure 1 were designed by William Gardner Greenleaf. Some were made with the wheel on the end and some were made merely with the sliding arm.

In the 1500s, John Feuilleuert (translated: Greenfoliage) came to England from France. His son Edmund, born in 1575, married and he and his wife came to America.

William Gardner Greenleaf was an eighth generation son in America.

He was born in Whitefield, NH in 1834. He married in 1860 and had two sons who went on to become shoemakers. After his wife died, he moved to Littleton, NH in 1868.

He remarried in 1869. He and his second wife Florence had two sons and two daughters. The fourth child Flossie was born in 1882 (more about her later).

Greenleaf was a carriage maker by trade, working for the C. F. Harris Co. in Littleton. He likely made the calipers in his spare time.

A very creative and talented person, Greenleaf built the first lookout or weather station (a log cabin), atop Mount Washington in New Hampshire. He also built a wooden bridge and a dam on the Connecticut River in Littleton which were still in use in 1930. He was a very accurate craftsman but never used a ruler.

However, he was a very eccentric individual. He was a tall man with a long white beard and wore a knee-length cape. I am told he would not have a button put on his cape even though his daughter offered to do it, but preferred to fasten it with a safety pin.

I purchased my first Greenleaf calipers with a wheel late in the 1960s. When displayed at one of our early meetings, it was the first any of our members had seen. One member, a very advanced collector of logging tools for a logging museum, told me he had seen one only in an early illustration. Mine was marked "W. H. Greenleaf". I sold it to him for his museum.

Fortunately, I had the opportunity to buy a second one in 1970. This one was marked "F. M. Greenleaf". It also was marked "Littleton, N.H.". (Figure 2)

This different marking aroused my curiosity. In attempting to trace the history of these calipers, my efforts began on a high note. I wrote to the town clerk in Littleton asking for information about Greenleaf.

Thank God there are super dedicated and cooperative people in this country. My letter wound up at the Littleton Area Historical Society. The group's curator, Mrs. Ella Nourse Carter, went to work on the project and passed along some of the above information about Greenleaf.

However, the "F.M." puzzle remained unsolved. There was a son named Franklin who died at birth. Possibly the initial "F" was in his memory I theorized.

Iron Horse Antiques pictured one of these calipers marked "F. M. Greenleaf, 62 Oak Ave.,Belmont, Mass." I wrote to Belmont.

There Mrs. Margaret McFarlane tried to trace Greenleaf and his calipers and an "F.M." on Oak. She did, in her third letter to me, uncover the fact that his widow Florence died at 62 Oak Ave. in 1928. Still no reference to calipers or "F.M." She gets three "Es" for her efforts!

Frustrated, I gave up. If he was so eccentric, maybe he used different initials as a whim.

After a period of time, Laurent Torno informed me that he had acquired a set of calipers marked "F.M., Jacksonville, Alabama". (Greenleaf apparently was a very ubiquitous person!) Here is where I struck the mother-lode.

My inquiry to Jacksonville ended up in the hands of Mrs. Margaret Greenleaf, the widow of his grandson. Mrs. Greenleaf had access to, and was considerate enough to pass along to me, a wealth of information concerning her eccentric, or as she put it, "more like weird!," relation.

With the information I had secured, it now became possible to ascertain the approximate date of "manufacture" of each of the calipers. Because Greenleaf moved to Littleton in 1868 and, as far as I can tell, never left there prior to his death in 1916, it is safe to assume that all calipers marked "W. H. Greenleaf, Littleton, N.H." and those marked "Wm. Greenlief" in the brass were made there by him between 1868 and 1916.

Greenleaf taught daughter Flossie how to make the calipers. She began making them while living in Littleton. Finally, this explains the "F.M." marking.

An accomplished violinist, she moved to Belmont to live with her sister while playing with the Boston Symphony. She made the calipers in Belmont while living with her sister, which explains why there was no listing in the directories.

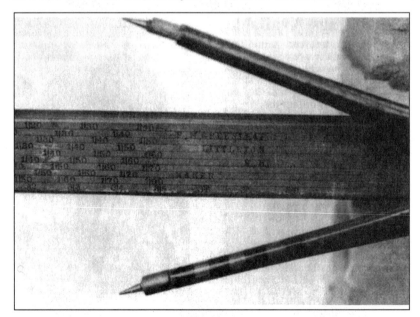

Figure 2

From *The Gristmill*, No. 28 (June, 1982), by permission

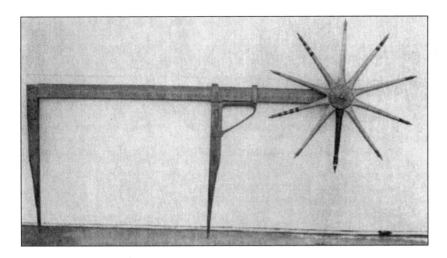

Figure 1

Assuming she was in her late teens or early twenties when she moved, all calipers marked "F.M., Littleton" must have been made between about 1897 and 1905. She lived at 62 Oak Ave. in Belmont until 1921. Thus, those calipers marked "Belmont, Mass." would date approximately 1905-1921.

She then moved to Jacksonville, AL. Because she moved from there to Denver, CO in 1926 to take care of a blind aunt, those calipers marked "F.M., Jacksonville" were made during this five-year period. I have no information of calipers marked "Denver" but if there are some in existence, they would have been made after 1926.

I have located eight different calipers which are marked "Greenleaf". There are a variety of markings even to the spelling of the name. According to Mrs. Greenleaf, he often spelled his name "Greenlief" as another example of his eccentricity. The eight examples are marked as follows:

• "Wm GREENLEAF, MAKER, LITTLETON, N.H." stamped on the wood, "Wm Greenlief" stamped in the brass.

• "Wm Greenlief" stamped in the brass. "Graduated in inches only" stamped on the wood. This one did not have a wheel.

• "Wm Greenlief" stamped in the brass. No markings on wood.

• "F.M. GREENLEAF, MAKER, LITTLETON, N.H." stamped on the wood. "128 ft to a cord" also stamped on the wood.

• "F.M. GREENLEAF, LITTLETON, N.H." stamped on wood. "Do not scale over 40 ft. in length" stamped on the wood.

• "F.M. GREENLEAF, MAKER, LITTLETON, N.H." stamped on wood. "Carey Standard, 100 Standards Equal 1000 ft. B.M." stamped on wood.

• "F.M. GREENLEAF, 62 OAK AVE., BELMONT, MASS." stamped on wood, "Greenleaf Cordwood Caliper" stamped on wood.

• "F.M." GREENLEAF, JACKSONVILLE, ALABAMA, MAKER" stamped on wood.

I have a feeling that the percentages indicated above, three made by the father and five made by the daughter, may well be indicative of the total numbers in existence.

As the father made calipers in his spare time, his daughter apparently used this job as her primary source of income prior to moving to Belmont and as an important supplemental income later on. Flossie died a wealthy woman in Denver, having inherited her uncle's estate.

E.S. Lane:
A Maine Rule Maker And Scaler

by Dale Butterworth and Ben Blumenberg

E. S. Lane – Upton, ME, Ca. 1930. Photo courtesy of Mrs. Eva Lane – Bethel, ME, widow of Lyman J. Lane

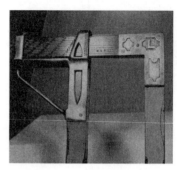

Caliper by "E. S. Lane, Upton, ME, Maker". Collection of Dale Butterworth. Metal castings are aluminum.

As part of our continuing research on the makers of hand tools in Maine for those trades that worked with wood, we have uncovered important information about a little documented manufacture, that of lumbermen's scales. Maine is the only state east of the Mississippi that still contains large coniferous forests that are logged commercially on a massive scale. The history of commercial lumbering in this state, however, goes back to the 18th century when the English Crown ravaged the Maine forests for 'tall timber' (huge pine trunks) to use for masts on their largest ships, particularly naval vessels. In the 1790's, commercial ship[1] building began in earnest in dozens of yards that sprang up along the Maine coast. Logging demands many tools, the most obvious of which is the ax. Scales are extremely valuable for rendering a quick determination of the quantity of lumber contained in newly cut logs. They were extensively employed in the lumber company yard where the bookkeeping needed to be continuously updated and loggers were often paid by the cord. Logging still continues today in the northern forests of the continental United States and the manufacture of wooden scales was an essential toolmaking endeavor that accompanied the industry. While portable electronic measuring devices are now available, the manufacture of wooden scales continued until very recently. We have documented one such tool maker who continued to make such 'rules' until the winter of 1942 (see below)! The exciting opportunity provided by these circumstances is that colleagues and relatives of these rule makers are still alive and are often most pleased by the opportunity to provide information needed by our research. We thus have an opportunity to delve into the lives of Maine's scalers with an information source at hand that is not available for planemakers. We will begin with two scalers who are father and son.

Several examples of lumbermen's scales are known stamped 'E. S. Lane/Upton, ME'. We illustrate his mark as it appears on a log caliper. Upton is tiny town on the New Hampshire border on state route 26, northwest of Grafton Notch State Park. It is too small to have its population included in the gazetteer of the well known Rand McNally road atlas. About 15 miles to the east is Rumford, Maine (pop. 6300) on the Swift River where the Boise Cascade Paper Company still has a large mill. Other nearby large rivers that could serve for floating logs are the Ellis and Androscoggin. The much smaller Bear River begins in Grafton Notch State Park and flows south into the Androscoggin. This area of Maine is still wilderness in which much commercial logging continues and it is dominated by what are termed 'minor civil divisions' and unincorporated townships. The tiny size of Upton, both in the past and now, raised hopes that uncovering the identity of E.S. Lane would not be too difficult and one of us (DB) 'hit the road' to do the field research.

An Ellsworth Lane, son of Charles H. Lane, born in Rumford Point, married to Rena Strickland and still living in Upton in 1942, is mentioned in Heywood (1973:88). Heywood also notes that Ellsworth had a son named Lyman. This is the only published reference to an E. S. Lane living in Upton Maine and confirmation was quickly obtained that he was, indeed, a scaler. A visit to the Upton cemetery located the gravestones of both Ellsworth and his wife. E.S. Lane was born in 1877 and died in 1945.

The information that follows about Ellsworth and his son was obtained from interviews with Lyman's widow Eva (now living in Bethel) and Mrs. Francis Dunn, now of Edgecomb, Maine, who was in the same class with Lyman at Gould Academy in Bethel, Maine. Mrs. Dunn proved to be a genealogist of considerable depth.

The Lane lineage in the New World begins with three sons of one James Lane who settled in Malden, Mass., about 1656. The grandfather of Ellsworth S. Lane was Jonathan Sewell Lane who was born in 1827 in Hanover, Mass and died in 1903 in Upton, Maine. We do not have enough information to reconstruct the exact family tree that leads from the 17th century Lanes to J.S.Lane Jonathan Sewell served two enlistments in the U. S. Navy before being married and went to California during Gold Rush days via the Straights of Magellan.[2] The first of Jonathan's three sons was Ellsworth's father, Charles H. Lane born in 1856 in Bethel, Maine.

Ellsworth was a tax collector and selectman in Upton for many years while making calipers and scales in a small shop behind his house. He continued to make scales until his death. Ellsworth was known by the nickname of 'Wert', the origin of which remained unknown to those who knew him. He apparently sold blank log calipers without wheels to F.M. Greenleaf and his daughter Flossie whose log calipers are well known and much desired by collectors. William H.

Original Wheel by "E. S. Lane, Upton, ME. Maker". Collection of Dale Butterworth. Wheel hub is brass.

From *The Chronicle of the Early American Industries Association*, Vol. 46, No. 1 (March, 1993), by permission

Greenleaf was born in Whitefield, N.H., in 1834 and after his first wife died, moved to Litttleton, N.H. where he worked as a carriage maker for the C.F. Harris Co. All of his log calipers were made in Littleton in his spare time between 1868 and his death in 1916. Interestingly, he taught the trade to Flossie (1882 - 1928), the fourth child of his second marriage. She began making calipers stamped with her initials while in Littleton, then moved to Belmont, Mass. because she was a violinist with the Boston Symphony! Several calipers are known with Flossie's initials and a Belmont, Mass. address (Holden 1982). Several log calipers have been observed with the marks of both E. S. Lane and either William or Flossie Greenleaf when they worked in Littleton, N.H.. The Greenleafs apparently stamped blanks bought from time to time from E. S. Lane and then fitted them out with a wheel of their own design.

Ellsworth's son Lyman was born in 1907 and attended Gould Academy. He learned scaling from his father and began to practice the trade at nineteen. He was a clerk and paymaster on the river drives on the Androscoggin in the 1930's and 1940's. As was his father, Lyman was also a tax collector and selectman in Upton. Lyman scaled for the Brown Company of Berlin, New Hampshire for 20 years and was also in charge of paying the crew and doing the clerical work in the woods. He also worked 22 years for the Boise Cascade Paper Company in Rumford. DB was able to interview Mr. Erlon Wentzell of Sebago, Maine whose father was woods boss for the Brown Company. He assigned the 11 year old Erlon to Lyman Lane as a recorder. Lyman would scale the cordwood that the choppers cut each week. In those days, 10,000 to 20,000 cords of wood were cut per season and hauled out in five to eight cord loads by pairs of horses. As soon as the tree was dropped and limbed, the bark was peeled off with spuds. Later it was cut to four foot lengths. Erlon told DB that some of the choppers would try to cheat the company by piling over stumps with short sticks or taking tallied sticks off the top of one pile and putting them on the next pile which had not yet been examined.

Lyman never made scales; therefore he offered his father's scale rule business to Leslie B. Sargent of Lincoln, New Hampshire c. 1949. who carried on the trade and tradition until his death in the winter of 1992! Leslie Sargent also bought out the scale rule business of Ken and Don Vineyard in Crystal (West Milan) New Hampshire in 1985 whose dies allowed him to increase production on rule orders from 24 to 100. He supplied such companies as Snow and Neally of Bangor, Maine; Weyerhauser Paper of Wisconsin; Boise Cascade Paper Co; Columbia Forest Products of Newport, Vt. and Presque Isle, Me and dozens of smaller firms. Leslie had kept several of Ellsworth Lane's 'master beams' without hardware which he showed DB.

The rule manufacturing system that he bought from Don Vineyard utilized a heating coil to burn the numbers and measurement

Beam made by "E. S. Lane, Upton, ME." on one side. "F. M. Greanleaf, Littleton, N.H., Maker" on reverse side. Collection of Tom Whelan, Marshfield, Mass.

increment lines into the face of the rule. Each rule was jacked up against the heating coil and the entire face burned in one operation. Leslie sometimes took old calipers and sanded off the old numbers. He ground off the numbers on the brass sliding jaw and then stamped his own without finishing off the coarse grinding marks.

Apparently E. S. Lane made only two types of scales judging from surviving examples with the mark of E.S. Lane. A simple slip stick was made in both eight and ten foot lengths. The foot mark was designated and the scales were in tenths of an inch. Some of these rules had a cull scale on one edge to detect defects such a rot or hollows in a four foot sticks. Choppers were paid by the cord and this rule was used in the woods to tally up the cords that each chopper cut, done by Lyman himself assisted by the young Erlon Wentzel. Lyman would scale the cordwood that the choppers cut each week. Erlon would write down the name of the chopper and the length and height of each pile. Back at the office, these two measurements would be multiplied together. This product multiplied by 4 and then divided by 128 would yield the cords of wood.

$\{[(\text{Pile Length}) \times (\text{Pile Height})] \times 4\}/128 = \text{Cords}$

A log caliper with wheel was made by E. S. Lane to measure long logs to calculate volume. Erlon Wentzell related that Lyman Lane would roll the wheel the length of the log, note the length, and then roll the wheel backwards to the center of the log length and then caliper the log. The number of units could be read from the caliper using the length column and diameter reading. Lane's design for his caliper wheel featured heavier spokes than usual and employed stove bolts to hold the spokes in position This is the only such use of stove bolts on a log caliper wheel known to DB and provides a quick and easy change over if a spoke became broken. Lane used both brass and aluminum cast fittings on the beam itself. The 'bullet-shaped' opening in the casting of the movable jaw is a E. S. Lane trademark. Wheels made by Lane were always signed 'E.S. Lane Upton Me Maker.'

"If it still works, don't buy a new one" is a well known axiom among mechanics with a traditional training and a respect for hand tools. The wooden scales employed by the lumber industry sufficed for the job at hand, assuming the user could understand some simple arithmetic and algebra. Therefore their manufacture continued in small shops almost until the present time. The recent demise of the scaler's trade is perhaps a comment upon the disintegration of public education in this country, a social phenomenon that continues to make weekly headlines. An electronic measuring device requires little knowledge except that of the correct sequence of buttons to push. Several Maine scalers are known from their marks on surviving rules and calipers. We hope that further research will uncover details of their lives and measuring instruments and force them to emerge from the obscurity that now cloaks them in the depth of the Maine woods.

ACKNOWLEDGMENTS

The information presented was graciously supplied by a number of people who consented to be interviewed and queried. Tom Whalen of Marshfield Hills, Massachusetts shared his extensive knowledge of the lumbering industry and its measuring instruments. Mrs. Arlene Bernier of Upton advertised our information search in the Bethel newspaper. Mrs. Francis E. Dunn (Edgecomb, Me) and Mrs. Joy Yarnell (Upton, Me) shared their experience and knowledge of the Ellsworth family. Erlon Wentzell of West Baldwin, Maine, and Mrs. Eva Lane graciously consented to be extensively interviewed. One of the unexpected fruits of this research was the contact it brought with Leslie B. Sargent, who was perhaps the last of the New England scale makers. Mrs. Sargent is presently allowing another forester to continue making scales.

REFERENCES

Heywood, C.E. 1973. *History of Upton, Maine*. Norway, Me: Oxford Hills Press.

Holden, B. 1982. "Wm. Greenleaf: Talented Eccentric". *Gristmill No.28, p* 8.

If you wish to read further about logging and lumbering in Maine:

Pike, R.E. 1967. *Tall Trees, Tough Men*. New York: W.W. Norton.

Wood, R.G. 1961. "A History of Lumbering in Maine 1820-1861". *The Maine Bulletin* XLIII (No. 15). Reprinted by the University of Maine Press in 1971.

FOOTNOTES

1. The term 'ship' is used here in the quasi-technical sense to denote a large vessel of more than a few tons deadweight.

2. Sounds bit like Abel Sampson!

A Most Unusual Rule

MORE FOR YOUR MONEY / *William J. Baader*

Although many multi-purpose tools have been devised to accomplish more than one function, a three-in-one rule would appear to be quite unlikely. The invention of Lewis M. Weaver and Zacharias C. Leiser was that very type of tool - a rule designed to be used as a marking gage, divider or circle scriber.

Here are two photographs of just such a rule from the author's collection. The only marking on this rule is PAT. SEPT. 20th, 92, and in my research to date I have been unable to determine where or by whom it was manufactured.

Accuracy is not achieved in the two extra uses of the rule, whereby the use of dividers would be more accurate and much quicker, and the same applies to its use as a marking gage.

The patent papers which follow and the patent drawings which accompany this article comprise all that is known of this rule. I would be interested in hearing from anyone having additional information. ■

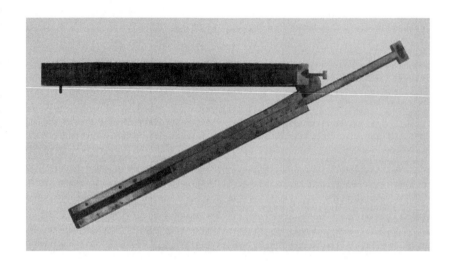

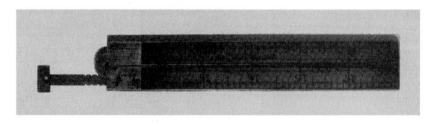

UNITED STATES PATENT OFFICE
LEWIS M. WEAVER AND ZACHARIAS C. LEISER, OF LOCK HAVEN, PENNSYLVANIA
COMBINED RULER, GAGE, AND CIRCLE-SCRIBER

Specification forming part of Letters Patent No. 482,886, dated September 20, 1892.
Application filed January 21, 1892   Serial No. 418,831 (No model)

TO ALL WHOM IT MAY CONCERN.
Be it known that we, LEWIS M. WEAVER *and* ZACHARIAS C. LEISER, *citizens of the United States, residing at Lock Haven, in the county of Clinton and State of Pennsylvania, have invented certain new and useful Improvements in Rulers; and we do hereby declare the following to be a full, clear, and exact description of the invention, such as will enable others skilled in the art to which it appertains to make and*

Our invention relates to improvements in rules; and the objects of the invention are to provide a simple, cheap, and durable ruler which can be used either as a rule, a gage, or a compass.

With these ends in view our invention consists in the peculiar construction and arrangements of parts, as will be hereinafter fully pointed out and claimed.

In the accompanying drawings, Figure 1 is a top plan view of a ruler constructed in accordance with our invention. Fig. 2 is a longitudinal vertical sectional view on the line xx of Fig. 4. Fig. 3 is a transverse vertical sectional view on the line yy of Fig. 4. Fig. 4 is a bottom plan view, and Fig. 5 is a detail view, of the extensible slide. Figs. 6 and 7 are detail views of the gages. Fig. 8 is a detail view of the plate for holding the movable slide in place.

Like letters of reference demote corresponding parts in all the figures of the drawings, referring to which —

AA designates the two sections or members of the ruler, which are pivotally connected together at a in a manner well known in the art. Each of these sections or members is composed of two parts connected by suitable hinges a.

The sections or members AA are preferably formed of a central longitudinal piece or block B, which is protected and strengthened by metallic side plates or strips B. (See Fig. 3)

The sections or members AA are graduated in the usual manner, and in a suitable chamber or recess formed in the member A and opening through the free end thereof is arranged a slide or member C. This slide or member C is pro-

From The Gristmill, *No. 40 (June, 1985), by permission*

vided at its outer end with and enlarged head D, in which is formed an aperture d, for a purpose to be hereinafter pointed out. This head D of the slide C is of greater thickness that the slide, and on one face thereof is formed an integral projecting lip or flange d. The sides of the slide C are provided with a series of notches c, extending throughout the length of the slide and preferably spaced apart one-sixteenth of an inch.

E designates a spring-plate which is arranged within the chamber in which the slide C normally fits. One portion e of this plate extends across said slide at right angles to the main portion of the plate E. The other two pins on the plate E are adapted to fit into aligned notches c on opposite sides of the slide C and hold said slide in any desired position with regard to the ruler. The spring-plate E is further provided with a thumb piece of stud F, which extends through a suitable aperture in the body of the rule and extends a slight distance beyond the surface thereof.

In a suitable chamber or socket formed in the member or section A and opening through the free end thereof is fitted a tapering pin G.

H, K, and L, designate three gages, which are pivotally connected to the side plates or strips B. These gages consist of a plate h, provided at one end with a disk-like portion h, through which the pivot-pin extends, and are preferably arranged on the section A at one-half inch, five inches, and ten inches, respectively, from the cuter free end thereof.

The gages H K L are adapted to be turned on their respective pivots, so as to extend at right angles to the length of the ruler, as shown in Fig. 2 of the drawings, but said gages are normally turned down and fit in suitable recesses formed in the upper surface of the central piece or block B, so that their expose surface lies flush with the edges of the side plates or strips b. The front end of each of these gages is cut away to form a projecting lip l. When the gages are forced down to contact with the body of the ruler, the lip d on the head D fits over the lip l on the gage ll and holds such gage in position. The other two gages K L, are held in their closed position by spring-pressed plates M, which take over the lips l on such gages. The plates M are arranged in suitable sockets, and each of said plates is provided with a longitudinal stem, around which is fitted a coiled spring N, and with a stem m, extending at right angles to the stem having the coiled spring thereon. The pins m extend through one of the side plates or strips B and are adapted to fit in sockets or recesses n, formed in the section or member A when the two sections A A are forced together.

O designates a series of apertures which are formed in the ruler at any desired distance apart. The aperture which is formed nearest the outer end of the section A is slightly out of line with the apertures in the other sections of the ruler to avoid the slide C.

When it is desired to use our improved ruler as a compass, the pin G is withdrawn from its socket and inserted in any desired one of the series of apertures O. A pencil is passed through the opening d in the enlarged head D of the slide C and a circle of the desired size is readily described about the pin G as a center.

The slide C can be withdrawn from its socket any desired distance by pressing the stud F to release the pins f from the notches in the side of the slide and measurements of less degree than the divisions of the ruler thus obtained.

We are aware that changes in the form and proportion of parts and details of construction of the devices herein shown and described as an embodiment of our invention can be made without departing from the spirit or sacrificing the advantages thereof, and we therefore reserve the right to make such changes as fairly fall within the scope of our invention. For instance, though we have shown but three gages, all on the section

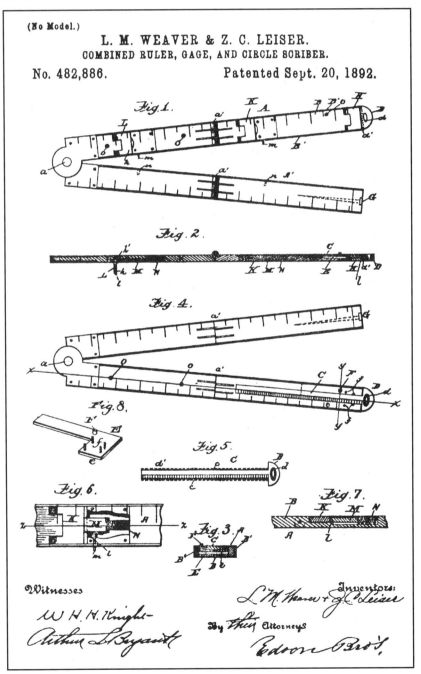

A of the ruler, we would not be understood as limiting ourselves to the number or exact location of the gage-plates herein shown and described, as the number and relative positions thereof can be varied at pleasure.

By turning up any desired one of the gages and inserting the pin G or a pencil in the aperture in the head of the extensible slide a gage of the desired size is obtained.

Having thus fully described our invention what we claim as new, and desire to secure by Letters Patent, is —

1. In a ruler, the combination, with the two pivotally-connected sections, of a notched slide fitted in a socket formed in one of the sections and a spring-plate arranged in said socket and provided with projecting pins adapted to take into the notches in the slide to hold the same in any desired position, substantially as described.

2. In a ruler, the combination, with the two sections pivotally connected together, each of said sections being composed of a central piece and metallic side plates or strips, of plates pivotally connected to the side plates and adapted to extend at right angles to the length of the ruler and means for normally keeping said plates in contact with the ruler, substantially as described.

3. In a ruler, the combination, with the two sections pivotally connected together, of a slide fitted in a socket in one of the sections and provided with an enlarged head having an integral lip or flange on one face, a plate pivotally connected to the ruler near one end and having a lip or flange formed on its free end and normally extending below the lip on the head of the slide, similar plates pivotally connected to the ruler at intermediate points of its length, and means for normally holding such plates against the ruler, substantially as described.

4. In a ruler, the combination, with the two sections pivotally connected together, of the gage-plates pivotally connected to one of the sections adapted to fit in suitable recesses formed therein, and spring-pressed plates for holding the gage-plates against the ruler, said plates being provided with outwardly-extending pins or studs which are adapted to fit in suitable sockets in the other section when the sections are brought together, substantially as described.

In testimony whereof we affix our signatures in presence of two witnesses.

Witnesses: A.W. Bittner
N.J. Leiser

LEWIS M. WEAVER
ZACHARIAS C. LEISER

THE STANLEY NO. 58½ RULE
PHILIP E. STANLEY

The Stanley #58½ Six-Fold Rule With Tables

Throughout its 140 year history the Stanley Rule & Level Company has constantly experimented with its line of rules, frequently introducing new ones in an attempt to address perceived needs in the workplace. Some of these added rules were eminently successful, such as the No. 98 desk rule, and the No. 32 caliper rule, and remained in the product line for years. Others, however, such as the Nos. 186 and 188 printer's rules, and the series of metric rules offered beginning in 1877, were less so, and were dropped almost immediately. This introduction of new products and cancellation of old ones was particularly common during the years before 1870, when the Company was still in the process of integrating the Hall & Knapp, Seymour & Churchill, and Seth Savage product lines, and trying to define itself more clearly.

One of the more interesting rules offered briefly during this period was the Stanley No. 58½ six fold rule with tables. This rule was first listed in the 1860 catalogue, described as follows:

> 58½: Two Foot, Six Fold. Arch Joint, Edge Plates, with Tables for ascertaining the Weights of all sizes of Iron, Steel, Copper, Brass, Lead, &c., ¾ inch wide, $24.00 per dozen.

It was offered twice more, with the same description, in 1862 and 1863, after which it was dropped from the catalogue. The number was later reused for another six fold rule, beginning in 1879, but that was a significantly different product, and does not concern us here.

Mechanically, the No. 58½ it is similar in construction to the No. 58 six fold rule introduced in 1859, having an arch joint and edge plates, and being ¾ inch wide. The difference between the two is in the markings: The No. 58 has the usual 8ths and 16ths scales; the No. 58½ has no scales, *per se*, but is instead entirely covered with the data tables mentioned in the catalogue description.

These tables occupy the entire surface of the rule on both sides, and are very densely packed. They run

From *Stanley Tool Collector News*, No. 5 (Spring, 1992), by permission

vertically down the sticks, and are exactly the full width of a stick (the two columns which make up each table are 7/64 and 17/64 inch wide, respectively). The figures in the tables are made up of 1/16 inch figures, and there are eight rows to the inch. Both the rows and the columns are separated by scribed dividing lines, as is usual. There are no markings on the edge of the rule except embellishment lines.

There are eighteen tables altogether, arranged six on one side of the rule and twelve on the other, as shown in the diagram to the right. Each table is identified by a one- or two-word title in its first row. These titles are not as informative as could be desired, but with some extrapolation, and analysis of the data in the tables, it is possible to identify the function of each table, as follows:

Table 1 - ROUND: The weight in pounds per foot of round wrought iron bars from ¼ to 5 inches in diameter.

Table 2 - SQUARE: The weight in pounds per foot of square wrought iron bars from ¼ to 5 inches on a side.

Table 3 - FLAT IRON: The weight in pounds per foot of flat wrought iron bars ¼ inch thick from 1 to 6½ inches in width.

Table 4 - FLAT STEEL: The weight in pounds per foot of flat steel bars ¼ inch thick from 1 to 6½ inches in width.

Table 5 - ROUND STEEL: The weight in pounds per foot of round steel bars from ¼ to 4 inches in diameter.

Table 6 - SQUARE STEEL: The weight in pounds per foot of square steel bars from ¼ to 4 inches on a side.

Table 7 - BAND: The weight in pounds per 10 feet of flat steel bands from ⅝ to 3 inches in width and from 11 to 21 gauge* in thickness. Only certain combinations of width and thickness are listed, presumably those which were in most common use in the 1860's.

Table 8 - SHEET: The weight in pounds per square foot of flat sheet steel from 1 to 30 gauge* in thickness.

Table 9 - BOILER: The weight in pounds per square foot of boiler (wrought) iron from ⅛ to 1 inch in thickness

Table 10 - SHEET COPPER: The weight in pounds per square foot of sheet copper from 1/16 to 1/2 inches in thickness.

Table 11 - ROUND COPPER: The weight in pounds per foot of round copper bars from ¼ to 2 inches in diameter.

Table 12 - SHEET BRASS: The weight in pounds per square foot of sheet brass from 1/16 to 1/2 inch in thickness.

Table 13 - SQUARE COPPER: The weight in pounds per foot of square copper bars from ¼ to 2 inches on a side.

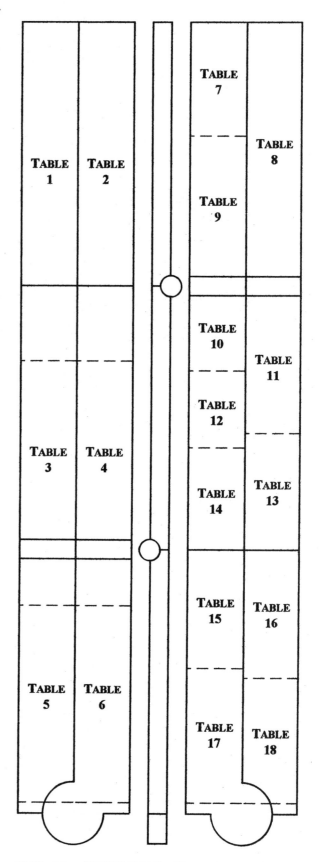

Table 14 - SHEET LEAD: The weight in pounds per square foot of sheet lead from 1/16 to 3/4 inch thick.

Table 15 - HAMR'D ROUND: The weight in pounds per foot of round wrought iron shafts from 5¼ to 8½ inches in diameter.

Table 16 - PROOF CHAIN: The breaking strain in tons, and the weight in pounds per foot, for chain made from wrought iron stock from ⅜ to 1⅜ inches in diameter.

Table 17 - HS.IRON: The weight in pounds per foot of square wrought iron bars from 5¼ to 9 inches on a side.

Table 18 - RC.IRON: The weight in pounds per foot of round cast iron bars/columns from 2 to 8 inches in diameter.

*: The gauge system used is not known; see below

These 18 tables, which are shown in the appendix to this article, certainly constitute the "... Tables for ascertaining the Weights of all sizes of Iron, Steel, Copper, Brass, Lead, etc. ..." described in the catalogue!

All this data was only packed onto the rule surface at a price, however. As mentioned above, the tables occupy the entire surface of the rule on both sides, to the extent that there is literally no space left for a common scale of inches. The makers were able to get around this, however, in a rather clever way. Table 1, located at the top of one of the tip sticks, has index values of from ¼ inch to 5 inches, by 8ths. The first row of the table is used for its name, so the indices begin starting in the second row. Each row is physically exactly ⅛ inch high, the same as the increments between the index values. Thus each index, in addition to its function of selecting a weight, also indicates the of the line below it from the end of the tip stick. This can be seen by comparing the illustration to the right of Tables 1 & 2 with the rule shown adjacent to them. This arrangement makes it possible to use the rule to make measurements up to 5 inches, using the lines of the table as a scale of 8ths, with the indices serving as the figures of the scale.

Nor is there space for either the Stanley name or the rule number on either face. They could have been placed on the edge of the rule, but this was not usual Stanley practice.

Fortunately, even in the absence of a marked name, it is still possible to identify the rule positively as a Stanley product on the basis of physical construction. The key to this identification is the shape of the head of the main joint pin. On boxwood rules this pin must, of necessity, be keyed, so that it is locked to the outer joint plates and turns with them when the rule is opened and closed. Most American makers (Stearns, Chapin, etc.) shaped the head like a circle with two small protrusions on opposite sides (like a stubby propeller with a large hub). Stanley, on the other hand, gave it a five-sided (pentagon) shape, and was apparently the only American maker to do so. The main joint pin on this rule has a definite five-sided shape, hence it is unquestionably a Stanley rule.

As corroborative evidence that this rule is the Stanley No. 58½, it should be noted that only Stanley, and no other American maker, ever offered a rule of this description.

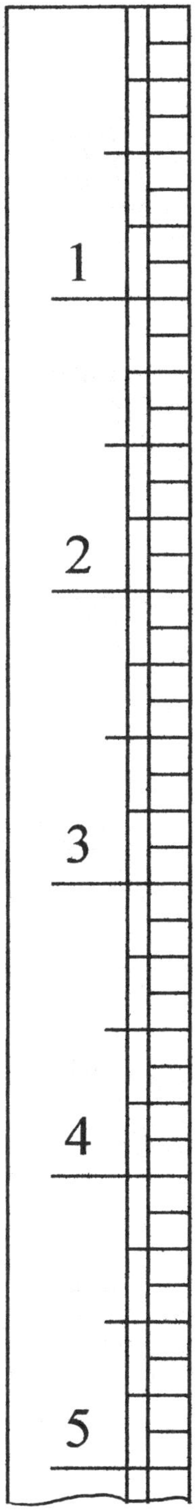

ROUND		SQUARE	
¼	.163	¼	.208
⅜	.368	⅜	.468
½	.654	½	.833
⅝	1.020	⅝	1.30
¾	1.472	¾	1.87
⅞	2.00	⅞	2.55
1	2.61	1	3.33
1⅛	3.31	1⅛	4.21
1¼	4.09	1¼	5.20
1⅜	4.94	1⅜	6.30
1½	5.89	1½	7.50
1⅝	6.91	1⅝	8.80
1¾	8.01	1¾	10.20
1⅞	9.20	1⅞	11.71
2	10.47	2	13.33
2⅛	11.82	2⅛	15.05
2¼	13.25	2¼	16.87
2⅜	14.76	2⅜	18.80
2½	16.36	2½	20.80
2⅝	18.03	2⅝	22.96
2¾	19.79	2¾	25.20
2⅞	21.63	2⅞	27.55
3	23.56	3	30.00
3⅛	25.56	3⅛	32.55
3¼	27.65	3¼	35.20
3⅜	29.82	3⅜	37.96
3½	32.07	3½	40.80
3⅝	34.40	3⅝	43.80
3¾	36.81	3¾	46.87
3⅞	39.31	3⅞	50.05
4	41.88	4	53.33
4⅛	44.54	4⅛	56.71
4¼	47.28	4¼	60.20
4⅜	50.11	4⅜	63.80
4½	53.01	4½	67.50
4⅝	56.00	4⅝	71.30
4¾	59.06	4¾	75.20
4⅞	62.21	4⅞	79.21
5	65.45	5	83.23

Table 1   Table 2

In the hope of identifying the gauge system used by Stanley in preparing tables 7 and 8, the thickness associated with each gauge number was calculated, working backwards from the weights and dimensions given, and taking the density of steel as 490 pounds per cubic foot. The results are tabulated in a table to the right.

The results do not answer the question, however. The Stanley gauge numbers do not correspond to any system known to the author which was in use during the 1860's. It is similar to the Birmingham (England) Wire Gauge, and to the Washburn & Moen Steel Wire Gauge (both included in the table for comparison), but differs from both by 2-3 mils or more in most gauges. It is probably some standard local to Connecticut which was originally derived from the Birmingham Wire Gauge or its predecessors.

The No. 58½ must not have been a very successful product. Its elaborate data tables and six joints made it extremely expensive to produce (at $24.00 per dozen it was one of the highest-pricced boxwood rules in the Stanley list). Additionally, its lack of a real inch scale made it not much more useful than a pocket book of data tables, which would cost less and contain more information. The fact that only three examples of this rule are known to exist would also indicate very limited acceptance. It is possible that only one or two batches were produced, and as soon as those were sold the rule was dropped.

Gauge Number	Decimal Equivalent		
	Stanley #58½	Birmingham or Stubs	Washburn & Moen
1	.307	.300	.2830
2	.275	.284	.2625
3	.256	.259	.2437
4	.234	.238	.2253
5	.212	.220	.2070
6	.204	.203	.1920
7	.184	.180	.1770
8	.163	.164	.1620
9	.154	.148	.1483
10	.135	.134	.1350
11	.116	.120	.1205
12	.105	.109	.1055
13	.089	.095	.0915
14	.079	.083	.0800
15	.073	.072	.0720
16	.064	.065	.0625
17	.054	.058	.0540
18	.047	.049	.0475
19	.042	.042	.0410
20	.035	.035	.0348
21	.032	.032	.0317
22	.028	.028	.0286
23	.024	.025	.0258
24	.023	.022	.0230
25	.021	.020	.0204
26	.019	.018	.0181
27	.018	.016	.0173
28	.017	.014	.0162
29	.015	.013	.0150
30	.014	.012	.0140

Stanley, Birmingham or Stubs,
And Washburn & Moen Wire Gauges

APPENDIX: TABLES ON THE NO. 58½ RULE

FLAT IRON	
1	.83
1⅛	.93
1¼	1.04
1⅜	1.14
1	1.25
1⅝	1.35
1	1.45
1⅞	1.55
2	1.66
2¼	1.87
2½	2.08
2¾	2.29
3	2.50
3¼	2.70
3½	2.91
3¾	3.11
4	3.33
4¼	3.53
4½	3.74
4¾	3.95
5	4.16
5¼	4.37
5½	4.58
5¾	4.79
6	5.00
6¼	5.20
6½	5.40

Table 3

FLAT STEEL	
1	.852
1⅛	.958
1¼	1.06
1⅜	1.17
1	1.27
1⅝	1.38
1	1.48
1⅞	1.60
2	1.70
2¼	1.91
2½	2.13
2¾	2.34
3	2.55
3¼	2.77
3½	2.98
3¾	3.19
4	3.41
4¼	3.62
4½	3.83
4¾	4.05
5	4.26
5¼	4.47
5½	4.68
5¾	4.89
6	5.11
6¼	5.31
6½	5.52

Table 4

ROUND STEEL	
¼	.167
⅜	.376
½	.669
⅝	1.04
¾	1.50
⅞	2.05
1	2.67
1⅛	3.38
1¼	4.18
1⅜	5.06
1½	6.02
1⅝	7.07
1¾	8.20
1⅞	9.41
2	10.71
2⅛	11.98
2¼	13.40
2⅜	14.92
2½	16.55
2⅝	20.00
3	23.80
3¼	27.96
3½	32.40
3¾	37.18
4	42.30

Table 5

SQUARE STEEL	
¼	.213
⅜	.479
½	.855
⅝	1.33
¾	1.91
⅞	2.61
1	3.40
1⅛	4.31
1¼	5.32
1⅜	6.44
1½	7.67
1⅝	9.00
1¾	10.44
1⅞	11.98
2	13.63
2⅛	15.37
2¼	17.22
2⅜	19.19
2½	21.21
2⅝	25.63
3	30.46
3¼	35.72
3½	41.38
3¾	47.48
4	54.00

Table 6

SHEET	
1	12.55
2	11.25
3	10.45
4	9.55
5	8.66
6	8.34
7	7.50
8	6.64
9	6.29
10	5.50
11	4.73
12	4.30
13	3.64
14	3.23
15	2.97
16	2.62
17	2.19
18	1.92
19	1.70
20	1.41
21	1.32
22	1.15
23	.99
24	.95
25	.84
26	.78
27	.72
28	.68
29	.62
30	.58

Table 8

SHEET COPPER	
1/16	2.88
1/8	5.76
3/16	8.65
1/4	11.53
5/16	14.42
3/8	17.30
7/16	20.19
1/2	23.07

Table 10

SHEET BRASS	
1/16	2.70
1/8	5.40
3/16	8.10
1/4	10.81
5/16	13.51
3/8	16.21
7/16	18.91
1/2	21.62

Table 12

BAND		
⅝	21	.685
¾	20	.885
⅞	19	1.24
1	18	1.60
1⅛	17	2.05
1¼	16	2.73
1⅜	15	3.40
1½	15	3.72
1¾	14	4.72
2	13	6.06
2¼	13	6.83
2½	12	8.96
2¾	11	10.85
3	11	11.84

Table 7

BOILER	
1/8	5.00
3/16	7.50
1/4	10.00
5/16	12.50
3/8	15.00
7/16	17.50
1/2	20.00
9/16	22.50
5/8	25.00
11/16	27.50
3/4	30.00
13/16	32.50
7/8	35.00
15/16	37.50
1	40.00

Table 9

SHEET LEAD	
1/16	3.70
1/8	7.40
3/16	11.10
1/4	14.81
5/16	18.51
3/8	22.22
7/16	25.92
1/2	29.62
9/16	33.33
5/8	37.03
11/16	40.73
3/4	44.44

Table 14

ROUND COPPER	
¼	.188
⅜	.424
½	.755
⅝	1.17
¾	1.69
⅞	2.31
1	3.02
1⅛	3.82
1¼	4.71
1⅜	5.71
1½	6.79
1⅝	7.94
1¾	9.21
1⅞	10.61
2	12.08

Table 11

SQUARE COPPER	
¼	.24
⅜	.54
½	.96
⅝	1.50
¾	2.16
⅞	2.94
1	3.84
1⅛	4.86
1¼	6.00
1⅜	7.27
1½	8.65
1⅝	10.15
1¾	11.77
2	15.38

Table 13

HAMR'D ROUND	
5¼	72.76
5½	79.85
5¾	87.28
6	95.03
6¼	103.12
6½	111.58
6¾	120.28
7	129.35
7¼	138.76
7½	148.49
7¾	158.56
8	168.95
8¼	179.68
8½	190.73

Table 15

PROOF CHAIN	
3/8	2TN9LB
7/16	3 11
1/2	4 14
9/16	5.5 18
5/8	7 24
11/16	8 28
3/4	10 32
13/16	12 38
7/8	14 44
15/16	16 50
1	20 56
1 1/16	22 63
1 1/8	27 70
1 3/16	32 78
1 1/4	37 88
1 3/8	44 106

Table 16

HS.IRON	
5¼	92.64
5½	101.68
5¾	111.13
6	121.00
6¼	131.30
6½	142.01
6¾	153.15
7	164.70
7¼	176.68
7½	189.07
7¾	201.89
8	215.12
8¼	228.78
8½	242.85
9	272.26

Table 17

RC·IRON	
2	9.81
2½	15.33
3	22.08
3½	30.06
4	39.27
4½	49.70
5	61.35
5½	74.24
6	88.35
6½	103.69
7	120.26
7½	138.05
8	157.08

Table 18

How Thick is "Extra Thick"?

by Cliff Fales

"Extra thick and strong" or "thick" rules were offered by at least two manufacturers—Stanley Rule & Level Co. and Belcher Brothers & Co. The available catalog listings for these rules do not specify how much thicker they were than the ordinary varieties.

In the two Belcher catalog listings which I have available (1853, 1860), there is some ambiguity in the terms: "Thin," "medium," "thick," and "extra thick." See TABLE below.

A comparison of the two Belcher price lists shows that, in the change from the 1853 price list to the 1860 price list, "thin" retained the same designation; "medium" received no designation; and "thick" was changed to "extra thick."

I am attempting to verify that the rule described below is an extra thick Belcher rule. The rule, as observed when laid out with several other 2-foot, 4-fold rules, appears noticeably thicker, but not extremely so.

round joint, center plates and brass tips. It is unsigned and unnumbered. The main (round) joint has three leaves rather than the usual two; they are steel and extend through the full width of the legs to the outer edges. The center plates of the knuckle joints are also steel. The rule is graduated in 1/16ths of an inch on both sides.

The embellishment lines display some unique characteristics. Not only are there double face embellishment lines on the inner ends of the graduations, but there is an additional embellishment line, very close to the edge, at the outer end of the graduations. The graduations terminate at this line and do not continue to the edge. This line is so close to the edge that it is mostly worn away on the outside but is clearly evident on the inside. Further, there is another embellishment line near the inner edges, which on the inside is a pair of lines. This totals four embellishment lines on the face and five on the inside.

round joint, the first embellishment line stops 1/2" from the end, the second continues to the end and the third stops 3/8" from the end.

Most unusual are the four edge embellishment lines, arranged as two pair, with each pair very close to the outer edge.

A statement in the Belcher Brothers & Co. price list reprint of 1860 states that their round joint rules "...being made of an inferior selection of Boxwood, although well seasoned, will not have the stamp of the firm upon them...." The 1853 price list includes a similar note: "The rules numbered 30, 50, 70, 80, 90 are not stamped Belcher Brothers & Co."

After considerable study, I am convinced that the figures on this

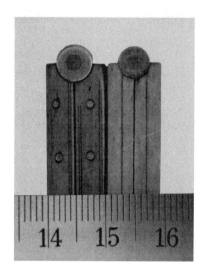

Edge view of Belcher rule compared to a "standard" thickness Stanley rule. Note the thickness (shown in mm) and edge embellishment lines.

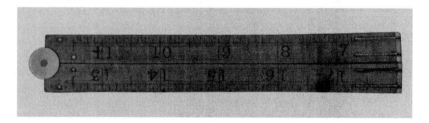

General view of the rule. Note the steel center plates and the embellishment lines which do not extend completely to the end.

Since no Belcher rules were stamped with numbers, and since their round joint rules were not stamped with their name, the verification necessarily must rely on other means, namely, comparison of the physical description with catalog listings, and comparison of the style of figures with other Belcher rules.

The rule has legs which measure a nominal 7/32" thick (compared to the usual nominal 3/16"). The thickness of the legs averages about .203" but at the round joint measures .215". I believe it likely that shrinkage would account for this difference.

It is of a common configuration: 2-foot/4-fold, one inch wide, with

The two principal face embellishment lines stop about 1/8" short of extending to the brass tips. At the

TABLE

First example from each series in 1853 & 1860 Belcher catalogs.

Rule Width	Catalog Number	1853 Catalog	1860 Catalog
3/4"	30	medium	—
	40	thin	thin
1"	50	thick	extra thick
	60	medium	—
1-3/8"	70	medium	—

From *The Gristmill*, No. 54 (March, 1989), by permission

rule were produced with the exact same stamps as those on a signed BELCHER BROTHERS MAKERS ship carpenter's bevel in my collection.

The construction style of the round joint (three steel leaves, full width of the legs) and the steel center plates of the knuckle joints are typical of examples by the earlier New York rule makers.

Based upon the fact that the figures on this rule match the signed BELCHER BROTHERS MAKERS ship carpenter's bevel in addition to the statements in the BELCHER BROTHERS & CO. price lists that their round joint rules were not marked with the firm name, I conclude that there is a high degree of probability that this example is a BELCHER BROTHERS & CO. No. 50 "extra thick and strong" rule.

Further information on rules similar to this one or comment by other collectors would be appreciated.■

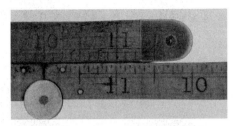

Comparison to Belcher bevel: note the damaged figure stamp of the first digit in the "11." It does not appear in the other double digits or the single 1s.

BIBLIOGRAPHY

Concordance of Major American Rule Makers. Philip E. Stanley. Westborough, MA: The Stanley Publishing Co., 1985.

Price List of Box-Wood and Ivory Rules, Belcher Brothers & Co. Reprinted in: *Introduction to Rule Collecting.* Ken Roberts. Fitzwilliam, NH: Ken Roberts Publishing Co., 1982. Plate IX.

Price List of Box-Wood and Ivory Rules, Belcher Brothers & Co. Reprinted by Ken Roberts Publishing Co., Fitzwilliam, NH: 1982.

Stanley Rule & Level Co., Catalogs of 1859, 1867, 1870, 1872, 1879, 1884, 1887, reprinted by Ken Roberts Publishing Co., Fitzwilliam, NH: 1975 through 1980.

The following article, submitted by William J. Baader, is taken from the book Cooper Industries 1833-1983, *by David N. Keller. The book covers the histories of the companies which Cooper Industries purchased. They are to be commended for compiling this information.*

We extend our special thanks to Mrs. Pat B. Mottram, Manager Public Relations, Cooper Corporation Office, Houston, Texas, for giving us permission to use this article. Also thanks to Roger H. Vaglia, Marketing Manager, Mechanical Drive, and Martin Miller, retired Works Manager of the Springfield, Ohio plant for their help in obtaining this information.

THE LUFKIN RULE COMPANY

When E.T. Lufkin began a board and log rule manufacturing company at Cleveland, Ohio, in 1869, lumbering had reached that general area in its slow sweep from New England to Middle America. Slightly north and west, across the thumb of Michigan, the city of Saginaw was a temporary center of the migrating lumber industry.

With virgin forests covering millions of acres of land, reforestation was not yet an American concern. As supplies were depleted, lumbermen simply moved farther west, selecting from seemingly inexhaustible resources those locations offering the best "lumber streams" where logs could be floated with minimum obstruction to sawmills. Such logs had to be measured for sale, as did the boards coming from the mill saws. Consequently, manufacturers of measuring rules popped up like bubbles in the wake of lumbering's westward movement. Most of them burst almost as quickly, lacking necessary knowledge and the skill to produce quality products, so despite their proliferation, suppliers were hard pressed to find reliable sources of measuring rules.

Quality Products

The E.T. Lufkin Board and Log Rule Manufacturing Company proved to be an exception. Using only the finest second-growth hickory butts available, it produced rules that were shaved down by hand, always following the natural grain, attached heads of carefully tempered steel, and burned figures into the wood, thereby avoiding the weakening caused by the less expensive stamping process used by most manufacturers. Handles were formed by skilled craftsmen and a protective finish came from varnish that the company prepared itself, following an exacting formula of selected gums and pure grain alcohol. Lufkin proclaimed that while its goods were not the lowest in price, they were "the most durable and therefore the cheapest." Business success amidst a cluster of faltering competition seemed to support its claim. The Lufkin log rule, made of flexible hickory, contained scales that provided quick board-foot calculations from measuring the length and varying diameters of a log. Because of natural log tapering and bark cover, the rule simply provided the means of making a feasible estimate.

It was produced with a variety of scales dictated by a strange situation that had developed with logging. Separate scales had evolved as standards in different geographic areas. Containing such names as Scribner, Doyle, Favorite, Baxter, Dusenberry, and Cumberland River, some favored the buyer, others the seller. As a

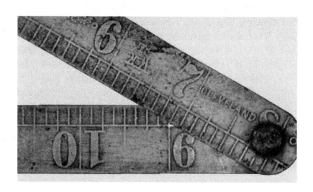

The rule pictured above, from the private collection of William J. Baader, is a three-fold, 24" metal blacksmith's rule in excellent condition. The close-up photographs give proof that it was manufactured between the years of 1885, when the company became known as the Lufkin Rule Company, and 1892, when the move from Cleveland to Saginaw was made. Cleveland-marked rules are the rarest of all Lufkin products, and, as you've guessed, this one is NOT for sale! *Tom Ward photographs.*

From *The Gristmill*, No. 49 (September, 1987), by permission

manufacturer, Lufkin supplied whichever variety a customer requested, avoiding controversy with a published statement of policy: "We are often asked which scale is the nearest correct. We leave that question for the lumberman to decide. We can furnish rules figured with any one of them." The less flexible board rule, also containing calculation tables along with the measuring scales, enabled lumbermen to figure exact board feet after sawing.

Within a few years, the company increased the size of its plant and added new manufacturing lines. It became the largest producer of forged boot calks - spikes that attached to the heels of logger's boots - in what was described then as "the West," and offered specialties ranging from large log calipers to tiny thickness gauges that could be worn on a finger.

New Name, New Manager

As Lufkin grew, it gained the attention of a measuring rule supplier, Morley Brothers Company, located in Saginaw, where logging and lumbering continued to flourish. Morley Brothers had many good customers. In 1877 alone, a billion board feet of lumber were cut and floated along the 26-mile Saginaw River, where there was about one saw mill per mile. Small wonder that Morley Brothers was disturbed by the difficulty in acquiring adequate supplies of quality measuring rules. Four men from the company decided to solve the problem by acquiring interest in a good log and board rule manufacturing concern. Acting independently of Morley Brothers, they contacted E.T. Lufkin, and in 1883 the group was able to purchase half of the Cleveland company. Within two years, they bought the remaining interest from Lufkin, who moved to Chicago, where he organized another business venture. Fred Buck, a 25-year-old Morley Brothers salesman, was hired by the new owners to manage the reorganized and renamed Lufkin Rule Company.

Tall and rugged, the young general manager looked more like a lumberman than a salesman. He moved to Saginaw from Adrian, Michigan, where his father was a hardware retailer. Seriously impaired vision had not deterred him from a successful career; declared totally blind at the age of twelve, he had somehow regained some of his sight, overcoming the handicap to such a degree that colleagues could never determine its true extent. Forceful leadership qualities and mechanical expertise made him a good choice to lead the Cleveland manufacturing operation.

Albert M. Marshall, a Morley Brothers executive who had provided most of the capital to purchase Lufkin, was named president although he did not take an active part in operational management. Soon after the purchase, a young man named Theodore Huss from northeastern Ohio was hired to head sales. A small man, he was a physical contrast to the 250 lb. general manager. But the men were well teamed as a production-sales combination that was to provide the company leadership for half a century.

New Lines Added

By 1890, the company had introduced the first steel measuring tape to be produced in the United States and extended other lines to include straight and folding steel rules, lumber marking implements, glass cutting boards, and a patented "magic pattern rule and chart" that reduced the time for elbow pattern cutting from half an hour to one minute. The still-popular

Lufkin employees standing in front of the company's first factory in Cleveland in 1869.

wooden rules were improved by company designed metal heads and end hooks. A company spokesman reported that "There is scarcely a lumber market in the United States or Canada where Lufkin products are not in daily use."

Saginaw Beckons

Meanwhile, logging continued its steady westward march, from eastern to western Michigan, then into Wisconsin and Minnesota. In the language of the industry, the Saginaw area was "all lumbered out." With its principal source of income vanishing rapidly, the city offered new business enticements in the form of free land and long-term tax exemptions. Lufkin Rule Company was receptive to such a lure. It needed a more modern manufacturing plant and it had management roots in the Michigan city. Constructing a new factory on twenty-three acres of land provided by the city, Lufkin moved in early 1892 to Saginaw, where it continued to expand into new markets requiring quality measuring instruments. Much of its specially designed equipment came from the fertile mind of Fred Buck, who visualized concepts, then described them to draftsmen who prepared the drawings.

New buildings were added with the introduction of new products over the years, and the single factory grew gradually into a manufacturing complex. Well before the turn of the century, the company established sales offices in New York and London. Then, in 1907, it opened a small plant in Windsor, Ontario, becoming Canada's first complete tape and rule manufacturer. (Later, the London office was closed in World War I.)

Although Lufkin had been incorporated in 1901, it was not until 1910 that Fred Buck succeeded Albert Marshall as president. Theodore Huss, continuing to head the sales organization, became vice president. (An interesting sidelight to the Lufkin history is Marshall's later move to Duluth, Minnesota, to form the Marshall-Wells Hardware Company, which became the largest wholesale hardware organization in the world before it was liquidated after World War II. Marshall and representatives of his family were members of the Lufkin board until the company was acquired by Cooper Industries, Inc., in 1967.) Among the company's new products were surveyor's chain tapes, made of steel, and an extensive line of measuring tapes combining woven linen with metallic warp, often contained in leather cases with brass winding handles.

Folding Rules, Precision Tools

Just before World War I, President Buck became interested in the folding wooden rule being manufactured in Europe. Convinced that the product would find wide acceptance in America, he made investigative trips abroad, and in 1915 Lufkin purchased a plant in an area that was to become part of Czechoslovakia when that country was formed from sections of Austria and Hungary three years later. Lufkin produced the zigzag folding wooden rules there for two years. Then the press of war prompted Buck to sell it and bring the process to the United States. In the years that followed, the folding "spring joint rule" became one or the best selling products in the company's long history.

As industrial advancement created increasing need for closer tolerances, Lufkin added a new division for the manufacture of precision hand tools. Starting slowly in 1919, it developed and produced such measuring instruments as micrometers, combination squares, precision scales, radius gauges, and related products. An engineer named Walter Schleicher was hired from an Eastern Seaboard company to develop the precision tools. Eugene Witchger, who had been with the company for several years, was head of engineering for other products, later taking over both segments when Schleicher left the company.

Entrenched competition made penetration of precision tool markets difficult during the 1920s, but the new Lufkin division survived early problems and became an important arm of the company, although it never challenged the leadership of tapes and rules. Product innovations spread Lufkin's reputation in that decade. Among the best known was the "push-pull" metal tape rule, with a concave configuration that held it rigid when extended. President Buck also invented a machine that greatly improved the printing of woven cloth tapes.

Several years later, the inventive Lufkin executive wavered temporarily from his established course of producing only quality products, by creating what he admitted to be a cheap product. Motivation for this uncharacteristic decision came from the nation's Great Depression.

The Lean Years

Layoffs from America's economic collapse of the 1930's spread through industry in a swath that left few companies untouched. There was no way Lufkin could escape, but Fred Buck was determined to preserve as many jobs as possible by developing a small, inexpensive measuring tape that wound like a telephone dial Sold almost at cost, it was created solely to maintain employment, and during the few years it was marketed it kept one hundred workers on jobs that otherwise would have been vacated. Many employees did suffer layoffs, however, as the company inched its way through the lean depression years. But the experience was not entirely new to Buck and Huss. They could remember similar economic panics of 1890 and 1907, and they remained optimistic.

Lewis Barnard, Jr., the president's grandson, who joined the company in 1932 just in time to experience the depression, remembered a story Fred Buck often told about the severe economic reversal of 1907. He recalled how his grandfather described walking along a Saginaw street with his close friend and business colleague, Huss. "Theodore," the president asked, "what have you got in your pocket?" Searching briefly, Huss replied, "Just fifteen cents." Buck thought a moment, then said, "I've got a dime; let's go have a beer."

Neither man lived to see the company recover completely from the Great Depression. Huss died in 1934 and was succeeded as sales manager by George Macbeth, a veteran company salesman, who later became a vice president. President Buck did lead Lufkin out of the worst part of the depression before he died in 1938 at the age of eighty. His successor as president and chief executive officer was Robert C. Thompson, who had headed the New York sales office and warehouse division as a corporate vice president for many years.

WW II Priorities

World War II brought a company commitment to defense industries. Precision tools were in great demand, as American industry tooled up for military production. Rules and tapes were needed also, and Lufkin went on a two-shift work schedule to meet its orders. As one of the company's fifteen hundred employees stated succinctly, "Everything that is built has to be measured." Obtaining material was a problem particularly since a special steel, most of which came from the Roebling Company of New Jersey, was needed for Lufkin products. Chrome, used as a rustproof coating for tapes and rules, was in scarce supply also. Even the high quality hard maple used in manufacturing spring-joint wooden folding rules was difficult to purchase; hard maple was used in the manufacture of army truck bodies.

When the war ended, Lufkin converted easily to peacetime marketing and began to search again for product improvements and new outlets. Lewis Barnard left the company briefly for further business study at Northwestern University, returning in 1947 as vice president. Canadian manufacturing operations, having grown during the war, were moved from Windsor to Barrie, Ontario, in 1948, occupying a new facility that in time was expanded fourfold.

Peacetime Prosperity

In 1950, Lewis Barnard was promoted to president and chief executive officer of Lufkin, Robert Thompson moving to chairman of the board until his retirement six years later. The company set goals of increasing its precision tool business and maintaining leadership in widening lines of metal, wood and woven cloth tapes and rules, to keep pace with accelerating American consumer and industrial needs.

The company increased its sales force, continuing an historical policy of using company salesmen, rather than manufacturers' agents, domestically, and directing foreign sales through New York export houses. The company's own New York sales and warehouse division was closed. Explained President Barnard, "Although we had operated it effectively for a great many years, times changed, and it became more economical to ship directly from Saginaw."

Sales doubled during the 1950s, and the company expanded into several new areas of operation. The first was the purchase of Master Rule Company, a producer of tape rules in Middletown, New York. "That company might have been small, but we learned something from it," said Barnard. "We had thought for years that it was necessary to apply tension on a tape using large, expensive machinery to assure accurate printing. Master Rule had a method of printing just as accurately and a whole lot easier without tension."

Later, the company acquired a plant in Maine, where an independent operator prepared lath, the strips of wood used in manufacturing folding rules. Previously, Lufkin cut its own lath out of hard maple board shipped from the lumber companies in the Saginaw factory. Since one-tenth of the board was usable, the rest became expensive sawdust, when viewed in the context of shipping weight. Purchase of the Anson Stick Company enabled it to prepare lath in Madison, Maine, near the source of hard maple, and send it waste-free to Saginaw.

A manufacturing and sales outlet for Lufkin products was then established as a joint venture, Lufmex S.A., in Mexico City. Later it was transferred to Guadalajara. Moving into Puerto Rico, the company established a dial indicator business, Lufkin Caribe. "We also took a few fliers on some things that didn't work out as well as we hoped, so we got rid of them quickly," the president said of the 1950s.

Lufkin Specialties, Inc.

Having dipped lightly into the manufacture of advertising specialties - products bearing a customer's name for promotional distribution - Lufkin in the early 1960s acquired half interest in Tigrett Industries of Jackson, Tennessee. John Tigrett, who had founded the company, had obtained a patent on a popular metal measuring rule that would roll up into itself. He also had patented a special spring in which Lufkin was interested. A few years later, the company bought the remaining half of the business, moved its own specialties production to Jackson, and Lufkin Specialties, Inc., became a wholly owned subsidiary.

Lufkin Rule Company of Canada began exporting to foreign markets in 1963. Separate arrangements also were made to have product parts shipped from Barrie to a plant in Australia, where they were assembled and marketed.

Cooper Takes Over

In 1964, Lewis Barnard moved to chairman of the board, and William C. Rector, president of True Temper Company in Cleveland, who had served as an outside director on Lufkin's board, joined Lufkin as president and general manager. Another True Temper executive, Harold A. Stevens, became vice president, in charge of Lufkin sales.

Following a feasibility study the company's precision tool line, which had reached more than three thousand items and accounted for approximately 20 percent of sales, was discontinued, with equipment, tools, parts, and patents sold to Pratt & Whitney, Inc. At the same time, a modern new manufacturing plant was constructed in Apex, North Carolina. Limited production was started at the Apex plant in May of 1967.

Five months later the company was acquired by Cooper Industries, Inc.

After the acquisition by Cooper, Lewis Barnard served as a consultant for a short time, then retired. William Rector remained president of Lufkin, with offices at the main plant location in Saginaw. Within a year, however, the antiquated Saginaw manufacturing facility was closed, and production was consolidated at the new Apex, North Carolina, plant, considered the most modern of its kind in the world. ■

ARTICLE SUPPLEMENT:

"Old catalogs dating back to 1885 offered a wealth of information on Lufkin products, and each contained historical company information pertinent to its time. Lufkin's early history was recorded in an updated brochure, in data compiled by the company in 1974, and in a brief sketch written by a company executive in the early part of the century. Details of the industry were outlined in the reprint of a 1916 magazine article, "Things Worth Knowing about Measuring Tapes," and in a booklet, *The Amazing Story of Measurement*. Post-World War II activities, acquisitions, and personal stories were related in great detail by former President Lewis Barnard, with supporting documentation he kept in the form of letters and newspaper clippings."

MEASURING RULES.

By J. RABONE, Jun.

Of the early history of this manufacture it may be sufficient to state that until the early part of the seventeenth century, at which time Edward Gunter invented the line of logarithms graduated upon a sliding scale, which solves problems instrumentally in the same manner as logarithms do arithmetically, the trade never assumed sufficient importance to cause it to be followed by persons who had no other occupation, and to make it worthy of being designated a craft. Up to that time the best measures had been made by the mathematical instrument makers; but this ingenious invention of Gunter, by reason of its universal applicability to measuring purposes, called into existence another class of workmen, superior to those who had hitherto chiefly made the notched sticks similar to those used in many rural districts at the present day, but still somewhat distinct from the opticians and makers of such instruments as quadrants, sextants, and the finer kind of optical and mathematical instruments. The first men who were worthy of the name of rule-makers were to be found only in London; but after a time the trade gradually extended itself to Wolverhampton and Birmingham. Under date of the 10th of August, 1664, Mr. Pepys records in his diary his experience of the scarcity of competent rule-finishers (as the men who make the rules are called) in London. He writes:—

"Abroad to find out one to engrave my tables upon my new sliding rule with silver plates, it being so small that Browne, that made it, cannot get one to do it. So I got Cocker, (the celebrated arithmetician and) famous writing-master, to do it, and I sat an hour by him to see him design it all; and strange it is to see him, with his natural eyes, to cut so small at his first designing it, and read it all over, without any missing, when for my life I could not, with my best skill, read one word or letter of it."

And then on the next day Pepys writes:—

"Comes Cocker, with my rule, which he hath engraved to admiration, for goodness and smallness of work; it cost me 14s. the doing."

From the fact that Browne, the mathematical instrument maker, was unable to find anyone to engrave Mr. Pepys's sliding rule, it is to be supposed that the art had not then attained to much perfection, or rather that proficiency in it had been made but by few. In 1755, nearly a century later, is recorded that even in London "it was a most difficult matter to get rules good, there being only one man who could make them perfectly well, and he had lately taken to other work." In July, 1755, James Watt found his first employment in London as a rule-maker, under Mr. John Morgan, of Finch-lane, Cornhill, of whom Watt writes to his father, that "though he works chiefly in the brass way, yet he can teach me most branches of the business, such as rules, scales, quadrants," &c.

From *Birmingham and the Midland Hardware District, the Resources, Products, and Industrial History*, edited by Samuel Timmins, pp. 628-632. London: Robert Hardwicke, 1866.

In less than two months after his arrival in London Watt says he had made a brass parallel rule eighteen inches long, "and a brass scale of the same length." By December he "could work tolerably well," and "he expected that by April he would understand so much of his business as to be able to work for himself." When June had again come round, and within twelve months of his first attempt at rule-making, Watt wrote to his father, that "he could now make a brass sector with a French joint, which is reckoned as nice a piece of framing-work as is in the trade." Watt, however, having doubtless acquired proficiency in the art of rule-making, did not "work for himself," as it appears he intended to do, but turned his attention to the improvement of the steam engine, and thus conferred greater benefits upon the world than he would, had he continued the manufacture of "parallel rules," or even "sectors with French joints." It may be interesting to know that the sliding rule, long used by James Watt, is still preserved in Birmingham, in the possession of an eminent founder; and it is still used as a check in testing calculations. Is it not probable that this rule may be the handiwork of its former great and famous possessor? Five and twenty years later the finer kind of measures were still made in London by the instrument makers; but the more ordinary kinds of carpenter's rules were manufactured almost exclusively in Wolverhampton. At the latter part of the past century only three or four rule masters, each employing a few apprentices and men, were to be found in Birmingham, and one at Harborne adjacent; but now the trade has almost deserted Wolverhampton, which numbers only four or five persons employed in it, while Birmingham affords employment to as many hundreds. With the exception of three or four makers scattered throughout the country the trade is now entirely confined to Birmingham and London. Many of the rules sold as London-made are produced in Birmingham, and many are framed in Birmingham and sold to the London makers, who mark or finish them themselves. The joints of the earliest made rules were cut from the wooden pieces which formed the legs, and brass plates were attached to the outside to strengthen the otherwise weak and wooden joints. Afterwards the joints were composed of metal, and attached to the wooden sides, instead of being made out of them. The joint, so well-known as the arch-joint, and which is seen upon so great a number of joiners', and other rules, was first made about sixty years ago; yet in a celebrated modern painting representing the early printers at their labours in Westminster Abbey, we find an accurate representation of a two-feet rule with this kind of arched joint, the artist being unaware that it was not invented till more than 300 years after those early printers had distributed their last form.

The general shape of rules has been but little varied, the convenient arrangement of two or four legs having been early introduced. Seventy years ago, the ordinary thickness of the boxwood carpenter's rule was not less than a quarter of an inch, but that size has been gradually reduced from time to time in order to give greater flexibility. Many rules have been made of only the sixteenth of an inch in thickness, this small size of wood having to be further cut away to admit of the insertion of the joint, which is inlaid so as to lie level with the surface of the wood. To remove this weakness of the joint, which was the common fault of all thin rules, a joint was produced by Messrs. John Rabone & Son, which being rivetted upon the surface of the boxwood legs instead of any portion of them being cut away to make place for it, adds strength to that part of the rule which had formerly been the weakest. These joints are also enriched by many artistic and ornamental designs stamped upon them, which render the rules, whether regarded for the greatest strength combined with extreme thinness, or for their elegant appearance, appropriately designated by the appellation of "nonpareil." The processes of sawing, rivetting, filing, and planing, are too well known to need reference here; but the numbering and graduation of rules is an operation which, to those unacquainted with it, is generally regarded as the most interesting of the processes. Since the invention of the dividing engine by Ramsden, and its improvement by Troughton, Adie, Reichenbach, Gambey, and others, all the most accurately and finely graduated astronomical and other instruments have been divided and marked by means of it; but the process is much too slow and costly to be employed upon the rules in general use by mechanics. Straight scales and rules are usually divided by placing the article to be divided and the original pattern side by side, then passing a straight-edge with a shoulder fixed at right angles to serve as a guide, along the original, and pausing at each division; then a corresponding line is made on the copy by the dividing knife. Segments of circles are also graduated in the same way, by making a straight-edge revolve on the centre of the circle, and marking off the divisions as on the straight scale. This method, in skilful hands, admits of much accuracy, and was applied to the graduation of theodolites and other instruments prior to the invention of the dividing engine.

About eighty years ago much labour was spent by a Wolverhampton rule maker in the construction of a pair of steel rolls, having upon their surfaces in relief, the various lines, divisions, and figures necessary to impress, at one operation, the two sides of a scale or rule. Its costliness may be guessed at, when it is stated that the scale it was intended to impress contains 2205 lines, divisions, and figures. But the attempt was, as might have been supposed, a failure. Owing to the inequality in the density of the wood, and the slight unevennesses of its surface, the work performed by it was defective, and its use was abandoned. Marking by pressure produces a thick, shallow, and ill-defined line; whereas percussion, when applied to the steel stamps which bear the figures or letters, gives a sharp, clear, deep and lasting impression. The filling up of the marks on the rules is easily effected by simply smearing over them a mixture of oil and charcoal, and wiping the surface clean.

In the year 1858, the late Dr. William Church, of this town, invented and patented a number of machines for the making and graduating of rules. These inventions displayed a vast amount of thought and ingenuity; but he was unaware of what had been done by a few persons in the trade before, and unfortunately for him all that was good and practicable in his inventions had been in use for many years, and his marking apparatus was too complicated ever to remain long in working order. And when it is known that the 540 divisions in the Gunter's line which may be seen on the brass slide of any carpenter's slide rule, are cut singly and accurately by hand in ten minutes by the process above described, it will not be surprising that in this instance manual labour should be cheaper and more to be relied on than machinery, designed as Dr. Church's was, to draw all those divisions at a single operation. In the working of several hundred teeth or cutters all connected together, the blunting or breaking of any one of them would render the work imperfect, and would consume more time in the filing up than the entire process, when performed by the usual method.

The box-wood, of which rules are made, is chiefly grown in Turkey, the English grown box-wood not being of sufficient size and quality. Formerly, large quantities of rules were exported to the United States of America; but, during the past thirty years, they have been manufactured in the States, a prohibitive tariff having prevented the introduction of English made goods. The way in which some nations—or rather—some Governments—regard the importation of rules into their territories is rather strange. Some allow, at fair and equitable rates of duty, the importation of English made rules, the measures being suitable to the use of their respective countries. But while one country will not admit them at all, unless the measure of the country is surrounded by other measures, so as to make the rule a scale of varied measures, instead of being only the legal standard of the country; another Government looks upon all rules, coming to its ports as illegal measures, because they have not the Government stamp upon them, and all so found are seized. The makers do not put a resemblance of the official stamp upon them; and, as there are no native rule makers, and all the rules are imported, measures, a little different from the legal standard are made, and being marked with the name of another country than that for which their use is intended, are allowed to be admitted; thus, the greater part of the people are necessarily using an incorrect measure, at the same time that the true standard might be supplied to them as cheaply and more readily. The various counties and states of the world have, at the present time, about 150 different measures; many small places, each containing only a few thousand inhabitants, have their own peculiar measures. At various times, many of these measures have been changed from the duodecimal to the decimal division, still retaining the original standard of length; but, within the past few years, the use of the French metrical measure has become much more common in all parts of the world. Many nations, now, have a large proportion of their rules marked with the French metre, in addition to their own standard; and are thus, doubtless, gradually tending to the general adoption of so convenient and perfect a measure of length as the French metre is admitted to be.

RULES.

THE DIFFERENCE BETWEEN SCIENTIFIC AND ORDINARY KNOWLEDGE. — MAN AS A MEASURER. — THE NEED FOR RULES. — AN ACCOUNT OF HOW THEY ARE MADE. — BOXWOOD. — ITS PRODUCTION. — THE FACTORY OF STEPHENS AND COMPANY. — THE PROCESSES OF RULE-MAKING. — THE MACHINERY USED IN THE BUSINESS. — MR. D. H. STEPHENS AS AN INVENTOR. — THE STEPHENS PATENT COMBINATION RULE. — THE VARIOUS USES TO WHICH IT MAY BE APPLIED.

THE ability to measure accurately, and thus obtain a definite and positive knowledge, instead of a general and indefinite knowledge of form, relation, distance, and the other phenomena of the existing condition of things in which we are placed, constitutes the difference between scientific knowledge and ordinary knowledge. By some philologists our term *man* is traced to a derivation from the Aryan root-word *ma*, to measure. Whether this derivation is true or not, certain it is that the most accurate and comprehensive definition of man, as classified at the head of the organic evolution of intelligence upon this planet, is that of a measurer, and, as symbols of his true domination of the world, a rule and a pair of scales would be much fitter and more expressive of his glory than a crown and a sceptre.

The use of the rule is so absolutely necessary in almost every mechanical or artistic pursuit, that the consumption is, of course, very great, and the manufacture is consequently a very important one. Rules are generally made of boxwood or of ivory, and are mounted and tipped with brass or silver. Boxwood is most extensively used, both on account of its being more plentiful than ivory, and also because it is less liable to expand and contract by variations of the temperature. This last consideration is the most important, since the accuracy of the rule depends upon the constancy with which it marks the fixed standard for lineal measurement.

The boxwood used by the chief manufacturers of rules grows in Turkey and Southern Russia. The forests in which it is produced are under the control of the respective governments, and are farmed out, or leased to special contractors, who pay for the privilege by a certain percentage of the income from the sale of their produce.

From *Great Industries of the United States,* by Horace Greeley et al, pp. 739-744. Hartford: J. Burr & Hyde, 1873.

The forests within the jurisdiction of Russia are leased by the government of that country to two persons, who have full control of the cutting and disposition of the boxwood. These forests occupy mountain ranges for the most part. The boxwood is cut, and after a suitable time brought down from the mountains to the market or depot for sale, on the backs of mules.

The tree producing boxwood, known botanically as the *buxus*, whence our name for it, is small in size, the average diameter of the logs which reach this country being from six to seven inches, and never more than fifteen. The boxwood imported by our manufacturers is now brought directly from the depots in Russia and Turkey to New York and Boston, by the way of Smyrna. Formerly it was taken to Liverpool, and there transshipped to this country. Boxwood is sold by weight, the prices varying from thirty to one hundred and fifty dollars a ton, the value depending upon the texture, color, and straightness of the grain; the color being an important consideration. The deeper the golden tint of the wood, the more valuable it is for rules.

The manufacture of rules is extensively carried on in the United States, and from its extent, and the multiplicity of interests which depend in a greater or less degree upon its accuracy, by which their own is regulated, it may be justly classed as one of the great industries of the land. The leading manufactory of the United States is that of Stephens and Company, at Riverton, Litchfield County, Connecticut. Their factory is situated on the Tunxis, a small river, which is one of the chief branches of the Farmington River, and which at this spot supplies a good water-power. The establishment of Stephens and Company has mainly grown up under the fostering care of Mr. DeLoss H. Stephens, who unites with the character of an able and indefatigable business man the genius of a first-class practical inventor, and who, by machinery of his own invention, has greatly promoted the manufacture of rules in this country. The most important inventions which Mr. Stephens has made have been wisely kept within the knowledge of Stephens and Company, and secured to their special control, not by letters patent, but by private use. By the aid of these machines, and improvements in others, which have been secured in the same way, Messrs. Stephens and Company have been enabled to manufacture the best quality of rules in the market, at less cost than many of their competitors have been able to. Some of Mr. Stephens's inventions have, however, been patented.

By the lessening of the cost of production brought about by the use of their machinery, Messrs. Stephens and Company are enabled to give the purchasers of their rules more perfect and more conscientiously made wares at lower prices than their competitors can well afford, and thus they have risen to their eminence in this branch of manufacture.

The making of rules is a nice art, and quite interesting in its details. The boxwood logs, on their arrival at the factory, are first "blocked up," or sawed into proper lengths or sections, which are then quartered, or split into four minor sections, which are then slabbed or cut into pieces about the width of a rule. These slabs are then slit into pieces about an eighth of an inch in thickness. The next process is to "dress off," or gauge, each piece as to its broader surfaces, or sides, and its edges, into the required shape and size. This is done rapidly and perfectly by an automatic machine of ingenious contrivance.

The "stuff," as the pieces of boxwood are called, is next fitted or adjusted to the kind of joints to which the pieces are finally to be united. These joints are respectively called "head joints" and "middle joints." The next step is to "tip" the boxwood pieces, that is, to fit the brass or silver caps upon their ends; brass being chiefly used for mounting the boxwood rules, and German silver, or real silver, for the ivory rules. Pure silver mountings are too expensive for the general demand; though on the occasion of our visit to the factory of Stephens and Company we saw several splendid ivory combination rules made to the order of Governor Claflin of Massachusetts, and other "republican sovereigns," which were mounted with pure silver.

The brass used in the manufacture of rules is brought to the factory from the rolling-mills, in sheets prepared for the purpose, and is slit by circular shears and saws into proper sizes, and then cut with dies into the forms needed for the construction of the joints, caps, bindings, etc., which are used in the rules. The joints of the rules made by Messrs. Stephens and Company are, by peculiar machinery, "scraped," or trimmed; this process doing away with the slower one of filing, milling, etc., and leaving the work more perfect. The machine which does this is the invention of Mr. DeLoss H. Stephens, and is patented.

Uniting the several pieces which form the common joint is done in this factory by a "driving machine," which performs this work at least one hundred per cent. more expeditiously than the hand method of driving which formerly prevailed. The "rolls," or collets, or the cylindrical parts of the rule joints, or shoulders over which the jointed parts of the rule turn, are made here by an ingenious automatic machine, with a great saving of labor and material. Everything in the establishment, even to the cutting of the pins, of which about twenty enter into the construction of a plain rule and forty into that of a bound rule, is done by machinery instead of by hand, as was formerly the practice. The heading of the rivets and the marking out of the "arches" to receive the joint-caps are also done by machinery.

After the work is put together and made ready for undergoing the process of "graduating," it is taken to the graduating room, where the lineal gauging is performed. The machines by which the surface of the rules is marked into inches and parts of inches are automatic in their operation, and quite complicated in their construction, and perform their work with more than human accuracy, and with almost living intelligence. These machines are the inventions of Mr. DeLoss H. Stephens, and alone would be enough to secure for him a place among the first scientific mechanical inventors of his time. Their work is neat, delicate, perfect, and rapidly performed. While these machines are necessarily so complicated in their construction, yet they are so simple in their action that they can be safely left to be operated by a boy.

After the rules are completed, they are then thoroughly inspected, any blemish or fault, however slight or trivial, condemning them. Stephens and Company manufacture over one hundred different varieties of rules, which are in demand all over the United States, in Australia, South America, and in Europe also. Some varieties of their rules are manufactured solely by Stephens and Company, among which is the celebrated "Stephens' Patent Combination Rule," cuts of which are here given. This rule, an invention of Mr. L. C. Stephens, the founder of the business and the father of the present owner, is made of boxwood or ivory, and combines in itself a carpenter's rule, spirit-level, square, plumb, bevel, indicator, brace-scale, draughting-scale, T square, protractor, right-angled triangle, and with a straight edge can be used as a parallel ruler. It has but one joint, and is bound with brass. When folded it is six inches long, one and three eighths inches wide, and three eighths of an inch thick. The cuts, which are just half-size, represent the rule in three positions: first, as a spirit-level; second, as a try-square level and plumb; when partially opened it serves also as a slope-level.

The angles formed by the blade and leg *decrease* just one half as

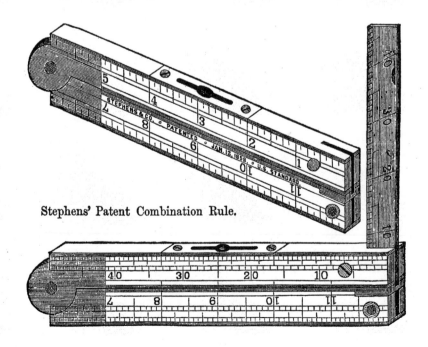

Stephens' Patent Combination Rule.

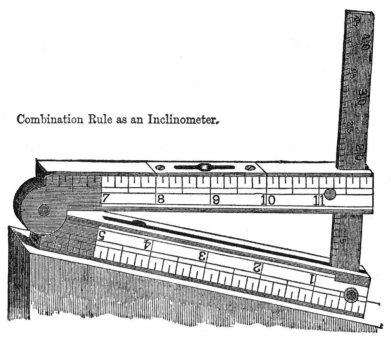

Combination Rule as an Inclinometer.

fast as the angles formed by opening the legs of the rule *increase*. The upper edge of the other side of the blade is graduated into inches and eighths, and thus shows the *pitch to the foot*. The inner edge of the leg which holds the glass is also graduated to measure the angles which are formed by turning the blade in the leg which holds it. These degrees show how much the right angle is reduced as the blade falls from that position. As a T square it is also a right-angled triangle. One side of the blade is divided into twelfths, also the inside edge of the leg which holds it, thus constituting a brace-scale.

The slotted screw which passes through the end of the leg is used in adjusting the square, should it require it. By this simple arrangement for measuring angles, this rule becomes invaluable to practical mechanics of all kinds, as well as to surveyors, draughtsmen, architects, and every one who knows the value of an instrument by which he can readily and reliably measure whatever he may wish.

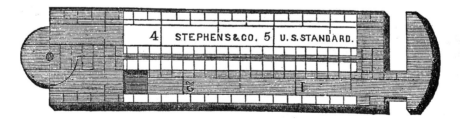

The above cut represents a caliper rule, also manufactured by Stephens and Company. The value of this convenient device for measuring accurately the diameters of round and cylindrical bodies is too well known to persons who have practically become acquainted with it to need more than the simplest mention here.

INTRODUCTION TO RULEMAKING
AT BIRMINGHAM
by Kenneth D. Roberts

Sketchley's 1767 Birmingham Directory, the earliest published, noted three rulemakers: Joseph Edwards at Bull Street, Frost & Whithnoll at Lichfield St. and Christopher Jacob at Small Brook. At this time the trade of rulemaking was principally carried out at London and was just beginning to assume importance at Birmingham. It steadily increased, peaking to 52 firms in 1876 and then declined to a dozen by 1900. An article, "Measuring Rules," by John Rabone, Junior was published in *Birmingham and the Midland Hardware District* [London, 1866, pp. 629-637]. At that date his father, who had been in the business since 1808, was still living and he undoubtedly contributed historical data. John Rabone, Jr. (b.1820; d.1892) spent his life in this business and died the year that this subject catalogue was published. At the time of his death, John Rabone & Sons was unquestionably the largest rule manufacturing firm in Birmingham and had the longest tenure, claiming establishment in 1784. [In the 1876 Birmingham Directory Francis Cox claimed establishment in 1780, but this firm dissolved in 1888.] This 1866 article is transcribed in entirety following this introduction and provides numerous details regarding the history of English rulemaking. Another excellent source concerning the historical development of rules in England appears in the *Carpenter's Slide Rule — Its History and Use*, originally published in 1880 by J. Rabone & Sons. This is being reprinted as a separate publication simultaneously with this 1892 Catalogue.

The 1861 Birmingham Directory was the first to note that the firm of John Rabone & Sons was "established in 1784." There were so many Rabones associated in different Birmingham rule enterprises that it is difficult to verify from the limited source of Directories the exact dates of their origins. The main problem is there are gaps of several years in the dates of publication of these early directories. The earliest listed rulemaker in this family was Michael Rabone, noted in 1800 as a "rulemaker and wood turner." Earlier noted Rabones were Richard, listed in 1767 as a buckle maker and Joseph in 1777 as a toy maker. [The importance of the trades of buckle and toy makers in the industrial history of Birmingham, 1750-1850, is discussed in my book *TOOLS FOR THE TRADES & CRAFTS (1976, pp. 14-20).*]

Michael Rabone was also listed as a rulemaker in 1803. Apparently soon after he died, and the business was succeeded to by his wife as Elizabeth Rabone & Son, noted in the 1808 Directory. Presumably the son was John (b.1795; d.1868). John went into the rule business under his own name at St. Paul's Square in 1825. Elizabeth Rabone continued in the rule business until 1845, then listed at Hockley Hill. Ephraim and Thomas Raybone, possibly other sons of Michael, were listed as rulemakers respectively 1816-1820 and 1829-1854.

In 1852 John, Jr. was taken into the firm, which then became John Rabone & Son. About this time, or shortly before, steam machinery was introduced as a power source for manufacturing rules. After the death of John, Sr. in 1868, the firm moved to Hockley Hill in 1871. About 1880 John, Jr. took in his two sons and the firm became John Rabone & Sons. After John, Jr. died in 1892, management passed to his two sons, H. J. and A. J. Rabone who respectively died in 1926 and 1932. Family successors were Eric Rabone (d.1953) and Arnold Godfrey who managed the firm until the merger with the Sheffield firm, James Chesterman Co. Ltd., when the new stock company Rabone Chesterman Ltd. was formed in 1963.

James Chesterman was first listed in the Sheffield Directory of 1837, noting he was a patentee of the spring tape measure and a window blind roller. In 1841 he was in partnership with James Bottom as patentees. They were issued a joint patent, #9214, on June 11, 1842 — "Improvements in Tapes for Measure." This was manufactured by I. P. Cutt (established 1828 in Sheffield), a firm that also made telescopes and opera glasses. John Bottom was granted a patent for an anti-friction brace (#10,526, Feb. 20, 1845), which he sold to Henry Brown. *[Some 19th Century English Woodworking Tools, 1980, p. 312.]* James Chesterman was also granted a patent for "Improvement in Carpenter's Braces" (#12, 843, Nov. 13, 1849) *[ibid, p. 332].* In 1849 Chesterman joined the firm of Cutts, Chesterman & Beddington which manufactured his patented tape measure (#11,962, Nov. 13, 1847). James Beddington left the firm and removed to Birmingham, entering the rule business there. The firm became Cutts, Chesterman & Co. *(see advertisement, SOME 19th CENTURY ENGLISH WOODWORKING TOOLS, Plate CLXXXVII, p. 336).* This was dissolved in 1851 and became James Chesterman & Co. After James died in 1867, the firm was managed by his son William until his death in 1930. His sons, Samuel and Gerald, succeeded in managing the firm until 1945 when it became a stock company, James Chesterman Ltd. On April 1, 1964 a merger was made with John Rabone & Sons Ltd. upon which the two companies traded by the name of Rabone, Chesterman Ltd. with offices at Birmingham.

At the turn of the 20th Century the next largest rule manufacturer at Birmingham was Edw. Preston & Sons, established as Edw. Preston, Jr. in 1864. His father was a planemaker, who began business in Birmingham in 1833. Initially Edw. Preston, Jr. started in 1864 as a spirit level manufacturer, but added boxwood and ivory rules to his line in 1867. His firm expanded, acquired his father's plane business and added other tools to their line. The firm became Edward Preston & Sons in 1888 when his three sons joined the business. The same year the firm acquired ownership of Thomas Bradbury & Sons, which had been established in 1834. A detailed history of Edward Preston & Sons has been published as a documentary in my reprint of their *1901 Illustrated Price List of Rules, Spirit Levels, Planes & Tools* [1979]. The line of rules appearing in that catalogue was equally as extensive as those in this *1892 J. Rabone & Sons Catalogue.*

A check list of 255 firms making rules during the period 1767-1900 has been compiled from the Birmingham Directories. Interesting insights of some of these firms from notations from various directory listings have been transcribed.

From *Fundamentals of Rule Collecting,* edited by Kenneth D. Roberts, pp. 36-39. Fitzwilliam: The Ken Roberts Publishing Co., 1997, by permission.

CHECK LIST OF BIRMINGHAM RULEMAKERS from the BIRMINGHAM CITY DIRECTORIES to 1900

ADLAM, George	1870 - 1900	BRITTLE, John	1882
ALLBUTT, I.	1845	BRITTLE, Joseph	1839
ALLEN, Thomas	1852 - 1856	BROCKHURST, Wm	1880 - 1886
ALLEN & ROWE	1849 - 1850	BROOKE, George	1861 - 1863
APPLELY, Edw.	1856	BROWN, William Henry	1835 - 1886
ASTON, J.	1854	BROWN, Wm. H. & Co.	1882 - 1886
ASTON, Jno.	1854 - 1870	BRUETON, John	1882
ASTON, John	1872 - 1890	BUTLER, Benjamin	1828 - 1866
ASTON, Samuel	1818 - 1820	BUTLER, Ben., jr.	1865
ASTON, Sampson	1833 - 1870	BUTLER, Mrs. H.	1866
ASTON, Sampson & Co.	1871 - 1872	BUTLER & POWELL	1812 - 1820
ASTON, T.	1845	CARR, Jno.	1856 - 1858
ASTON, Thos.	1818 - 1871	CARR & WISEMAN	1861
ASTON, Thos.	1845 - 1890	CHARLES & Co.	1876
ASTON, Thos. & Sons	1876 - 1895	CLARK, Chas.	1882 - 1900
BAKER, Arthur	1880 - 1886	CLARK, Wm.	1829 - 1852
BAKER, Henry	1839 - 1876	COOPER, Geo.	1863 - 1900
BAKEWELL, Richard	1719 - 1825	COPE, Thomas Henry	1852
BAKEWELL, W.	1854	CORBETT, Thos.	1849 - 1876
BARRATT, Jno.	1856 - 1858	COX. A.	1800 - 1802
BARRINGTON, Philip	1845 - 1900	COX, A. & Geo.	1808
BARROW, F.	1863	COX, Francis B.	1828 - 1882
BEDDINGTON, Mrs. E.	1854	COX, George	1812 - 1825
BEDDINGTON, E.	1856 - 1863	COX, Thomas	1775 - 1798
BEDDINGTON, James	1849 - 1876	DARLING, J.	1863
BEDDINGTON, James & Son	1886 - 1890	DAVIES, John Richard	1890 - 1895
BEDDINGTON, John	1845 - 1852	DAVIS, George	1880 - 1886
BEDDINGTON, PHILIP	1863	DAVIS, G. H.	1845 - 1866
BEDWORTH, Thos.	1863	DAVIS, John George	1876
BEES, HIGGISON & Co.	1888 - 1900	DAWSON, William	1818 - 1820
BELCHER & HASSELL	1798	DENCHFIELD, James	1876
BESWICK, W.	1862	DIXON, Benj.	1856 - 1867
BETTS, Edwin	1849 - 1886	DUWELL, Wm.	1865 - 1870
BETTS, James	1825 - 1849	EDWARDS, Joseph	1767 - 1777
BETTS, John	1825	EDWARDS, Wm.	1880
BETTS, Peter	1860 - 1865	FARNOLL, John	1818 - 1845
BETTS, Richard	1870	FLOYD, John	1863 - 1876
BETTS, Thos.	1812 - 1863	FLOYD, W.	1845
BETTS, William	1829 - 1835	FLOYD, Wm.	1856 - 1867
BIRD, Joseph	1882	FLOYD, Wm. Henry	1876 - 1900
BLUNT, T.	1860	FLOYDS, Wm.	1839
BOLTON, Fred.	1852 - 1858	FRANKLIN, E.	1856 - 1858
BOLTON, Thos.	1825 - 1845	FROST,	1777
BOLTON, T. & H.	1849 - 1850	FROST & WHITNOLL	1767 - 1774
BRADBURN, G.	1858	GABRIEL, W.	1845 - 1852
BRADBURN, George	1876	GARBETT, Henry Joseph	1870 - 1872
BRADBURN, Geo. & Thos.	1841 - 1852	GEAR, Henry	1876
BRADBURN, Robert	1862 - 1863	GERREDD, James	1835
BRADBURN, Thos.	1852 - 1861	GERRETT, J.	1856
BRADBURN, Thos. & Son	1862 - 1890	GIBBONS, Mrs. E.	1845
BRADLEY, J.	1829 - 1866	GRANGER, James	1876
BRADSHAW, Alex.	1854 - 1876	GREATOREX, Thos.	1863 - 1882
BRAILEY, John	1880 - 1882	GREEN, Thos.	1865 - 1882
BRAMPTON, Fred.	1863 - 1866	GRIFFITHS, Francis Henry	1876
BRAMPTON, F. & Co.	1882	GRIFFITHS, T.	1861 - 1870

Name	Dates
GRIFFTHS, Thos.	1841 - 1863
GROSVENOR, Richard	1882
HAND, George	1860 - 1876
HARLOW, Isaac & Co	1833
HARMER, James	1880 - 1895
HARPER, Alfred	1888 - 1900
HARPER, Nathan	1849 - 1850
HARRIMAN, J.	1858 - 1865
HARRIMAN, John	1829 - 1835
HARRIMAN, Thomas	1812 - 1852
HARRIS, Chas.	1865
HARRIS, T.	1825
HARRISON, Wm.	1839 - 1850
HASKINS, Danl.	1863 - 1882
HAYCOCK, Edward Samuel	1852 - 1856
HAYCOCK, E.S. & Co.	1845 - 1850
HAYCOCK, Samuel	1829
HEMMING, Edwin	1845 - 1856
HEMMING, H.	1854
HIDSON, Frank	1860 - 1895
HIGGISON, John	1833 - 1890
HILL, Abraham	1830 - 1831
HITCHEN, Thos.	1849 - 1861
HODGETTS, Thos.	1829
HOLLOWAY, John	1856 - 1895
HOLMES, Thomas	1856 - 1858
HORTON, Edward	1880
HUDSON, C.	1861
HUGHES, James	1860 - 1865
HUGHES, Jno.	1852 - 1865
HUGHES, William	1876
HUGHES & Co.	1871
INMAN, Chas.	1849 - 1876
INMAN, Wm.	1839
INMAN, Wm. & Son	1841 - 1842
JACOB, Christopher	1767 - 1770
JACOB, William	1770
JOHNSON, George	1882
JOHNSON, Wm.	1882
JONES, J.	1845
LANGFORD, R.T.	1845
LANGFORD, Thos.	1845
LAPWORTH, Wm.	1852
LAWLEY, G.	1833 - 1842
LEE & HAYWARD	1861
LOGIN, John	1876
LONGFIELD & CHAPLIN	1870
LUCAS, Wm.	1841 - 1870
LUDLOW, T.	1825 - 1842
LUDLOW, W.	1845 - 1854
LUNT, Joseph	1835 - 1861
MALLIN, Abraham	1845 - 1882
MALLIN, W.	1845 - 1854
MALLIN & Co.	1842
MAN, T.H.	1849 - 1850
MILLWARD, Catherine	1870
MILLWARD, Henry	1852 - 1867
MORRIS, Joshua	1876
MORRIS, William	1801 - 1833
MOTTERAM, R.	1862
MOVITTA, Wm.	1882
NELSON, Joseph	1839 - 1845
NICHOLLS, John	1867 - 1880
NICHOLLS, John Merritt	1852
NICHOLLS, T.M.	1858
NISBETT, J.	1845 - 1850
NORRIS & FLETCHER	1882
NORRIS & Son	1870 - 1876
ONIONS, Benjamin	1825 - 1852
ONIONS, Thos.	1785 - 1798
PALSER, WM. Hands	1866 - 1876
PARKE, Henry	1835
PARKES, Jas. & Son	1856 - 1876
PARKES, Samuel Edward	1871 - 1876
PARKES, Mrs. Sarah Eliza	1880 - 1882
PERKS, Wm.	1856 - 1882
PERKS, Wm.	1856
PHILLIPS, Geo.	1829
PHILLIPS, Wm.	1861 - 1870
POWELL, Robert	1825
PRESTON, Edw., jr.	1867 - 1888
PRESTON, Edw. & Sons	1888 - 1933
PRICE, Benj.	1865 - 1866
PRICE Brothers	1867 - 1870
PROBIN, W.	1845 - 1850
PURSER, G. & W.	-1876
QUINTON, Jno.	-1856
QUINTON, Wm.	1856 - 1863
RAYBONE, Elizabeth	1808 - 1835
RAYBONE, Mrs. E.	-1845
RAYBONE, Elizabeth & Son	1808
RAYBONE Ephraim	1816 - 1820
RAYBONE, John	1825 - 1842
RAYBONE, John & Son	1852 - 1877
RAYBONE, John & Sons	1880 - 1963
RAYBONE, Michael	1800
RAYBONE, Thomas	1829 - 1854
REEVES, D.	1860
RICHARDS, Daniel	1876
RICHARDS, John	1815 - 1825
RICHARDS, Joseph	1876
RICHARDS, Wm.	1785 - 1798
RICHARDS & LOCKYEAR	1828 - 1831
ROBERTS, J.	1858 - 1861
RODGERS, Walter	1876
ROTTON, Ambrose	1829
ROWE, James	1858 - 1900
ROWE, Thos.	1862
RUSSELL, Edward	1852 - 1858
RUSSELL, Edward & Co.	1849 - 1850
SALT, Abraham & Son	1801 - 1825
SALT, Isaac	1829 - 1833
SHAW, Thos.	1870
SHAW, Wm.	1825
SMALLWOOD, Josiah	1863
SMALLWOOD, John & Daniel	1862 - 1900

SMITH, Mrs. Ann	1867	TURNER, Joseph	1839 - 1845
SMITH, Samuel Abbot	1845 - 1872	TURNER, Thomas	1870 - 1872
SMITH, T.	1845	WALTON, E.	1845
SNELSON, Jospeh	1829	WARWICK, Henry	1890 - 1893
STARMER, Thomas	1871 - 1882	WESTON, G.	1845
STINTON, Horatio	1856 - 1886	WHITTINGHAM, T.	1845
STUBBS, Geo.	1866 - 1882	WILSON, Alfred	1876 - 1882
SUMMERS, John	1890	WILSON, W.	1845 - 1865
TAPLIN, W.	1854	WITHNOL, Wm.	1777
TERILL, Edmund Henry	1856 - 1882	WOOD, Henry	1777
TINSLEY, J.	1866	WOOD, John	1777
TITTERTON, Thomas W.	1852 - 1866	WRIGHT, John	1888 - 1900
TROW, Isaac	1835 - 1858	WRIGHT, Wm.	1785 - 1798
TRUEMAN, Thomas	1861 - 1866	YATES, Hy.	1856 - 1861
TURNER, John	1854 - 1865		

SOME LISTINGS OF RULEMAKERS from BIRMINGHAM DIRECTORIES, (1815-1886)

1815 - BAKEWELL, Richard - Loveday St., manufacturer of mathematical instruments, brass compasses, plated and brass dog collars, ivory and box rules, land surveyors' measuring tapes, &c.

1845 - COX, F. B. (successor to G. Cox & Co.) - 50 Camden St., New Hall Hill

1856 - TAPLIN, Wm. M. (Late of 50 Hatton garden, London) - Poole St., Aston Brook

1860 - BEDDINGTON, James [Steel] (late Cutts, Chesterman & Beddington of Sheffield) - 10 Russell St., Steelhouse Lane

1861 - ASTON, Sampson (woods cut to any size and prepared for engravers) - 23 Masshouse Lane, near Dale end

1862 - RAYBONE & SON - 61 St Paul's Square - Box, ivory and brass rules suitable for the home, colonial, continental and South American markets. The only manufacturers of the celebrated nonpareil joint rulers.

1865 - STINTON, Horatio (parallel and rolling) - 69 Lovejoy St.

1867 - RABONE, John, & SON (the oldest house in the trade, established in 1784) manfrs. by steam machinery of every description of English and foreign warranted box-wood, ivory and metal rules, suitable for the home, colonial, continental, Indian and South American markets - 61 & 62 St. Paul's Square

1867 - SMALLWOOD, John & Daniel, also makers of iron & steel rules, spirit levels, squares, &c. - Leopold St.

1870 - PRESTON, Edward jun. - 26 Newton St. - manufacturer of box, ivory, wood and brass rules; spirit levels, brass and lead plumb bobs; trammel heads, steel squares, plane maker's stops and plates; improved carpenters' bench stops; inventor and sole manufacturer of the patent spring punch sett; straight and spiral spill machines.

1876 - BRADBURN, Thomas & Sons - 111 Great Charles St. - manufacturer of rules, measuring tapes, joiners' and edge tools, general hardware factors, Established 1834

1876 - BROWN, W. H. & Co. - 10 Fisher St. - manufacturer of box and ivory rules, chain scales, surveyors' dipping rods, plain ivory edged and metal parallels, cattle gauges, clinometers, size sticks and most of the articles connected with the trade

1876 - COX, Francis Blakemore - 50 Camden St. (Established 1780) - manufacturer of box and ivory rules, Carrett's, Hawthorn's and Routledge's engineers' rules, Slater's improved rules for cotton spinning, foreign rules with their correct measure.

1886 - HIGGISON, John - 270 Bishop St. (Established 1831) - manfr. of all kinds of English and foreign rules, mathematical scales, &c., in boxwood and ivory - quotations invited

Rule Manufacturing
H. E. Drake

This article was written on December 20, 1910, by Harold E. Drake, then a student at Worcester Polytechnic Institute. At that date his father, Wilbur Drake, was Superintendent of the Rule Shop of Chapin-Stephens Co.

One of the most peculiar and interesting, as well as the least known of the industries of this country, is the making of rules. By rules I mean measuring instruments of wood or ivory with one or more joints. Little does a person think that whether he picks up a cheap rule in a five and ten cents store in Boston or an ivory rule in a first class hardware store in Galveston, thay are both made within a radius of fifteen miles in western Connecticut. There are only three factories in existence in America which make them.[1] These, supplemented by some in Germany supply the world.[2] These are, in order of size: The Stanley Rule and Level Co. of New Britain; The Chapin-Stephens Co. of Pine meadow and the Upson Nut Co. of Unionville. The second named firm, which is to be the basis of my article, has a capacity of about five thousand rules a day. The secrecy with which the work is carried on, as well as the few concerns manufacturing, is due to the fact that the machinery, most of which is automatic, cannot be bought, and therefore must be invented and constructed at great expense within the shop itself.

Since the Spanish-American War the boxwood of which every rule is made is no longer profitably bought in the West Indies, but must come from the north coast of South America.[3] This is shipped in logs about fifteen feet long to New York by boat and then by rail to the factory. It must be handled by crane, as the weight of a single log is enormous. This wood is chosen for its fine grain, hardness and capability of taking a fine finish. The boxwood logs are piled loosely to allow free circulation of air in timber sheds to start the process of curing. When about half dry the logs are rolled in great chutes to the cross cut saws and cut up into six inch lengths. From there a belt carries them to a splitting saw and then to sizing saws. The pieces, slightly larger than the finished product, are piled in a brick kiln, where for weeks stifling hot air is forced through them by huge fans. Now they are ready to "burr", that is, to take down to the required size. The work is now inspected and any dark colored grain thrown out. It is next slotted for the tips, joints and binding, if it has the latter.

The ivory is imported in sticks about a foot long and three inches square, and after being sun cured,[4] is prepared the same way as boxwood, but with greater care, owing to its brittleness.

The making of the joints may be said to be the most interesting part, with the exception of marking, in the whole process. The sheet brass for the joints, which must be a certain percent mixture,[5] comes in huge rolls weighing one fourth ton each.[6] This is cut and sheared by a machine into strips about eight feet by two inches, for joints; and one foot by six inches for binding, other sizes and widths being used in other cases. Now it is ready for the presses. The dies and punches of these presses are considered so valuable that they are kept in a stone vault. Without them the company is lost. They are very hard to keep in repair. The presses punch out the tips, plates, and binding from the sheets of brass. The plates are countersunk, milled and scraped to make the pin hold and joint work easy. The soft iron pins are made from wire coils by automatic machines. The rolls are also the product of a single process. The machine takes large brass wire from rolls, and drills, turns down and cuts off with mechanical precision. Now the joint may be "driven up." A roll, two plates, another roll, two more plates, a third roll and a pin comprise the most common joint. These are driven together into a joint by air pressure. The tips are sunk and squared by a turret roller. Now the work is ready to be put together.

German silver[7] and ivory, and brass and boxwood are in most cases drilled, countersunk and riveted in one operation, and delivered to the power files, which in turn deliver them to the scrapers who

From Wooden Planes in 19th Century America, Volume II: Planemaking by the Chapins at Union Factory, 1826-1929, by Kenneth D. Roberts. Fitzwilliam: Ken Robert Publishing Co., 1983, by permission.

finish them ready for inspection. After they are "passed" they go to the marking room. The graduations on a rule are not stamped on as one would expect, but scratched on by accurate little hardened steel knives, thus eliminating the denting of the brass work. The machines which do this are worth their weight in silver at least. Many years and men are needed to build one machine, as they must be absolutely without fault. Linseed oil and lampblack are rubbed over it, giving the dark color to the lines and figures. After this is rubbed off they are shellacked, rubbed down, re-shellacked and after another inspection are ready for the market.

NOTES

1. At the date of this writing, Dec. 20, 1910, two other firms, the Lufkin Rule Co. Of Saginaw, Michigan and R. B. Haselton of Contoocook N.H. were making wooden rules (maple and hickory) for lumber measuring, but not folding rules with joints. After 1920 Lufkin began making boxwood folding rules.

2. The writer, H. E. Drake, was obviously not aware of the large number of rules manufactured then at Birmingham, England at that date.

3. Turkey boxwood grows very slowly in a dry climate at the rate of about 2½ inches diameter increase in 10 years, and thus has a very dense structure. Some time about 1900 it became in short supply, and so-called "Maracaibo" boxwood, from Venezuela, was substituted.

   This statement regarding the purchasing of Turkey boxwood in the West Indies is questionable. While mahogany and rosewood came from the West Indies and South America, Turkey boxwood principally came from southern Russia and Turkey, and would probably have been purchased on the London market.

4. Often ivory was stored in the cupola of the building to expose it to sunlight, thus bleaching out stains. See chapter XII of *Wooden Planes in 19th Century America; Volume II - Planemaking by the Chapins at Union Factory, 1826-1929*, by Ken Roberts

5. Leaded high brass (66% copper, 0.5% lead, balance zinc)

6. This appears to be in error. At that date an ingot from which the strip would have been rolled was limited to about 125 pounds, and there was no way to continuously join coils together. It is quite possible that the author meant to state that a single order delivered to the factory, consisting of several rolls, was 500 pounds.

7. German silver, now commonly known as nickel silver contains nominally 66% copper, 17% nickel, balance zinc. Its color resembles silver.

RULEMAKERS' PLANES

By
Kenneth D. Roberts

Fig. I

Rulemaking, like planemaking, was a trade developed to a high degree in England and subsequently introduced into America by immigrants. Several years ago I acquired a collection of planes used for rulemaking, together with other tools, marking jigs and fixtures used in one shop. These were purchased from Winthrop L. Carter, Jr., of Portsmouth NH, who had obtained them from Tony Barwick in London. Reportedly they had belonged to Jephthah Corbishley, whose forbears had been spirits rulemakers for at least three successive generations, working in the Elephant and Castle district of London into the 1960's. I had hoped to make a study of these tools and the processes, but with other commitments I have not been able to attend to this project. Recently I decided to turn over the collection in its entirety to a younger person, who indicated that he would in the near future complete such a study.

These special planes are illustrated herewith in four photographs. Figure I shows a jack and two fore planes with their irons set at 90° for planing boxwood stuff to size. I assume that boxwood, with its extremely close grain, required a scraping process, rather than a shear cut—hence the 90° angle. In any event, there were very fine chips wedged in the mouths from last use.

Figure II shows another jack and an iron plane. The latter was used on brass, and is engraved on the side: "Wm. Gillumuy (? or Galloway) / Engineer / -7 Fleet Street & / 59 High Holborn / London." This may refer to the maker.

Figures III and IV show four different special plow type planes for cutting grooves in rules for slides. The irons in three of these are set at about 70°, while the fourth is at about 45°.

The rule business in America appears to have become mechanized at Hermon Chapin's *Union Factory* at Pine Meadow, New Hartford, Connecticut. There the stuff was processed by circular saws rather than hand planes. If such rule planes as are illustrated herewith were ever used in America, they are not known to have survived.

From *The Chronicle of the Early American Industries Association,* Vol. 36, No. 2 (June, 1983), by permission.

Fig. II

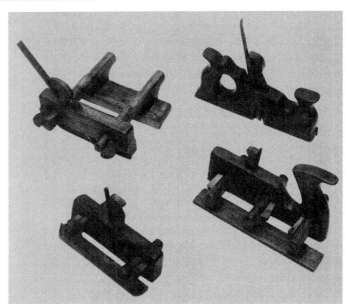

Fig. III

Fig. IV

Crafting Scale Rules by Hand – An Historical Note

David Goodson [1]

During a chance encounter with an engineer from Blundell Harling Ltd.(UK) the fact emerged during our conversation that, back in 1963, I had visited the old manufacturing works of the W.H.Harling company in Clapton in the East End of London. In retrospect the timing of my visits appear to have been fortuitous, as shortly after this date the W.H. Harling company, who had been manufacturing high quality drawing instruments in London from 1848, amalgamated with the slide rule manufacturer Blundell Rules Ltd, a rather younger company celebrating its 50th Aniversary in 1998. The successful merging of these precision instrument makers had some fairly immediate effect in 1964 and in particular Blundell's use of modern plastics rapidly displaced a traditional, but very labour intensive, method of engraving boxwood and celluloid edged scales.

The following decription is extracted from a study of "Hand made drawing instruments" made by the author during 1963/4. The study attempted to record a number of craft skills, now lost, and in particular describes production methods employed for engraving scale rules which seemed quite anachronistic, even at the time of writing.

Chapter 6 – Scale Making

Formerly scale makers relied on boxwood and ivory for their base material. Now, of course, ivory has been replaced by the cheaper and more readily obtainable plastics, principally celluloid, although other plastics are used. The first requirements for the base material are:

1. Dimensional stability and resistance to warping
2. A degree of hardness sufficiently to withstand fair wear and tear
3. A strong colour contrast with the graduation filling
4. A surface which will take clean clear-cut division lines
5. Flexibility to absorb punch marks without chipping or splitting.

With any wood the risk of shrinkage due to incomplete seasoning is always present and to reduce this risk to a minimum all cutting and shaping operations are separated by long periods of natural air seasoning. The change from log to scale takes several years, without which the maker could not be sure that the scale would remain stable after sealing. For boxwood scales the wastage factor is very high - above 50% in the initial stages. This is due to the costly method of cutting, which has to be radial to avoid warps, numerous splits in the wood, bad color changes and small knots and twists in the grain.

Celluloid edged boxwood blanks are stacked carefully to ensure stability before finishing

When the seasoned, shaped and sanded blanks arrive in the engraving department, the machine and mass produced methods hitherto employed give way to the art of the scale maker. He has one machine to assist him, namely the engine dividing machine which is used to prepare brass master plates for open divided scales and also for engraving fully divided scales. The machine, as can be seen from the photographs, consists of an oscillating arm carrying two knives, the travel of which is limited by a rotating stepped cam. The drive is taken up from an electric motor (a new adaptation) direct to the headstock which at one end operates the knife arm through a crank and at the other, through a Geneva motion gear train and lead screw, moves the table (and hence the scales) the required spacing per stroke of the knife. The Geneva motion is required to ensure that the movement of the table only occurs when the engraving knives are clear of the scale.

The stages of engraving and finishing a scale are as follows:

1. The edges are trued with a jack plane, a process which also strengthens them, for they are supplied from the shaping process with razor-like edges which would be easily damaged.
2. The scales are then rubbed down with a flour grade of garnet before the limit lines are scratched,

---

[1] Electronic & Electrical Engineering Department, Loughborough University, LE11 3TU, UK (a.d.goodson@lboro.ac.uk)

The dividing engine set up ready to cut a pair of fully divided, bevel-edged, boxwood scales.

Scribing the limit lines along the length of the scale using a simple marking gauge.

The dividing engine ready to go.

Close-up view of the dividing engine cutter head. The cuts are made on the back stroke, towards the scale's edge, and extending into a supporting wooden buffer strip. A cam then lifts the cutter blade clear of the scale surface while repositioning for the next division.

Punching in the individual numbers and letters along the bottom of the scale.

The final touch – Stamping on the BS Kite mark and the W.H. Harling logo.

using a special version of a woodworker's marking gauge.

3. The actual engraving or dividing of the scale, which can be done in either of two ways. First by hand, using a special board and square working from a brass master plate, will be described. Secondly by clamping to the table of the engine dividing machine. The essential point in either case must be to ensure that no movement takes place while dividing is in progress. Hence the numerous brass clamps which may be seen in the photographs and which serve to hold the scale both stationary and flat.

4. When the divisions have been cut, the numbers, letters, British Standard Kite mark and trade marks are punched in by hand.

5. The division lines and figures are practically invisible at this stage, so a filling which consists of a paste made of charcoal, linseed oil and whiting is rubbed on with a hard cloth pad.

6. The filling is wiped off the surface with another cloth, allowed to dry in the divisions and then rubbed down with fine garnet paper. This cleans the surface and sharpens the division lines by removing any surface roughness left by the knife.

7. The final stage is merely to restore the natural colour of the surface by a wipe with linseed oil. The scale is then sealed by spraying with white polish. Until a few years ago, French polishing was used. The scales are now packed in polyethylene sleeves which offer protection from dust, damp and abrasion in storage.

Except for some very minor changes, the methods used in scale making have remained the same from the very start. The dividing engine was developed over century ago and has been little changed since. In many cases the machines in use today have been in constant use for fifty years or longer. As with a lathe, its accuracy is not affected by wear, provided that all the backlash is taken up before the first cut and provided that the sequence of operations is not interrupted from first to last. The machine in the photograph is at least 60 years old but is still capable of cutting a scale on celluloid with divisions graduated as fine as 1/10 mm. This cannot be done on boxwood, of course, for the wood would break up between the divisions.

Open divided scales are nearly always cut by hand owing to the difficulties inherent in setting a dividing machine to alter its pattern of action without introducing inaccuracies. The apparatus required is very simple. It consists of a small board, the required brass master plate, three or four thin brass clamps and a small brass-stocked engineer's square with a small steel hooked cursor. The knife is very similar to a woodworker's marking knife with a very oblique cutting edge. In use it is drawn along the edge of the square from the scribed or pencilled limit line to the edge of the scale. The cursor on the square engages with the graduations on the master plate with an audible click when the square moves after each division has been scribed. The sequence of operations is as follows:

1. First the trued blank, with the limit lines scratched and pencilled on, is clamped against the master plate, which itself is screwed to the edge of the board.

2. The cursor on the square is adjusted to engage with the required scale on the master, on which there are usually four different scales.

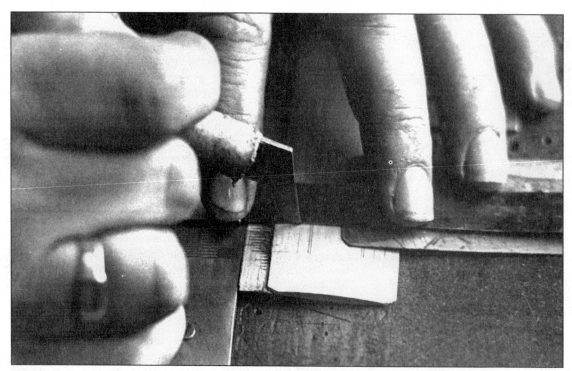

Cutting the graduations on an open-divided rule by hand using a square and graduating knife

3. The end line is now cut, and working from right to left slowly the graduation marks are cut click-by-click. Considerable concentration is required for this operation; if a mis-count results in a long graduation being made where a short one is required, the scale will be rejected at inspection.

End Notes

Detailed information about the hand methods employed when making traditional design drawing instruments was obtained from the senior staff employed at the W H Harling company's East London works at Clapton during 1963. Boxwood scale making as described above continued into 1964. However, the W H Harling company merged with Blundell Rules during that year and this labor intensive method of scale making ceased soon afterwards. The Clapton works in London were closed and the amalgamated company, now known as Blundell Harling, continued to develop in Weymouth on the South coast of England where they are still trading.

Bibliography

1. Anon. (1958) British Standard Specifications B.S.1709, pub. HMSO, UK
2. Goodson A. D. (1964). *Hand Made Drawing Instruments*. Unpublished craft study dissertation - London University Institute of Education Teacher's Certificate, Shoreditch College, pp. 58-66.
3. Soole P. (1998). 'Blundell - Blundell Harling - W.H.Harling Slide Rules, A brief history of Blundell Harling Ltd', J.Oughtred Soc. v7(1). ISSN 1061-6292, pp. 17-22.
4. Vaughan F.E. (1950). 'Historical note on drawing instruments' Booklet published by the V&E Manufacturing Co. Pasadena.

A finished Harling rule. It is made of boxwood and was cut by hand. Shown actual size.

A. D. Goodson is in the Department of Electronics and Electrical Engineering at Loughborough University, where he serves as Chief Experimental Officer.

IV
Mensuration

In 1987, three years after the publication of *Boxwood & Ivory*, the author, in association with James Hill, a collector in Maryland, began publication of a newsletter for rule collectors entitled *Mensuration*, dedicated to promoting the collection and study of measuring devices and related objects. The first issue appeared that summer, and three more issues were published at irregular intervals over the next three years.

Publication was discontinued after the fourth issue; the author discovered that he had underestimated the time commitments associated with a full-time job and the raising of three teenagers. Consequently, he found it impossible to continue producing a quality publication in a timely fashion.

The four issues which were produced, however, contained a number of useful and interesting articles and commentary (see the Tables of Contents in the next column), and we are reproducing the first three issues on the following pages so that this material will be more readily available to collectors.

The fourth issue has not been reprinted here. That issue was dedicated entirely to William Gilliland's excellent article "Some Measures of History," a study of pre-metric European measures. Chapter VIII of this book deals even more extensively with this same subject, however, and reprinting the fourth issue of *Mensuration* seemed redundant.

MENSURATION

CONTENTS:

Vol. I, No. 1 Summer, 1987
  E. A. Stearns & Co
  The E & M Scales
  A Proposed System for Classifying Rules As To Scarcity
  Six Fold Rules.

Vol. I, No. 2 Fall, 1988
  Common Graduation
  The Stanley No. 133 Rule
  Cleaning Rules ...
  ... and Storing Rules
  A Relic

Vol. I, No. 3 Fall, 1989
  Barrel Mensuration & Ullage
  A "Take-Down" Extension Stick
  Why 12?
  "Full and Complete Instructions ..."

Vol. II, No. 1 Spring, 1990
  Some Measures of History

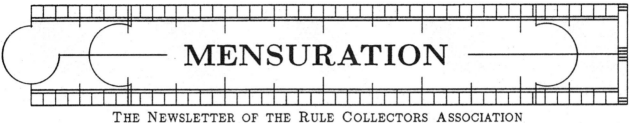

MENSURATION

THE NEWSLETTER OF THE RULE COLLECTORS ASSOCIATION

VOL I NO 1   SUMMER 1987

FIGURE 1: Brattleboro, Vermont in 1856. A portion of a lithograph drawn by J. H. Bufford and published by John Batchelder in that year. The probable location of the E. A. Stearns Co. is indicated by the arrow. (Courtesy of Jeff Barry)

E. A. STEARNS & CO.

Jim Hill

Edward A. Stearns was born in Warwick, Massachusetts in 1806, and moved to Brattleboro, Vt. in 1831. He married Elizabeth Salisbury of that city in 1834. Some time before 1836 he had begun to work for S. M. Clark in his recently-established rule shop. He learned rulemaking while working for Clark, and became familiar with the the newly-invented rulemaking and graduating machinery which Lemuel Hedge, a Vermont inventor, was making for Clark. By 1836 Stearns had become "... the principal workman in the factory, ..." (1), and when Clark was forced to close by the business crash of that year Stearns bought him out, and recommenced the manufacture of rules under the name of E. A. Stearns & Co.

Clark's rules had been somewhat unusual in their design and style of marking, but Stearns were more

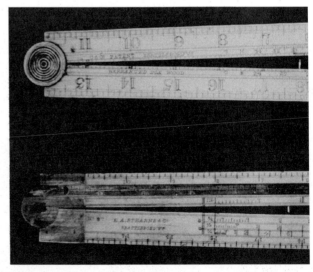

FIGURE 2: TOP: A 2 foot, 2 fold carpenters' rule marked CLARK & CO. BOTTOM: A No. 9 2 foot, 2 fold carpenters' rule with extension slide, marked E. A. STEARNS & CO.

VOL. I NO. 1 SUMMER 1987

MENSURATION is published three times a year by the Rule Collectors Association, a non-profit organization dedicated to promoting interest and research in measureing devices and related subjects. To this end, we solicit and will publish serious articles dealing with the history, use, and manufacture of measuring instruments, and provide a forum for discussion of issues of interest to collectors and others interested in early rule technology.

Editor:
   Philip E. Stanley
   40 Harvey Lane.
   Westborough MA 01581-3005
   (617) 366-9442

Associate Editor:
   Jim Hill
   13933 Wayside Drive
   Clarksville MD 21029
   (301) 854-3170

Annual Subscription/Membership: $7.00 U.S. and Canada, $10.00 elsewhere. Make checks payable to Philip Stanley, 40 Harvey Lane., Westborough, MA 01581-3005.

Copyright © 1987
The Rule Collectors' Association

CONTENTS:

   E. A. Stearns & Co. 1
   The E & M Scales 6
   A Proposed System for Classifying
      Rules as to Scarcity 9
   Six Fold Rules. 10
   Recent Rule Prices . . . . Insert

Stearns (continued)
conventional. This change was probably caused by a fire in 1842 which totally destroyed the rule shop and all of the machinery (2).

After the 1842 fire Stearns reestablished himself in a building formerly occupied by the Typographic Co. (papermakers). This new location supplied adequate water power and (probably) some of the overhead shafting from the former owners, but Stearns had to have all new manufacturing and graduating machinery made. To accomplish this he retained the technical advice of Hedge, who by that time was living in New York, and hired Edwin Putnam, a well-known machinist, to build the machines. Putnam, born in Gilford, Vt. in 1820, had been apprenticed to the machinists firm of Hines and Newman before joining Stearns in 1842.

The new equipment developed by Hedge and built by Putnam during this rebuilding process created a generation of graduating equipment that remained in constant use for almost one hundred years. In an article written about the Hedge in 1923 by Guy Hubbard for *American Machinist* magazine, Hubbard refers to a Hedge dividing engine still in operation, "... used at the Stanley Works to divide its high-grade ivory rules, in preference to the modern types of machines." (3) (see photo)

FIGURE 3: A Hedge graduating machine designed for E. A. Stearns & Co. in 1842. Used by them from 1842 through 1863, and subsequently by the Stanley Rule & Level Co. into the 1920's. (Courtesey of the Vermont Historical Society)

Mr. Putnam's contribution played an important part in the reputation for high quality which the Stearns Co. achieved in the mid 1800s. This rebuilding period was indirectly mentioned in an 1847 Stearns broadside which referred to the use of "improved machinery" in the manufacture of the some fifty-seven rules and measures then offered for sale.

The following is a list of confirmed employees who worked at the E. A. Stearns firm from 1838 to 1863: Edwin Putnam, Deacon Augustus Stearns (22 years), John Stearns, Lewis S. Higgins (18 years), George H. Bond, Horatio S. Noyes, F. C. Edwards,

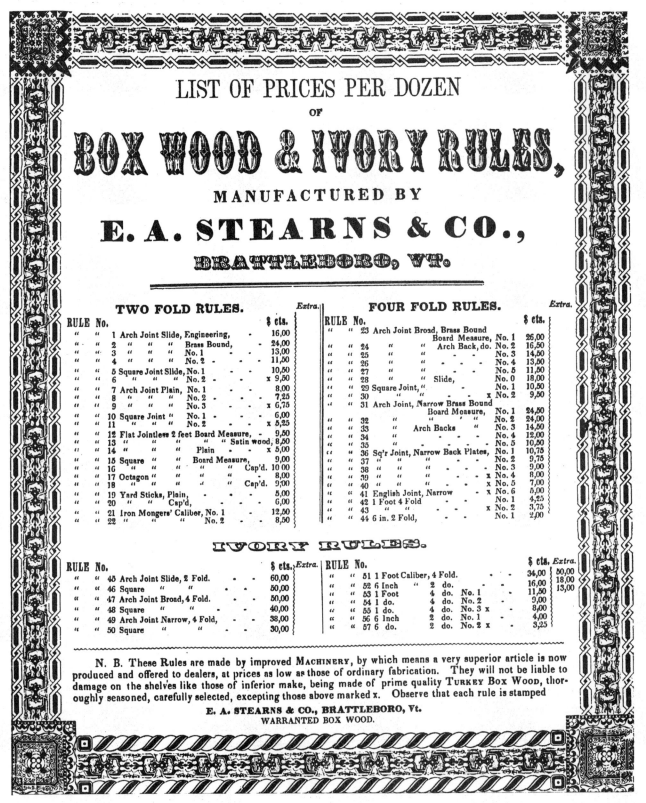

FIGURE 4: An early E. A. Stearns rule price list, found pasted in the front cover of a Stearns account book from the period 1847-1849. (courtesy of Paul Kebabian)

Stearns (continued)

Madison Ranney, Dr. Charles Stratton, Theodore Cole, Asa Marshal, Lewis M. Burditt, and Charles L. Mead.

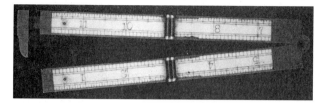

FIGURE 5: A 1 Foot, 4 fold ivory caliper rule bound in German silver marked E. A. Stearns & Co. Note the early style of caliper and number stamps.

The company's success was due, in part, to the great care which was taken in fitting, graduating, and finishing their rules. An indication of production volume is the statement in The Annals of Brattleboro that in 1853 the E. A. Stearns Co. advertised eighty patterns of rules and measures, and employed thirty hands, with annual gross sales between $15,000 and $18,000. (4) Their prosperity was also due to the marketing and organizational abilities of Charles L. Mead, who was hired by Stearns during the early fifties. Mead's father, Larkin J. Mead had been the president of the Typographic Co., the building's former tenents. Mr. Mead gained controlling interest in the E. A. Stearns Co. in March of 1857, eight months after Stearns' death on July 29, 1856.

Charles Levi Mead was born in Chesterfield, New Hampshire on January 21, 1833, moving with his family to Brattleboro, Vermont in 1836. He spent a good deal of his adolescence working with his father and keeping the books at the Typographic Co. He returned to this same building in 1850 as an employee of E. A. Stearns.

In September, 1857, only six months after Mead took control, the company was again nearly destroyed by fire, when the Typographic building was burnt to the ground in a conflagration involving several entire blocks in the center of Brattleboro. Fortunately, the advance of the fire was slow enough to allow the removal of most or all of the precision machinery, including the Hedge engines, before the end, and Mead

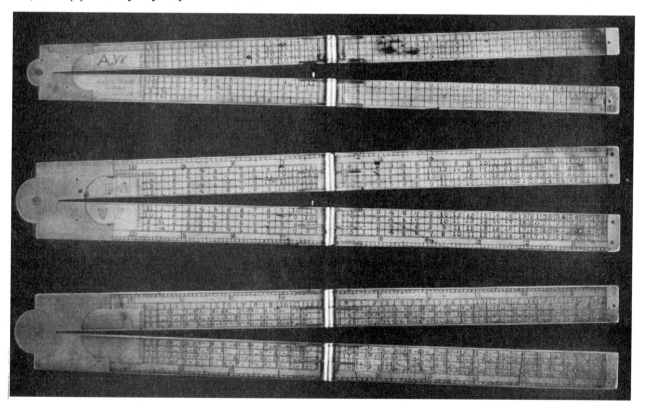

FIGURE 6: Three 2 foot, 4 fold boxwood rules with board tables by E. A. Stearns & Co. TOP: A No. 24, with arch joint and edge plates; MIDDLE: A No. 18, with arch joint, full bound, broad; BOTTOM: A No. 14, with arch joint and edge plates, broad.

PRICE LIST.

No.	DESCRIPTION.	Width. Inches.	Length Inches.	Price per doz.
	TWO FOLD RULES.			
1	Broad Arch Joint, Slide, Engineering, Bound,	1½	24	$29 00
2	Broad Arch Joint, Slide, Engineering,	1½	24	19 00
3	Broad Arch Joint, Slide, Full Scale, Bound,	1½	24	25 00
5	Broad Arch Joint, Slide, Scales, - - -	1½	24	14 00
6½	Broad Arch Joint, Thin, Full Scale, - -	1¼	24	10 00
7	Broad Arch Joint, Scales, - - - -	1½	24	7 50
9	Broad Square Joint, Slide, Scales, - -	1½	24	10 00
11	Broad Square Joint, - - - - - -	1½	24	4 75
12	Medium Iron Mongers, Caliper, Brass Leg,	1¼	6	9 00
13	Medium Iron Mongers, Caliper, - - -	1¼	6	8 00
13½	Narrow Iron Mongers, Caliper, - - -	13-16	6	7 00
	FOUR FOLD RULES.			
14	Broad Arch Joint, Board Measure, Bound,	1½	24	24 00
15	Broad Arch Joint, Full Scale, Bound, - -	1½	24	22 00
16	Broad Arch Joint, Arch Back, Board Measure,	1¼	24	17 00
17	Broad Arch Joint, Arch Back, Full Scale, -	1½	24	15 00
18	Broad Arch Joint, Board Measure, - -	1½	24	14 00
19	Broad Arch Joint, Full Scale, Slide, - -	1½	24	15 00
20	Broad Arch Joint, Full Scale, - - -	1½	24	12 00
21	Broad Arch Joint, Scales, - - - - -	1½	24	9 00
22	Medium Arch Joint, Board Measure, Bound,	1¼	24	20 00
23	Medium Arch Joint, Full Scale, Bound, -	1¼	24	18 00
24	Medium Arch Joint, Arch Back, Board Measure,	1¼	24	15 00
25	Medium Arch Joint, Arch Back, Full Scale,	1¼	24	12 00
26	Medium Arch Joint, Board Measure, - -	1¼	24	11 00
27	Medium Arch Joint, Scales, - - - - -	1¼	24	7 00
28	Medium Arch Joint, Yard Sticks, - -	1¼	36	9 00
28½	Narrow Arch Joint, Full, Six Fold, - -	¾	24	13 50
29	Narrow Arch Joint, Bound, Full, - -	13-16	12	13 00
29½	Narrow Arch Joint, Caliper, Bound, Full, -	13-16	12	20 00
30	Narrow Arch Joint, Full, - - - - -	13-16	12	8 00
30½	Narrow Arch Joint, Caliper, Full, - - -	13-16	12	12 00
32	Broad Square Joint, Scales, - - - -	1½	24	8 00
34	Narrow Square Joint, Back Plates, - -	1	24	9 00
34½	Medium Square Joint, Scales. - - -	1¼	24	7 00
35	Narrow Square Joint, Scales, Bound, - -	1	24	15 00
37	Narrow Square Joint, Edge Plates, - -	1	24	6 50
38	Narrow Square Joint, Middle Plates, - -	1	24	5 00
41	Narrow Round Joint, Middle Plates, - -	1	24	3 25
39	Narrow Square Joint, Edge Plates, - -	13-16	12	6 00
40	Narrow Round Joint, Middle Plates, - -	13-16	12	3 00
43	Round Joint Pocket, - - - - - -	¾	12	2 50

No.	DESCRIPTION.	Width. Inches.	Length Inches.	Price per dozen. Un-bound.	Price per dozen. Bound.
	IVORY RULES.				
45	Broad Arch Joint, Slide, Engineering, 2 Fold,	1½	24	$72 00	$92 00
46	Broad Square Joint, " " 2 "	1½	24	66 00	84 00
47	Broad Arch Joint, Four Fold, Full Scale,	1½	24	60 00	72 00
48	Medium Arch Joint, Four Fold, Full Scale,	1¼	24	50 00	60 00
49	Broad Square Joint, Four Fold, Full Scale,	1½	24	54 00	66 00
50	Narrow Square Joint, Four Fold, Full Scale,	1	24	42 00	56 00
50½	Narrow Arch Joint, Six Fold, Full,	¾	24	66 00	
51	Narrow Arch Joint, Four Fold, Full,	13-16	12	25 00	36 50
52	Narrow Square Joint, Four Fold, Full,	13-16	12	22 00	32 50
53	Narrow Arch Joint, Four Fold, Caliper, Full,	13-16	12	36 00	48 00
54	Narrow Square Joint, Four Fold, Caliper, Full,	13-16	12	32 00	44 00
55	Narrow Square Joint, Two Fold, Caliper,	13-16	6	15 00	
55½	Narrow Square Joint, Silver Leg, Caliper,	13-16	6	20 00	
57	Narrow Square Joint, Pocket, Four Fold,	¾	12	14 00	24 00
59	Round Joint, Pocket, Four Fold, Brass,	¾	12	8 50	
61	Round Joint, Pocket, Two Fold, Brass,	¾	6	8 50	
	Miscellaneous Articles.				
62	Bench Rule, Board Measure, Boxwood,		24		8 50
62½	Bench Rule, Board Measure and Slide, Boxwood, - - - - - - -		24		10 25
63	Bench Rule, Board Measure, Satin Wood,		24		5 50
63½	Bench Rule, Boxwood, Bound, - -		24		11 00
64	Bench Rule, Plain, - - - - - -		24		3 50
65	Board Measure, Square, Capped, - -		24		5 50
68	Board Measure, Square, Capped, - -		48		14 00
69	Board Measure, Octagon, Capped, - -		24		8 00
71	Yard Sticks, Capped, - - - - - -		36		3 50
73	Ship Bevels, Prime Boxwood, Double or Single Tongue, - - - - - - -		12		5 50
77	Ship Bevels, Prime Boxwood, Double Tongued and Tipped, - - - - - - -		12		7 00
78	Guage Rod, Maple, - - - - - -		36		7 50
79	Guage Rod, Maple, Wantage Tables, - -		48		10 00

EXPLANATION.

Rules designated FULL SCALE, are graduated 8ths, 10ths, 12ths and 16ths of inches, with Drafting Scales, (and Octagonal Scales on Broad Rules,) also 100 parts to the foot on edges of Unbound Rules. Those marked Scales, bear Drafting and Octagonal Scales, with 8ths and 16ths of inches. FULL signifies graduations of 10ths, 12ths and 16ths of inches, with 100 parts to the foot. All other Rules are graduated 8ths and 16ths of inches. Slide Rules (except No. 9, which has simply a slide graduated 8ths of inches,) are designed for Instrumental Calculations, in Engineering and Mechanics.

The SLIDE RULE GUIDE, a book explaining in full their use, will be sent to order. Price, cents.

The Joints and Bindings of all Ivory Rules will be made of good quality GERMAN SILVER, excepting those numbers marked BRASS. The Boxwood Rules are made of prime quality Turkey Boxwood,—strict attention being given to selecting and seasoning.

Discount on Boxwood Rules, 40 per cent. Discount on Miscellaneous Articles, 35 per cent.
" Ivory " 35 " " for Cash, 10 "

NEW BRITAIN, CT., April 1st, 1863.

FIGURE 7: An E. A. Stearns rule price list issued by Stanley at the time of Stearns acquisition by that company in April 1863 (courtesy of the Worcester Historical Museum)

THE E & M SCALES

Phil Stanley

The carpenter of more than 250 years ago may have been a skilled woodworker, but like most of his contemporaries he had little, if any, mathematical education beyond the most basic. He could add and subtract, and (perhaps) multiply and divide, but that was about all. Anything beyond that, such as simple geometry or trigonometry, was beyond his ken. It is no wonder that over the years a number of aids had been developed to deal with the technical aspects of his work that might require skills beyond basic arithmetic. Some of these have been described in the literature: the use of a stick or pair of sticks to record and transfer dimensions from one place to another without having to write the numbers down, for instance, or the quarter-girth formula for estimating the volume of a log without recourse to any "exotic" constant such as π.

Another aid of this type are the two scales marked E and M which are so commonly found on 2 foot, 2 fold rules, the so-called E & M Scales. Properly used, they enabled a carpenter, without recourse to arithmetic, to determine the correct points at which to cut the corners off a square to turn it into an octagon, or how long to make a diagonal framing brace.

Stearns (continued)
resumed business about 5 months later in a new building erected on the site by Woodcock and Vinton.

FIGURE 8: The Stearns "Rule Manufactury" as rebuilt in 1857. This picture was taken in 1869 (Courtesy of Jeff Barry)

Mead skilfully promoted the "state of the art" Stearns rules which continued to bear that name under his own leadership. By 1863 the line of rules and measures had grown to eighty-one.

During Mead's tenure as a soldier in the Civil War, Henry Stanley of the Stanley Rule & Level Co. persuaded him to sell the E. A. Stearns Co. This transaction took place on April 1, 1863. In return, Stanley offered Mead stock and an executive position as Treasurer and Sales Manager of the Stanley warehouse at 57 Beekman Street in New York City, a position in which he would later excel.

The Stanley Rule and Level Co. continued to produce the Stearns rules at the Brattleboro location, trading on the excellent reputations E. A. Stearns and C. L. Mead had gained.

By 1870, the machinery and a number of the Stearns employees had been moved into a separate building at Stanley's New Britain, Conn. plant site. Rules produced there continued to be sold under the Stearns name but were no longer stamped "BRATTLEBORO, VT." The extra care which was taken in the manufacture of these rules lessened their competetiveness and by 1884 all but four of the remaining forty-nine boxwood rules had been discontinued. These four were renumbered as Stanley rules. The Stearns ivory rules were also greatly reduced. By 1902 only twelve remained and these were dropped shortly thereafter.

Mr. Mead moved to the New York City warehouse location and there focused his talents on sales. He bacame a driving force in the Stanley Rule and Level Co., becoming president in 1884, a position which he occupied until his death in 1899.

NOTES:
1. Burt, Henry M. *The Attractions of Brattleboro*. Brattleboro, Vt.: D. B. Stedman, 1866. p. 80
2. Cabot, Mary R. *Annals of Brattleboro, 1681 to 1895*. Brattleboro, Vt.: D. B. Stedman, 1921-1922. Vol. I, p. 415
3. Hubbard incorrectly assumed that the graduating equipment being used at the Stanley Works was built in 1836 when in fact, this equipment was developed by Hedge after the Stearns factory fire of 1842.
4. Cabot, *Op Cit.* p. 415

The author wishes to thank Jeff and Peg Barry of Brattleboro for their assistance in locating a great deal of the original materials upon which this article is based. □

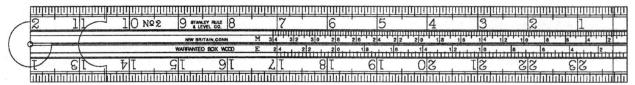

FIGURE 1: The E & M scales on an American 2 fold rule

These two scales were marked on one side of the rule on the inner edges of the two legs (see Fig. 1), each beginning at its respective tip and ending a little more than half way down the rule. They were used with the rule closed, in conjunction with the first 7 inches of the common inch scale, and their figures were oriented the same as those for that section. The divisions of the E scale were a little less than 1/3 (1/3.414... more exactly) the size of common inches, and those of the M scale about 1/5 (1/4.828... more exactly); thus the E scale was about 24 divisions long and the M scale 34.

Assuming the carpenter wished to measure in from the edges on each side of a square piece of stock to mark the places where the diagonal cuts should begin and end to turn it into an octagon, he would would hold an ordinary try square against his closed rule so that the edge of the blade would cross the E scale at the division which represented the width of the square stock (say 15-1/2 inches). Then he would read off the division where the blade crossed the common inch scale (4-17/32 inches in this example), and that would be the distance from each edge where he would make the marks on each side (see Fig. 2).

These scales were sometimes called Mastmakers Scales, from their utility in one of the steps used in shaping timbers into ships' masts. The usual method of making a mast consisted in first hewing a timber until it was square in cross section and tapered as desired along its length. The four corners would then be hewed and planed off until it was octagonal in cross section (This was the step where the E & M Scales came in; by their use the mastmaker could determine where to mark the four surfaces of the mast to guide him in cutting off the corners). This procedure was then repeated, with the eight corners of that octagon

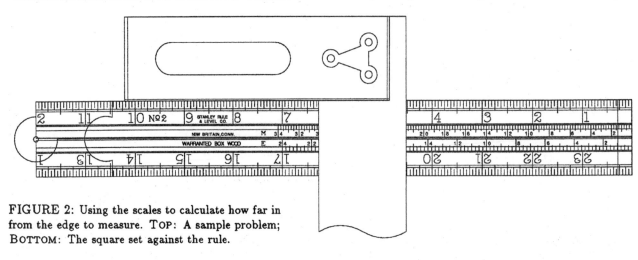

FIGURE 2: Using the scales to calculate how far in from the edge to measure. TOP: A sample problem; BOTTOM: The square set against the rule.

E & M Scales (continued)

being hewed and planed off to turn it into a "sixteenagon", and then, finally, the 16 corners rounded off with round-bottomed spar planes to achieve the desired cylindrical or conical shape. Sometimes, on large masts, an extra step would be taken to reduce the timber to 32 sides before rounding, but the principal remained the same.

It was not always possible to measure from the edges of the stock, however. Perhaps one or more corners was damaged or cut off, or, in the case of a mast timber, the log selected was large enough to make the desired mast, but still too small to be fully squared at the beginning. In such a case the carpenter would use the M scale instead. After marking a center line on each of the four sides of the stock, he would hold the try square against the rule as before, but with the edge of the blade crossing the M scale at the division which represented the width of the stock (say 15-1/2 inches, as before). Then he would read off the division where the blade crossed the common inch scale (3-7/32 inches in this example), and that would be the distance from the center lines where he would make the marks on each side (see Fig. 3).

Another application where this rule was useful was in calculating brace measure, that is, figuring how long a diagonal brace had to be to exactly span the distance from the post to the beam when framing a house or barn. To perform this calculation he would use both the E and the M scales, but ignore the common inch scale. Suppose he wished to ascertain the proper length for a 45 degree brace whose two ends were 22 inches out from the meeting point of the post and beam. As before, he would hold his square against the closed rule, but with the edge of the blade this time crossing the E scale at the 22 inch point, and would read off the corresponding point on the M scale to get the correct length for the brace (31-3/32 inches, see Fig. 4, opposite).

On English and very early American rules these scales were usually longer, extending almost the full length of the folded rule to within an inch of the joint. This resulted in the E scale being about 38 divisions long and the M 53 divisions.

It is probable that the E & M scales were introduced some time in the early 18th century. They are described in Coggeshall (1767, Ref. 1), but not in Bion (1758, Ref. 2), two contemporary authoratative sources, suggesting that they were not uni-

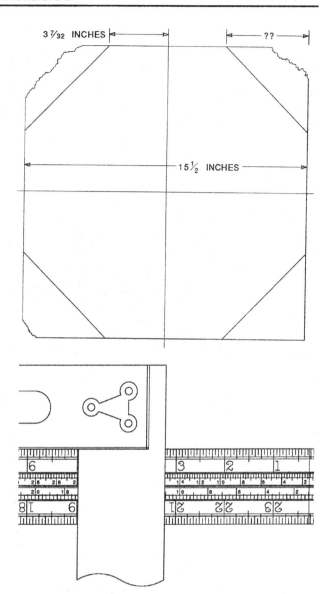

FIGURE 3: Using the scales to calculate how far out from the centerline to measure. TOP: A sample problem; BOTTOM: The square set against the rule

versally accepted during this period. Later sources, such as McKay (1815, Ref. 3) all describe them, however, indicating that by about 1800 they had become commonplace.

Whatever their exact date of introduction, by the middle of the 19th century these scales were being placed on all but the least expensive 2 foot 2 fold rules, and sometimes on the higher quality 2 foot 4 fold rules as well. Different makers called them by various names, such as "E&M scales", "Octagonal scales", "4-square lines", or, as mentioned above, "mastmakers' scales". At least one maker simply de-

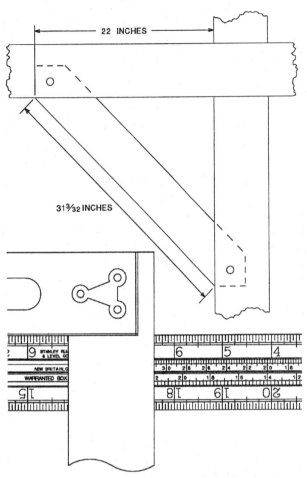

FIGURE 4: Using the scales to determine brace length. TOP: A sample problem; BOTTOM: The square set against the rule.

scribed a rule which had these scales as "lined", with no further description.

These scales became less and less available and thus less used after the turn of the century. Most rules having them were discontinued at about the time of the First World War, and the last rule to be so marked, the Stanley #5, was offered for the last time in 1932. Today, a carpenter would have no idea what these scales were for, let alone how they were used

BIBLIOGRAPHY

1. Coggeshall, Henry. *The Art of Practical Measuring, by the Slide Rule: Shewing How to Measure Round, Square, or Other Timber, ... Whereunto Is Added, In a Short Method, the Use of Scamozzi's Lines for Finding the Lengths and Angles of Hips, Rafters, &c. ...by John Ham. Seventh Edition.* London: Edward and Charles Dilly, 1767.

2. Bion, Nicholas. *The Construction and Principal Uses of Mathematical Instruments,* Paris, 1723. Translated by Edmond Stone; ...*To Which Are Added, the Construction and Uses of Such Instruments As Are Omitted by M. Bion,* by Edmond Stone; 2d. ed. London: J. Richardson, 1758

3. MacKay, Andrew. *The Description and Use of the Sliding Rule, in Arithmetic and in the Mensuration of Surfaces and Solids. Also the Description of the Ship Carpenter's Sliding Rule, Its Use Applied to the Construction of Masts, Yards, &c., Together With the Description and Use of the Gauging Rule, Gauging Rod, & Ullage Rule. 2d ed., ...Improved, Enlarged, and Illustrated With an Accurate Engraving of the Different Rules.* Edinburgh: Oliphant, Waugh, and Innes, 1811 □

A PROPOSED SYSTEM FOR CLASSIFYING RULES AS TO RARITY

Phil Stanley & Jim Hill

The experienced tool collector, on seeing an item for sale, usually has a pretty good idea of how rare or common it is, and thus whether the price asked is high, fair, or low. He has acquired this knowledge through years of attending auctions and flea markets, reading tool lists, and seeing the collections of others.

The beginner, on the other hand, lacks this body of experience. He will sometimes pay too much for a relatively common item which he has never happened to have seen before, or, even worse, pass up some real rarity because the price looks too high.

To help these novice collectors, and to provide an objective standard for describing the relative availability of rules and measuring instruments, we are proposing the scale shown in Fig. 1, Page 12 for classifying rules as to scarcity:

The numbers in the third column represent the probable number of existing examples of the rule being rated, as estimated by the editors of this newsletter and/or any other experienced collectors specializing in measuring instruments they may consult. This estimate will include all known examples in their own and other collections, plus a guess as to how many more may be in collections not yet seen. Admittedly,

Continued on Page 12

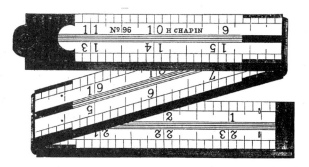

SIX FOLD RULES

Phil Stanley

The 2 foot, 6 fold rule, measuring only 4 inches when folded, was very easy to carry in the pocket, and had much less tendency to drop out when the workman leaned over. It was also an attractive rule, with pleasing proportions and plenty of nice-looking brass or German silver joints.

The number and arrangement of these joints posed some problems, however. The rolls for the four middle joints protruded on both sides of the rule, as compared to one side only on 4 fold rules; thus the 6 fold rule could only be laid flat on one edge while measuring or marking. Also, because these middle joints folded both ways, this rule was somewhat more difficult to handle; it had a tendency to sag when held unsupported in a horizontal position. A third problem, pointed out by Crussel (Ref. 1), was that some carpenters, instead of using the figures on the rule, calculated the distance by counting inches from the nearest joint. It must have been confusing to such a user to use a 6 fold rule, where the joints were not at 6, 12, and 18 inches, but at 4, 8, 12, 16, and 20 inches instead.

It is not known when this type of rule was first offered commercially in the United States. Its 5 separate joints made it an even more labor-intensive product than the 4 fold rule, which required only three, and in all probability it was not economically practical to make commercially until the introduction in the 1850's and 1860's of machinery to stamp out and file joint plates.

The earliest known price lists offering these rules were those of the Stanley Rule & Level Co. in February 1859 (Ref. 2), and of Hermon Chapin in July of the same year (Ref. 3). E. A. Stearns & Co. began making them some time prior to 1861 (Ref. 4). It is not certain when Stephens or Belcher Brothers first offered them; the scarcity of catalogues or price lists for these makers makes it very difficult to date the introduction of any product exactly. Suffice it that from about 1870 on these rules were commonly available from all of the major makers.

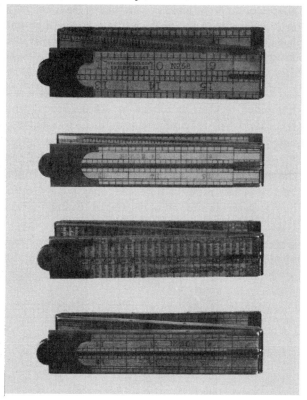

FIGURE 1: Four examples of 6 fold rules by various makers. TOP TO BOTTOM: A Standard Rule Co. #58 [9] (unusual in that it is 7/8 inch wide instead of the usual 3/4 inch); an E. A. Stearns #60 [9] (ivory & German silver); an early Stanley #58½ [9] (completely covered with various weights of metals tables); a late Stanley #58½ [7] (no tables, but full bound).

During the period when they were offered, these rules were available in at least three different construction patterns, in both boxwood and and ivory. The three patterns differed only in the construction of the four "middle" joints; the main joint was always an arch joint. The most widely offered, and the most frequently encountered today, are the boxwood rules with edge plates. They were also available full bound, however, and one maker (Belcher Brothers & Co.) also offered them with middle plates. As was customary, the wood rules were trimmed with yellow brass, and the ivory (except for the Belcher #S146) with German silver (white brass)

Almost invariably these rules were graduated simply in 8ths and 16ths of inches. The one exception was the Stanley #58½ offered in 1860-1862. That rule, apparently an experimental product, was completely covered with 18 tables giving the weight in pounds (per square foot, per inch, etc.) of various shapes of different metals (iron, brass, lead, etc.). (an article describing this rule and its tables will be published in a future issue of **MENSURATION**)

Because of the extra work involved in their manufacture, these rules always sold at a premium. In 1879 (Ref. 5), for example, The Chapin 6 fold #96 was listed in their catalogue at $13.00/dozen, while the 4 fold No. 17, otherwise similar, was listed at only $8.00.

These rules continued to be made right up until the eve of World War I, when, like many other low-volume labor-intensive products, they were dropped due to labor and material shortages. Stanley offered them for the last time in 1915; presumably Chapin-Stephens and the Upson Nut Co. discontinued them at about the same time.

BIBLIOGRAPHY

1. Crussell, E. H. *Jobbing Work for the Carpenter.* New York: The David Williams Co., 1914

2. *Price List of Boxwood and Ivory Rules, Levels, Try Squares, Sliding T Bevels, Gauges, &c.* Hartford: The Stanley Rule and Level Co., February, 1859 (reprinted by the Ken Roberts Publishing Co.)

3. *Price List of Rules, Planes, Gauges, Hand Screws, Bench Screws, Levels, &c.* Hartford: Hermon Chapin, July, 1859 (reprinted by the Ken Roberts Publishing Co.)

4. *Price List of Box Wood and Ivory Rules Manufactured by Charles L. Mead, Successor to E. A. Stearns & Co., Brattleboro, Vt.* Brattleboro: E. A. Stearns & Co., January 1, 1861. (reprinted in *The ACTIVE Scrapbook*)

5. Roberts, Kenneth D. *Wood Planes in 19th Century America, Vol. II, Planemaking by the Chapins At Union Factory, 1826-1929.* Fitzwilliam: The Ken Roberts Publishing Co., 1983 □

	STANLEY	STEARNS	CHAPIN	STEPHENS	CHAPIN-STEPHENS	BELCHER	STANDARD	UPSON
BOXWOOD								
Middle Plates						S45 [10]		
Edge Plates	58 [4]	28½ [9]	96 [10]	33 [9]	58 [10]	S46 [10]	58 [9]	58 [9]
Edge Plates, Weights of Metals Tables	58½ [9]							
Full Bound	58½ [7]		96¼ [10]		58½ [10]		58½ [9]	58½ [10]
IVORY								
Edge Plates		60 [9] (=50½)		34 [10]		S246 [10]		
Edge Plates, Brass Joints & Trim						S146 [10]		
Full Bound		60B [9]		35 [10]				

(All rules had an arch main joint)

FIGURE 1: The different patterns of six fold rules and their makers

---

FROM THE MEMBERS:

One of the things we hope to achieve with this newsletter is increased communication and a lively exchange of ideas among the membership. To this end, we will regularly publish in FROM THE MEMBERS: relevant letters on rules and/or rule collecting. We're looking forward to some interesting discussions.

To kick things off, here are a few questions which should elicit some spirited responses for our next issue: How thoroughly should rules be cleaned, and what is the best/worst method? How should rules be stored? Is there any way to bleach ivory without damaging it? Why were American rules figured right to left, when British rules were the opposite?

Let's start those letters coming in! ...

Classifying Rules as to Scarcity (continued)

this is only an estimate, and not as exact as one might wish, but rule collecting is a still relatively small community, and an "experienced specialist collector" can usually come pretty close.

Rules classified [1] through [10] are rules which can reasonably be assumed to have been produced in more or less volume, based either on catalog listing or on the existance of one or more examples which appears to be a production product. Rules in class [10] are the rarest, known to have been made but with no examples known extant; rules in class [1] are plentiful to the point of surplus.

We have also included an 11th class, [U] (for Unique), for describing a rule where something special about it makes it impossible that another will ever be found. One example could be an otherwise standard production rule which has been engraved with the owner's name or initials or some sort of presentation message, or has been modified for some special use. Another could be a nonstandard pattern or obviously handmade rule which is clearly one-of-a-kind. Catalogued variations in standard production patterns, such as metric graduations in the case of Stanley rules, would not qualify for this class.

Except in the case of rules classified [U], any rule could be moved to a lower class if new examples turn up. As an example, a Standard #113 ivory & ebony caliper rule currently has a scarcity of [9], since there are only two known examples; however if a couple more were to turn up it would probably be downgraded to an [8].

This scale attempts only to classify rules as to scarcity, and makes no pretense at establishing selling price. What a rule should be bought or sold for depends on much more than just rarity. Charisma, material, the buyer's/seller's degree of enthusiasm, condition, and the type of selling environment all influence the price profoundly and should be considered.

From this time forward, all rules mentioned in MENSURATION will be classified as to scarcity in the terms of this table, by putting the scale value in square brackets after the rule name or number, thus: "Stephens #38 [8]". Members advertising tools for sale are encouraged to do the same, or if they do not have sufficient experience may request the editors to do so for them.

It is hoped that this system of rating rules will be of assistance to both new and long-term collectors; their comments are solicited, and will receive appropriate editorial response. □

Relative Scarcity	Scale	Number of Known Examples	Representative Rules
Unique	U	1	Rule Engraved with Owner's Name, Special or Proven Provenance, Custom Rule, Etc.
No Known Examples	10	None	Stanley #28, Belcher #S146
Extremely Scarce	9	1-3	Standard #113, Lufkin #2072P, Stanley #58½ (1860-1862)
Very Scarce	8	3-6	Stephens #38, Stanley #97
Scarce	7	7-12	Stanley #48½, Stanley #66, Stanley #036
Rare	6	13-20	Stanley #48, Stanley #62C
Exceptional	5	20-40	Stanley #87, Stanley #31½, Upson #32
Unusual	4	40-80	Stanley #40, Belcher #956, Stanley #54
Uncommon	3	80-200	Stanley #53½, Stephens #36, Chapin-Stephens #62½
Common	2	200-500	Stanley #84, Stanley #63
Extremely Common	1	500 up	Stanley #18, Stanley #68

FIGURE 1: Descriptions of the eleven classes and examples of each.

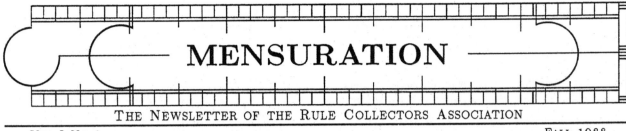

MENSURATION

The Newsletter of the Rule Collectors Association

Vol I No 2 — Fall 1988

COMMON GRADUATION

(From the article *Graduation*, in Tomlinson's Cyclopedia of Useful Arts & Manufactures, ca. 1858)

Common graduation is the method of taking copies from a pattern which has been already laid down by original graduation; but, as generally practised, it consists of taking copies of a copy. It also includes those cases of original graduation where the usual patterns do not apply, and where the utmost possible degree of accuracy is not required.

The apparatus used with certain tools for common graduation, consists of a *dividing-plate*, which is either a complete disc, or a broad rim connected with the centre by radial arms, and made inflexible by circular rings or edge bars beneath. It may vary from 14 to 30 inches in diameter. The extreme border is divided into degrees and quarters, and just within this is another circle, divided into degrees and thirds. Within are usually engraved such numbers as are required for the dial of the perambulator, Gunter's line of numbers arranged in a circle, and other logarithmic lines. Also tangents in hundredth parts of the radius, and the difference of the hypotenuse and base as applied to the theodolite; also the equation of time for dialling, the points of the compass, &c. In the centre of the plate is a circular hole, made truly perpendicular to the surface, into which is nicely fitted a circular pin or arbor, which also fits the centre hole in the circle or arc to be divided, and is the principal connexion between it and the dividing-plate while the work is being done. In Fig. 1 the dividing-plate is shown with a compass ring attached, in the process of graduation; this ring is prevented from turning round by means of a couple of holdfasts, two being used, since it is necessary to remove one of them when its position obstructs the work. An index of tempered steel, with a very straight edge, is attached at one end *A* to a plate of brass, furnished with an angular notch exactly in a line with the straight edge, which notch receiving the arbor of the plate directs the straight edge to the centre. The length of the index is equal to the radius of the plate. At, and below the exterior end, is fixed a secondary index *B*, reaching as far inward as the lines of the plate extend, its edge being also directed to the centre, but usually placed a little to the right of the other edge, as in the figure. By an arrangement of nuts and screws, the distance of the two parts can be adjusted according to the thickness of the work. For instruments which are required to be divided on feather edges, such as protractors, a flexible index is sometimes used, so that the pressure of the hand may bring it in contact with the inclining plane; but a secondary index is preferable, if it allows of its position being adjusted to the plane which is to receive the divisions.

FIGURE 2: A dividing-knife

The dividing-knife, Fig. 2, consists of a blade of good steel and a handle of beech-wood. The cutting edge should be quite straight, in a line with the handle, and of the same thickness as the intended divisions. The edge must not be

FIGURE 1: A circular dividing-plate

| PAGE 2 | **MENSURATION** | VOL I NO 2 |

VOL. I NO. 2   FALL 1988

MENSURATION is published three times a year by the Rule Collectors Association, a nonprofit organization dedicated to promoting interest and research in measureing devices and related subjects. To this end, we solicit and will publish serious articles dealing with the history, use, and manufacture of measuring instruments, and provide a forum for discussion of issues of interest to collectors and others interested in early rule technology.

Editor:
> Philip E. Stanley
> 40 Harvey Lane.
> Westborough MA 01581-3005
> (508) 366-9442

Associate Editor:
> Jim Hill
> 13933 Wayside Drive
> Clarksville MD 21029
> (301) 854-3170

Annual Subscription/Membership: $7.00 U.S. and Canada, $10.00 elsewhere. Make checks payable to Philip Stanley, 40 Harvey Lane., Westborough, MA 01581-3005.

Copyright © 1988
The Rule Collectors' Association

CONTENTS:
> Common Graduation 1
> The Stanley No. 133 Rule 4
> From the Members 4
> Cleaning Rules 6
> A Relic 7
> ... and Storing Rules 7
> Recent Rule Prices . . . . Insert

sharp, but rounding, so as to present to the surface which is to be divided a small semicircle, whose radius is equal to half the breadth of the line which it is to make. The back of the blade is about $\frac{1}{15}$ inch thick. The left side should be flat, but the opposite side chamfered in a faint curve from back to edge. The extreme end of the blade makes with the line of the edge an angle of about 70°. A small chamfer on the side to the right, broad at the back, but vanishing at the edge, reduces the end to an equal thickness. A semicircular recess is made in the edge of the blade near the handle, which affords a relief when the tool is sharpened, and is useful for receiving the inner side of the end of the middle finger. For the further accommodation of the finger, a part of the ferrule of the handle is cut away, as show in the figure. The curved back of the blade enables the operator to see his work better. The knife is held very much like a pen, but the handle must be quite home between the thumb and forefinger, which, being placed upon the ferrule directly over the back, is, by its pressure, the chief agent in giving depth to the divisions, the thumb and middle finger acting as supporters, while the other two fingers, as in writing, prop the hand. The knife is held at an angle of about 45° with the plane to be divided, and is used with the flat side in contact with the index of the dividing-plate. The action of this knife is the reverse of the graver, which is *pushed* outwards, and cuts away a fibre in the line of its course, leaving the rest of the metal undisturbed; whereas the knife is *drawn* inwards, and without producing chips cuts a furrow, while the metal displaced rises in a bur on each side, which is afterwards removed by rubbing with willow charcoal and water.

Before making the divisions, the circular lines which limit the lengths of the strokes are marked out by means of a beam-compass and made of sufficient breadth and depth by the dividing-knife. The compass-ring to be graduated is attached to the dividing-plate, its zero, or north point, being placed so that the index when set to it may also agree with the zero of the plate. The operator then drops the point of the dividing-knife into the line on the plate, and pressing the index to prevent its moving, cuts the corresponding stroke upon the ring. The index is now drawn forwards something more than the value of a division, and the knife being fixed in a second line, it is then pushed back

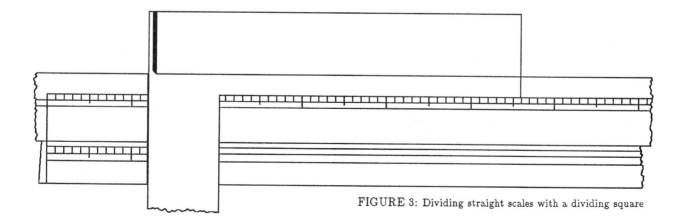

FIGURE 3: Dividing straight scales with a dividing square

into contact with it, and a second stroke cut as before. In this way he proceeds from right to left until the circle is completed. In dividing upon metals this work is laborious for the hands, and there must be frequent intervals of rest, during which the work is examined with a lens. In dividing upon ivory and wood, the work is so light that an operator will keep pace with a skilful seamstress, a division for a stitch, for any length of time.

Common dividing, as applied to straight lines, is similar to circular dividing. The pattern, of course, is straight, and the scale on which the copy is to be laid is placed beside it; instead of the index above described, the lines are ruled with what is called a *dividing square*, Fig. 3; this consists of a straight-edge made of thin tempered steel, with a shoulder at right angles, like a carpenter's square. It is made to slide along the original, stopping at each division, when a corresponding stroke is cut by the dividing-knife on the copy.

When box, or other wood, has been divided upon, the bur is first well rubbed off the divisions, and the whole surface polished with a dry rush; the surface is next burnished by rubbing it hard both ways, in the direction of the grain of the wood, with a clean piece of old hat, whereby an agreeable gloss is produced. The divisions are blackened by a mixture of powdered charcoal and linseed oil, laid on quickly, rubbed hard, and cleared away. This finishes the process. In ivory the divisions are first filled in with a composition of lamp-black and hard tallow, or bees'-wax and olive-oil; when this has been hard rubbed into the strokes, the whole surface is well rushed, and then polished with chalk and water, laid upon a linen rag. In finishing brass, after the bur has been taken off with charcoal and water, the surface should be finished off with wet blue-stone, which is a very soft slate. Divided gold and silver are best finished with charcoal; but all metals are improved by being rubbed with the hand after a little oil has been applied.

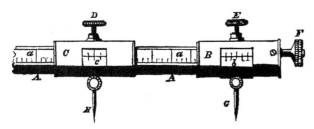

FIGURE 4: The beam-compasses

The *beam-compasses* consist of a beam, AA, Fig. 4, of any length required, generally made of well-seasoned mahogany. Upon its face is inlaid throughout its length a slip of holly or box, $a\,a$, upon which are engraved the divisions or scale. Two brass boxes, $B\,C$, are adapted to the beam, of which boxes the latter may be moved by sliding to any part of its length, and is fixed in position by tightening the clamp screw D. The two points $G\,H$ are attached to the boxes, and these may be made to have any extent of opening by sliding the box C along the beam, the other box, B, being firmly fixed at one extremity. The nice adjustment of the points $G\,H$ is attained by means of the two verniers b,c, fixed at the side of an opening in the brass boxes to which they are attached, and afford the means of minutely subdividing the principal divisions $a\,a$ on the beam, which appear through these openings. E is a clamp screw for a similar purpose to the screw D, viz. to fix the box B, and prevent motion in its point after adjustment to position. F is a slow motion screw, by which the point G may be moved a minute quantity for perfecting the setting of the instrument after having been set as nearly as possible by hand alone.

FIGURE 5: A Pair of spring dividers

For some descriptions of dividing, the compasses called *spring dividers*, Fig. 5, are useful. They consist of a circular steel bow and two legs, all in one piece: the bow allows a motion of the points from the distance of about an inch, or 2 inches, to their contact. The points may be brought near together by means of an adjusting screw attached to one of the legs by a pin passing through it, and upon which it turns as in a centre, the elasticity of the spring exerting a pressure against the screw, (or rather against a sharp angle attached to the screw,) and thus keeping the points at the distance to which they are set. The legs are bored to receive the cylinders or points, which are ground to the requisite fineness and brought very near the inner extremity of the diameter in order to measure the shortest possible distance; but at the point they are made round, and in every direction the sides must make equal angles with the perpendicular, otherwise a distance set off with them would be altered by pressure. In using this instrument the forefinger is pressed upon the bow, the thumb and middle finger keeping it upright, while the other fingers prop the hand; but where a distance is to be set off many times in succession, the dividers are to be twirled round in the same direction, making a dot at every half turn. For accurate bisection of a distance, the instrument must be held by the leg near the point, which is lodged in one extremity, while with the other a faint arc is described; the same thing being done from the other end of the distance, the middle point is secured by making a dot with a fine conical pointril. In using the dividers a magnifying glass held in the left hand must always be employed.

In dividing a common thermometer, several points, 12 or 15 degrees apart, are marked off in accordance with a standard instrument; these, which are always unequal, are filled up with equal parts. Suppose the distance from one mark to the next is 15 degrees; instead of first dividing the space into 3 or 5, the operator guesses, or estimates the distance of 1 degree, and running the divider over the space almost as quickly as he can count its steps, he ascertains how much he has erred; a second or third trial is sure to give him the proper distance. The dots in these trials, two of which should never be in the same line, are scarcely to be recognized by the lens, and he requires the last set only, so that by repeating the steps with greater pressure he may make the dots sufficiently large to receive the point of the dividing knife.

Original written material describing the tools and processes of hand graduation are extremely rare. This article from Tomlinson is one of the most complete we have ever found, and we are happy to be able to share it here with the members.

Figures 1, 2, 4, and 5 are those which accompanied the original article in Tomlinson; Figure 3 is an addition, copied from a similar article in the *Edinburgh Encyclopedia*.

This article has been reprinted in its entirety, with much material dealing with with circular graduation left in. These sections, of lesser interest to rule collectors, could have been omitted and replaced with ellipses (...), but to do so would have reduced the value of this reprint as a reference for others in the future. If any member finds them tedious and unnecessary, we apologise.

– The Editors

The Stanley No. 133 Rule

We have recently found a catalogue reference to a 3 foot, 4 fold Stanley rule not listed by Stanley in **Boxwood & Ivory**.

This rule, called the Number 133, was only offered one time, in the newly-introduced tool supplement appended to the third (1933) reissue of the 1929 Stanley #34 catalogue. It had not been mentioned in the previous reissue the year before, nor was it listed in the #134 dealers catalogue when that came out a year later.

As can be seen from the catalogue illustration reproduced here, the No. 133 was similar to the No. 66-1/2A, offered since 1923, but had a square, instead of an arch joint, and was made of maple instead of boxwood. The figures and graduations were printed, and were identical to those on the No. 66-1/2A.

Because of its very short period of manufacture, examples of the No. 133 are relatively scarce [8]. Only a 3 or 4 have surfaced, and those in only in FAIR to GOOD+ condition.

A similar inexpensive 2 foot rule, the No. 27, had been introduced in 1929, and this may have been a companion piece to that. The number assigned to this rule does not seem to be consistent with that of any other 3 foot 4 fold rule offered by Stanley; it is possible that the number was just derived from the year it was introduced. If this is the case, then perhaps the No. 27 was actually introduced two years earlier than the catalogues listings would indicate, and was assigned its number for the same reason.

The Stanley No. 133, 1933   Source: 1933 Stanley Catalogue #34

FROM THE MEMBERS:

(Relevant letters from our membership relating to rules and/or rule collecting)

It was a first for me when I found on a summer acquisition (a Keen Kutter No. K660½) a small brass plug with the dimple (like a "zero point"?) at the one inch mark on the inside. The only thing I could figure was for setting compass or divider to avoid the wear of the point. Then when I remembered the KK catalog reprint I found I was on the right track.

Are the Keen Kutter rules the only ones that incorporated this feature? This is the only one which I have seen.

Clifford Fales
Lakewood CO

We were not aware, until we recieved this letter from Cliff, that any 20th century rulemaker equipped his rules with these so-called "gauge points", as these brass plugs were called. They were a standard feature of nonfolding rules intended for use with dividers, such as Gunter's scales, etc., but we thought their use had died out when these instruments had ceased to be made sometime before 1900.

These plugs served three purposes: they identified significant locations in the scales they were set into (usually, but not always the zero point), they provided an accurately located mark for setting one point of the dividers, and they protected the wood from being damaged by repeated piercing with that point during use.

Folding rules so equipped are extremely rare; we know of only one example, a 2 foot, 2 fold carpenters' sliding rule by Adlam/Boston (pre-1850) which has them in the four architects' scales on the lower leg.

The reprint which Cliff mentions is the 1930 Simmons Hardware (owners of the Keen Kutter trademark) catalogue which was reprinted in 1984 by the M-WTCA. On page 106, at the beginning of the boxwood rule section, it says:

"For convenience in setting Compass and Dividers, a Small Rivet with Countersunk Center is inserted at the One-inch Mark, in which one point of Compass or Divider can be placed and the other point extended the desired distance on the Rule; this distinctive feature will be found on all *KEEN KUTTER* Boxwood Rules except No. K680; It insures Absolute Accuracy in setting Compass or Divider and also prevents marring the Rule."

We checked our collection, and were able to find two Keen Kutter rules equipped with these plugs, a No. K610 [5] and a No. K320½ [5]. Another, a No. K680 [4], did not have them; this agrees with the catalogue note above. A photo of the No. K320½ is shown here to illustrate this relatively rare feature.

The Editors

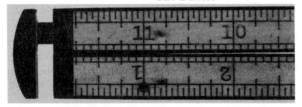

Gauge points on a No. K320½ rule

Enclosed is a half scale drawing of a Hook Rule [U] in GOOD to GOOD+ condition that I bought in January in an antique store in Orange, California for $20.00. The store clerk did not know its origin so we'll have to do some guessing. I thought it most interesting because it has been made to be used by the left hand (the numbers start on the right and end on the left – ED) My guess is that it was used at a lumber mill to sort boards by width. As it is only 18" long I would tend to think that it is of eastern origin. The wood looks like maple, but I am far from an expert. The inch and half inch lines as well as the numbers were done with a knife. The lines are not bad, but the numbers are rather crude. The blacksmith that made the hook did an excellent job, and it is fastened with four pointed flat head screws. Was it made entirely by the smith, or in cooperation with the user? Interesting indeed.

In July at the flea market in Moss Landing, California I found a 2 foot 3 tier board rule in GOOD+ condition for $20.00. It is also marked from right to left for left handed use. It is professionally made, but not signed. It has three minor errors, perhaps due to working backwards.

Were these two pieces made for left handers, or for people who wanted to free their right hands for tallying?

Which brings us of English and American rule markings: American (right to left graduations – ED) rule markings seem to be almost exclusively woodworking, at least among the major crafts. The English, as well as most other American trades, probably used left to right as a natural extension of reading. There are exceptions, of course. I have three Gunter's scales and five sectors. All but one are English, and all have inch scales that are right to left. I have exceptions in rules of other American trades, but two trees do not a forest make.

Henry Aldinger
Carlsbad NM

As far as we know, the difference in figuring between American (right to left) and English (left to right) carpenters' rules goes all the way back to the very beginnings of rule-making in this country. Commercial rule manufacture in significant quantities began in America about 1800-1825, and the few dateable examples extant from that early period show that the distinction existed even then.

There are a few exceptions to this generalization, of course. A few American rules, usually the early production of rulemakers just come to this country from England, follow the English pattern. Except for these, however, until after 1900, right to left was the standard.

Why the difference existed, we don't know. The primary cultural heritage of the American colonists was English, and they had a tendancy to use English (as opposed to Continental) tool patterns and methods. It has been suggested that American woodworkers adopted this difference at about the time of the American Revolution or the War of 1812, as a symbolic "rejection" of all things British. This could be true; Anglophobic sentiment ran high in 1775-1815 and, indeed, persisted until well after the Civil War. However, no documentary evidence exists to support this assertion, and it must be viewed only as an interesting theory.

Only the common carpenters' 2 fold and 4 fold rules were figured in opposite directions. As Henry has pointed out, some trades used one direction or the other irrespective of nationality. Gunter's scales and sectors, whether English or American, seem to have always been graduated right to left, while patternmakers' shrinkage rules were always left to right. It's all very inconsistent and confusing.

The distinction between the rules of the two nations began to break down with the gradual replacement of boxwood rules with other types after 1900. "Blindman's" rules, with the extra large printed figures were always graduated left to right, in America as well as in England. Zig-Zag rules and measuring tapes, when they were introduced in 1880-1900, were also usually graduated left to right in both countries. Finally, when Stanley, the only remaining American manufacturer of boxwood rules, switched to vertical figures in the 1940's, the last vestige of the national distinction was gone.

The Editors

I have a few cleaning related comments:

I suspect you'll receive some cautions against cleaning rules too vigorously and thoroughly for fear of damaging any painted numbers, lines, etc., on them. I acknowledge that as a theoretical possibility, but I've yet to see any evidence of it in the course of cleaning some 125+ rules via the quite vigorous use of turpentine and #0000 steel wool. Accordingly, I have strong doubts about the ugliness of ground-in dirt being worth tolerating due to any fear of that nature.

On some rules, a chipped finish can present quite a blotchy appearance. Most rule finishes are alcohol soluble, and a light wipedown with alcohol can greatly improve this appearance. I dont particularly advocate completely removing the original finish, just smoothing and redristributing it.

Some people don't like "overshined" brass on rules. I'm not exactly one of them, but Simichrome metal polish is the only type of polish that gives an "overshined" look that I do like. Simichrome is a pink paste that turns black in use. I suggest precleaning heavily oxidized brass with #0000 steel wool. (NOTE: This is especially in cases where the brass has oxidized under the original finish.) Use Simichrome sparingly, both out of respect for its cost and to avoid spreading too much black onto the wooden parts of the rule; what black does spread can be wiped off with turpentine.

Some ivory rules will have rust stains acquired while lying in contact with iron tools or spreading out from an iron pin. I've found a product named ZUD, sold as a rust removing cleansing powder, to be quite effective in removing such stains. Besides an abrasive powder, ZUD contains rust dissolving chemicals. Since the abrasive part of the ZUD can damage lines, numbers, etc., it should be used in a way that puts more reliance on the other chemicals than on the scouring action if the stain is in a marked area of the rule. Mix some ZUD and water to form a paste, put that on the rust spots, and allow it to sit for a few minutes. Then, using a cue-tip or a rag over the blade of a small screwdriver, rub the paste around gently – limiting the scouring action to the immediate area of the stain. Rinse off, check, and repeat if/as necessary. I've had no particular success with ZUD on any other types of soaked in ink, paint, etc., stains.

Robert Nelson
Cheverly MD

The issue of cleaning and caring for measuring instruments (or antique tools in general) has been much discussed in the past, and will surely continue to be so in the future. We feel strongly about this, to the point where a brief reply to this letter would not represent our views adequately. Accordingly, we have addressed ourselves to the subject at some length below in the articles: **Cleaning Rules ...**, and ... **and Storing Rules**

The Editors

Cleaning Rules, ...

Jim Hill

An ongoing controversy regarding restoration has been conducted by museums, historical societies, and private collectors for decades. The PRO's maintain that antiques should be returned to as close to their new condition as possible, and that the owner has the right to clean and refinish the objects in his collection as he sees fit; the CON's that ageing, patina, and wear are as much part of an antique as its origins and should never be removed. They worry about the gradual deterioration of potential national treasures through well-intentioned but uninformed restoration, and assert that we are only caretakers of these objects of antiquity,

These practices, as they pertain to rules, measures, and other instruments, can be broken down into four general categories: cleaning or polishing with abrasives, cleaning with solvents or chemical cleaners, passive cleaning, and the application of new surface materials. I would like to present my views on each of these.

Cleaning or Polishing with Abrasives

The long term effects of abrasive cleaning must be carefully considered before it is attempted. This process irreversibly removes surface material, and should be avoided.

In some instances, the decision to perform abrasive cleaning/polishing is made based upon the assumed desireability of brightly polished brass. It may enhance the aesthetics of an object in the view of some, but I believe this view is short-sighted and needs to be examined more carefully. We must consider that the patina aids in dating an object and can, in some cases, identify the type of environment it has been subjected to and how it has been used, and can actually increase its value.

In 1981 I purchased the contents of the factory of the defunct Buff & Buff Instrument Co. Various people had acquired objects from the Buff inventory prior to my purchase, and one of them offered to sell me a number of unassembled Buff sextants. When I inspected these instruments, I found that he had buffed each part to a mirror finish (which by itself was disconcerting), but still worse, had overlooked the fact that the graduated arcs were silver inlay, much softer than the surrounding brass. The results of his buffing was to obliterate most of the fine graduation lines on those arcs, rendering them useless! I declined the purchase.

Relating this incident reveals my thoughts on polishing and buffing any object of antiquity. Over the generations, repeated polishing can result in severe loss of metal and the destruction of the tool marks which are vital to historical research and authentication. A very soft cleaning with #0000 steel wool on the wood parts only of an object may be acceptable but even this mild cleaning can be easily abused. Careful abrasive cleansing and removal of corrosion or verdegris is justifiable, if leaving it unattended would result in the rapid deterioration of an object. In general, however, my recommendation is to do as little as possible or nothing at all to an object.

Solvent or Chemical Cleaners

With chemical cleaners it is important to realize that they have side effects other than the primary one of removing surface material from an object. This problem is compounded when there are rivet joints on wood adjacent to the metal being cleaned which could allow the cleaning chemicals to be absorbed into the wood, where they might continue to react with the metals. This could shorten the life of an object significantly.

Rust spots on ivory are ugly, and their removal by ZUD (as suggested by Bob Nelson) seems attractive. However, ZUD's active ingredient is oxalic acid, making it ZUD Vader (brother of Darth), the arch-enemy of ivory rules. I shudder to imagine cleaning around a steel pin on an ivory rule, trapping the oxalic acid below the surface where it continues, eternally, acting on the metal pins.

A RELIC ... A curious relic of the rebellion has lately been shown to us, in the shape of a carpenter's rule, which from its appearance has borne an active, if not all the time an honerable part in the late war. The rule was made at the rule factory in this village, having the trade mark of "E. A. Stearns & Co.", and from some peculiarities in its manufacture is known to be ten or a dozen years old. Mr. Charles L. Mead, agent for the Stanley Rule and Level Co., the present owners of the rule factory, accidentally fell in with a returned soldier at New Britain, Conn., a few days ago, who was the possesor of this rule. The soldier was an artificer in the 1st Conn. Artillery, and accompanied Gen. Butler on his expedition against Fort Darling. As our troops fell back on the second day, they found many of the rebels who had been killed in the first day's advance, and from the pocket of one of these - an artificer in a Louisiana artillery regiment - the rule referred to was partly visible. Our Conn. soldier (it being right in the line of tools wanted by an artificer) seized upon the rule, and confiscated it for the service of "Uncle Sam". The rule was carried through the remainder of the war by the soldier, and until now he has used it in his trade as a machinist. The utility of the thing had about gone, its "war worn" appearance rendering it valueless as a tool. The owner exchanged it for a new rule, and Mr. Mead now has this stray production of our peacable village. It has been left for a few days at the periodical store of E. J. Carpenter, where it may be seen by all who have a curiosity to see it.

From the *Vermont Phoenix*, Brattleboro, Vermont, April 26, 1867

Liquid cleaners contain many ingredients inherently detrimental to the materials with which rules and measures are usually constructed. Here is a list of a few common cleaners and their ingredients:

- ZUD   Contains oxalic acid.
- SIMICHROME 409   Contains petroleum distillate and ammonia. States that it is "not recommended for painted surfaces"; this means it could be harmful to remaining varnish or shellac, or to the compounds used to black in the graduations and figures.
- SOFTSOAP/SOFT SCRUB/IVORY (liquid soaps) Contain a long list of substances that may be good for your hands and dishes, but whose effect on shellac, lacquer, wood, brass, ivory, etc. is totally unknown: Sodium C14-16, olefin sulfonate, sodium laurel sulfate, lauramide DEA, glycol stearate, ... etc.

My recommendation is to stay away from chemical cleansers, whatever their form.

Surface Restoration

Restoring the surface finish of a rule or instrument is very problematical. As Bob Nelson explains, alcohol will work to re-distribute original shellac resulting in a satisfactory look. It not likely that this process will work with many rules, though, and one must be careful that the alcohol does not dissolve the blacking used to fill the graduation lines and figures.

Re-shellacking or varnishing rules is in my opinion not a good practise either, as this often results in over-finishing, leaving the object with an unauthentic look.

In Summary (Passive Cleaning)

The safest approach to restoration is to do as little as possible to disrupt the original/remaining finish while removing surface dirt. I recommend careful wiping with a moist cloth immediately followed by gentle buffing with a dry cotton cloth.

Developing an appreciation for "as found" condition of our rules, measures and other instruments will lengthen the life of these treasures and may enable us all to learn more from them.

... and Storing Rules
Phil Stanley

We have a tendancy, almost all of us rule collectors, to store our treasures in bulk. That is to say, we don't have all our rules spread out in a drawer or case, or set out on a shelf, or in a display frame on a wall, but instead usually keep most of them packed into some container or other somewhere. I know of one fine collection stored in a couple of machinists' chests, and another, one of the largest in the country, in four cardboard cartons in a closet. This is not surprising when you think of it; if you have more than 50 or 100 rules, the cost of cases or whatever to display them is not inconsiderable, and you quickly find yourself running out of wall, shelf, and/or table space.

Storage in bulk has one disadvantage, however. It is hard on the rules. Every time you pull the box out to, say, compare a rule with another newly acquired, or to show your collection to a friend (we all do that from time to time, don't we?) there's a chance that one or more will bang against another and pick up a nick or scratch. Even if you are just moving the box without opening it, to get at whatever is stored behind it, the rules, stacked 4 or 5 deep, can rub against one another.

I have found that, for me, the best way to prevent this kind of damage is to put each rule in a clear plastic bag before putting it in the box. This completely prevents the possibility of one rule scratching or nicking another no matter how much they are moved around or taken out and put back, and has the additional advantage of keeping each rule clean; it is impossible for rust on one rule to stain another (as once happened to me), or for fingerprints from handling to gradually soil a mint surface.

The bags I use are 4 mil polyethylene bags just wide enough to fit the rule, about 2" longer than its length, sealed at one end and open on the other.

I get them without any "zip-lock" or other closure seal on the open end; its easier to simply fold over the top and hold it down with a short length of transparent tape (see Fig. 1, below). The tape adheres well enough to the bag that it won't come off by accident, but not so tightly that it cannot be peeled off, if desired (it helps to fold about 1/4" of the tape back on itself to provide a grip tab). With this type of seal, the bag can easily be reopened as often as necessary.

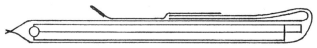

FIGURE 1: Recommended method of sealing

I get these bags from a local plastic bag company (listed under BAGS-PLASTIC in the Yellow Pages). They are sold in almost any combination of width and length, but I find that that in practise I only need three sizes: 1½ by 9 inches, 2 by 9 inches, and 2 by 16 inches. The smallest of these is perfect for 2 foot, 4 fold rules 3/4 and 1 inch wide, and for 2 foot, 6 fold and 1 foot, 4 fold rules as well with a little trimming. The next is suitable for broad 1 foot, 2 fold and broad 2 foot, 4 fold rules. The third is ideal for 2 foot, 2 fold and 4 foot, 4 fold rules, and can also be trimmed to fit 3 foot, 4 fold rules as well.

All my rules are stored in these bags now (except for large or awkward pieces such as tailors' squares, shrinkage rules, etc.). I recommend them highly; they do and ex-

cellent job of preventing damage, but do not significantly increase the amount of storage space required. I have even started taking a few with me to antique shows, flea markets, etc.; if I find a rule I can put it right in a bag when I buy it and prevent damage while walking around or on the way home.

Such bags are not expensive. The 1-1/2 X 9 inch size costs around $2.25/100, the 2 X 9 inch around $2.75/100, and the 2 X 16 around $4.75/100. It would only cost about $5.00 to put a collection of 150 rules safely in bags, and if doing so prevented damage to just one rule it would justify the expenditure.

The only problem with using these bags for storage is the minimum order size, typically 2000 pieces for each size ordered. Few of us want to lay out $200.00 for something we only need $5.00 worth of, so as an accomodation, the editors of **MENSURATION** are offering to purchase these bags in bulk and then resell them to the members in lots of 100. We have enclosed a sample of each size with this issue; any member wishing to buy some should let us know by February 1 how many 100 of each size he or she requires. The price to each member will be the bulk cost prorated over the quantity ordered (probably fairly close to the figures given above) plus shipping. We will purchase the bags shortly after February 1, and ship & invoice the individual orders shortly thereafter.

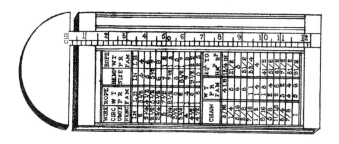

Errata – Vol. I, No. 1

Six Fold Rules: This article stated (incorrectly) that these rules were "... almost invariably graduated simply in 8ths and 16ths of inches", the only exception being the #58½, which was marked with numerous tables." We are indebted to both Paul Kebabian and Bill Gustafson for pointing out that of the five six fold rules illustrated, two are also graduated in 10ths of inches and one in 12ths. The offending sentance should be changed to read thus: "... usually graduated with the same selection of scales offered on narrow 2 foot 4 fold rules: 8ths and 16ths of inches, and often 10ths and/or 12ths as well. The one exception to this was the early Stanley No. 58½, which was so completely covered with tables that it bore no proper scales whatever."

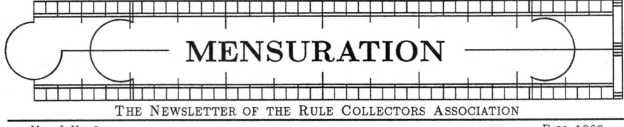

MENSURATION

THE NEWSLETTER OF THE RULE COLLECTORS ASSOCIATION

VOL I No 3   FALL 1989

BARREL MENSURATION & ULLAGE

Robert E. Nelson

Stanley (Ref. 1) provides one of the more lucid explanations to be found of the theory and use of the barrel gauging rod in establishing the liquid capacity of a barrel by taking a single diagonal measurement through the bung hole. It is not certain when or where the first such rod was created, but it is known that they were in use in 1615 in Austria. In that year, Johannes Kepler (one of the foremost mathematicians of his time) put out a paper which formally examined the subject of measuring barrels in general and the accuracy of barrel gauging rods in particular.

Kepler's work was reportedly triggered by his observing an Austrian wine merchant calculate a barrel's volume by using a diagonal rod. He intuitively sensed an error potential associated with this technique and proceeded to disprove its validity in a manner somewhat as follows: To avoid the complexity created by a barrel's compound form, imagine a straight sided cylinder whose volume is easily computed by the formula $\pi r^2 l$. Take two cases: one a cylinder whose length is 60" and diameter 40", the other a cylinder whose length is 80" and diameter 30". Now picture, for each case, a gauging rod inserted diagonally through a central bung hole to the opposite chine (juncture of side and head). In each case the reading will be 50". However the volumes of the cylinders in the two cases will be very different: the first holding 326.23 gallons, and the second 244.67 gallons.

Although he had successfully disproved the general mathematical validity of the gauging rod, Kepler finally concluded that the method was viable in practice due to the different wine barrels being measured all falling within a narrow range of sizes and proportions. This conclusion is reflected in Stanley's statement regarding a barrel's proportions always being (approximately) the same.

The diagonal rod was useful as a working expedient, but was not accurate enough to satisfy all governmental and mercantile requirements. Kilby (Ref. 2) indicates that as late as the middle 1900's British brewers who sold beer in casks did not sell a specified number of gallons — just a particular size of cask. Kilby's description of the barrel making process cites only a very crude measuring technique being used by the cooper to establish a cask's general size and indicates that the capacity was also varied by the amount the staves were shaved without changing any of the basic dimensions involved. The fact that the British weights and measures laws are quite strict probably accounts for the brewers' reluctance to specify gallonage.

FIGURE 1: Measuring casks with a gauging rod in 16th century Austria

Other mathematicians after Kepler also studied the problem of barrel measurement, but it was not until about 1790

VOL. I NO. 3   FALL 1989

MENSURATION is published three times a year by the Rule Collectors Association, a nonprofit organization dedicated to promoting interest and research in measureing devices and related subjects. To this end, we solicit and will publish serious articles dealing with the history, use, and manufacture of measuring instruments, and provide a forum for discussion of issues of interest to collectors and others interested in early rule technology.

Editor:
  Philip E. Stanley
  40 Harvey Lane.
  Westborough MA 01581-3005
    (508) 366-9442

Associate Editor:
  Jim Hill
  13933 Wayside Drive
  Clarksville MD 21029
    (301) 854-3170

Annual Subscription/Membership: $10.00 U.S. and Canada, $13.00 elsewhere. Make checks payable to Philip Stanley, 40 Harvey Lane., Westborough, MA 01581-3005.

Copyright © 1989
The Rule Collectors' Association

CONTENTS:

  Barrel Mensuration and Ullage 1
  A "Take-Down" Extension Stick 5
  Why 12? 6
  "Full and Complete Instructions . . ." 7
  Recent Rule Prices . . . . Insert

First Variety—Casks of the ordinary form, being that of the *Middle Frustrum of a Prolate Spheroid*. Rum puncheons and whiskey barrels are fair exponents of this form, which comprises all casks having a spherical outline of stave.

Second Variety—Casks of the form of the *Middle Frustrum of a Parabolic Spindle*. Wine Casks are exponents of this form, which comprises all casks in which the curve of the stave quickens slightly at the bilge.

Third Variety—Casks of the form of the *Middle Frustrum of a Paraboloid*. Brandy casks and provision barrels are exponents of this form, which comprises all casks in which the curve of the staves quickens at the chime.

Fourth Variety—Casks of the form of two equal *Frustrums of Cones*. A gin pipe is an exponent of this form, which comprises all casks in which the curve of the staves quickens sharply at the bilge.

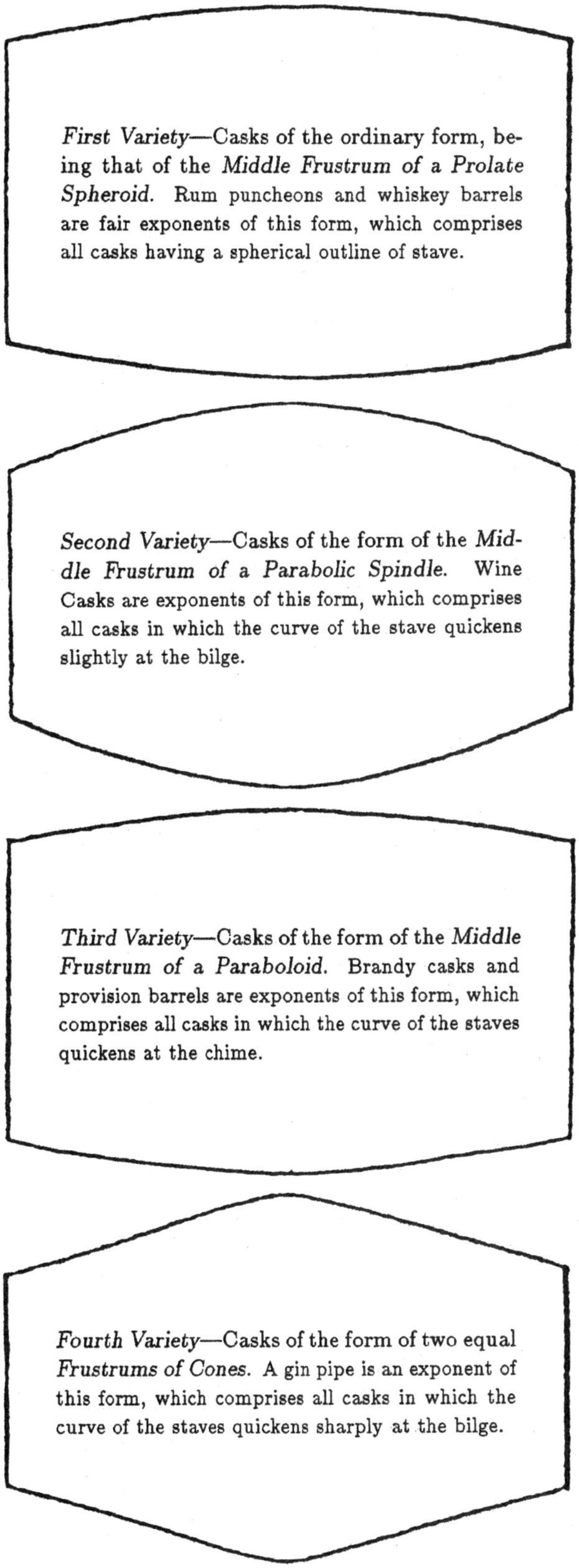

FIGURE 2: The Four Varieties of Barrel

that a Nicholas Pike devised the simple method that eventually became commonly accepted. Pike divided barrels into four varieties (see Fig. 2). The first had uniformly curving staves, the form commonly used at that time for whiskey barrels; the second a more bellied out form with the staves curving most in the middle, corresponding to wine casks; the third relatively flat sides with staves which curved most near the ends, as was used for brandy and provisions; and the fourth almost V-shaped staves, nearly straight except for a sharp bend in the middle, as in gin "pipes". For each variety Pike developed a multiplier to apply to the sum of the head and bilge diameters to approximate a mean diameter which could be used in the $\pi r^2 l$ volume formula cited earlier. Using the two most extreme of these multipliers, those for the fourth and third varieties, two barrels 50" long, with bung diameters of 30" and head diameters of 25" would have volumes of 130.96 and 138.13 gallons (note that a gauging rod would give the same volume reading for both barrels).

FIGURE 3: Catalogue illustration of a Stanley gauging rod

As can be seen, Pike's multipliers were quite critical to the accuracy of the gallonage obtained. Subsequent adaptations by other writers altered these multipliers slightly and/or created expanded versions of the basic volume formula to otherwise accomodate the essentially nonlinear dimensions involved.

Haswell (Ref. 3) included a discussion of cask gauging in which he explained the use of four mathematical formulas that updated Pike's work, two additional formulas for determining the capacity of any style barrel, and two formulas for calculating the contents of partly filled barrels. He also identified the five different measuring tools used for cask gauging:

One tool cited by Haswell is the "gauging or diagonal rod" (see Fig. 3), and his explanation of it and its use is essentially the same as is provided by Stanley. A second tool, called a "bung rod", is described as being alike to the diagonal rod and customarily combined with it in a single tool. The bung rod is simply used to measure the inner diameter of a cask through the bung hole; accordingly, the scale(s) of inches on the diagonal rod render it entirely adequate for this purpose. The distinction Haswell makes is that when using the tool as a bung rod a sliding collar is run down it to mark the reading on the inner side of the bung hole.

A third tool described by Haswell is the "callipers" (*sic* – with two l's and a singular form s) This is "a sliding rule adapted to project over the chimes of the cask to measure their inner length, when it is adjusted to just touch the outside surfaces of the heads of the cask,the inside distance between them may be read off, an allowance of one inch for the thickness of each head being included in the divisions of the rule." Such a rule is pictured in Hicks (Ref. 4) where it is called a "long calliper" (*sic* – two l's, but

no s). The one pictured there (see Fig. 4) and one in my collection (made by Belcher Brothers) are identical; two square wood pieces are interlocked to slide along one another with, on their outer ends, short wood pieces braced by brass supports that descend down 90 degrees and then back in toward each other. The scale on the tops of the interlocked pieces on my tool (and most others I have seen) only read 1-1/2" less than the separation of the two ends, as compared to the 2" cited by Haswell.

This illustration also shows a tool not mentioned by Haswell: a "cross calliper". This tool had straight, long jaws, and could measure the diameter of very large circles. One of Haswell's formulas calls for the use of a diameter reading taken midway between the bung and the head, but none of the tools he cites could provide such a measurement. Perhaps the cross calliper was intended for this purpose (it is assumed that it included an allowance for stave thickness in its graduations).

Haswell's fourth tool is the "gauging slide rule" (see Fig. 5, TOP), described as having a linear inch scale extendable by its slide to permit its use in measuring the head diameters of casks (note the short projecting metal "jaws" on the pictured example, designed to rest against the inside surface of the staves while measuring). This rule also had the normal slide rule scales, plus various special scales adapted to calculating barrel capacities. The example he describes (in detailed, if confusing terms) is cited as having been made by Belcher Brothers.

Haswell's final tool is a "wantage rod" (see Fig. 6), which is essentially the same as described by Stanley. Hicks describes, but does not picture, a 24 inch "vacuity rule ... for ascertaining the number of gallons out or in a cask." This may be a wantage rod (he does not list any rod which he calls a "wantage rod") by another name, but its price

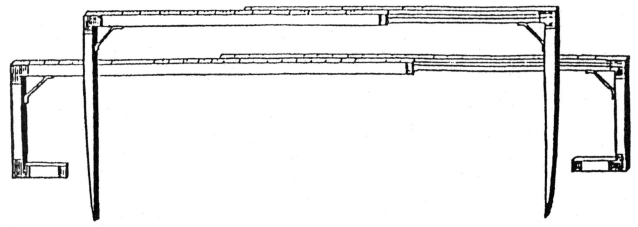

FIGURE 4: Catalogue illustration of a long calliper and a cross calliper

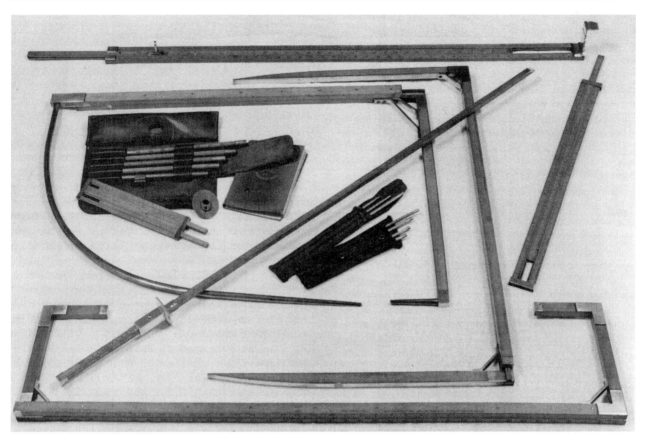

FIGURE 5: Some gaugers instruments sold at an English auction in 1987: TOP: Gauging slide rule. CENTER: Cross calliper and head thickness calliper. BOTTOM: Long calliper. DIAGONAL: Bung rod.

Photograph courtesy of David Stanley Auctions

relative to other rules listed seems somewhat higher than such a rod should cost. Additionally, Hicks includes two "bar cask rules" which give readings of the remaining contents of bar casks (vs. their "wantage" of being full) on a dip-stick basis and two similar, but shorter (2' vs 3'), "bottle rules" which could easily be confused with barrel rules. In observations of such "bar cask rules" in the marketplace, it has been noted that they are frequently called "wantage rods" despite the technical difference between the two types.

To summarize the apparant current availability of these barrel measuring tools: wantage rods and their dip-stick variations are quite common. Gauging rods with collars for use as bung rods are rarely seen, but plain rods are only slightly scarcer than wantage rods; long callipers, are fairly scarce, and cross callipers are extremely scarce – I have never seen one. It is not certain whether gauging slide rules are also extremely scarce or are just difficult to identify for what they are; slide rules whose slides include an extensible linear measuring scale are to be found, but I have never been able to correlate one with the specific description provided by Haswell.

Haswell's formulas for Pike's four types of barrels are as follows: For the first type: twice the square of the bung diameter plus the square of the head diameter; multiply this sum by the barrel's length and multiply that product by 0.2618. For the second type: twice the square of the bung diameter plus the square of the head diameter, less 40the two diameters; multiply the result by the barrel's length and multiply that product by 0.2618. For the third type: the square of the bung diameter plus the square of the head diameter; multiply this sum by the barrel's length and then multiply that product by 0.3927. For the fourth type: the square of the difference in the head and bung

FIGURE 6: Wantage Rod

diameters plus three times the square of their sum, multiply this sum by the barrel's length and then multiply that product by .06566. All of these formulas give a cubic inch figure which is divided by 231 for gallonage.

Note that, compared to the standard formula for volume used today, Haswell's formulas do not cite π, use diameter instead of radius figures, and include what I'll call "magic numbers." These magic numbers can all be shown to represent the results of combining π with various other acceptable mathematical processes. For example, a squared diameter divided by four equates to a squared radius (e.g., if a radius is 20", d^2 equals 1600 which, divided by four, equals the same 400 that r^2 equals); this division by four represents part of Haswell's magic numbers. As another example, if a number is a divisor, its reciprocal (1/n) can be used as a multiplier (e.g., dividing by four produces the same result as multiplying by four's reciprocal, 0.25); Haswell's magic numbers include such reciprocals. Finally, he had to supply additional divisors (or their reciprocals) to average out his multiple squared diameters or to otherwise inject the factoring needed to establish a mean diameter. In another part of the book, where he cites their use in connection with the gauging slide rule, it is clear that his factorings were 0.7 for the first type barrel, 0.63 for the second, 0.56 for the third, and 0.52 for the fourth. In more normal usage (as by Pike), the difference in bung and head diameters would be multiplied by these factors and the result added to the head diameter. Haswell's factoring figures for the same 4 varieties of barrels are, respectively, 0.7, 0.65, 0.6, and 0.55.

The fourth type of barrel is the easiest to visualize on an intuitive basis and was used for some comparisons of the various formulas. If purely straight lines were involved, the mean diameter of this type barrel would simply be the average of the head and bung diameters and the $\pi r^2 h$ formula would give the capacity. Pike's use of a 0.55 factor and Haswell's use of 0.52 come very close to providing such a simple average; the difference would represent the nominal curvature deviations involved in reality.

Haswell (Ref. 3) gives an example of a barrel of the fourth variety with a 24" bung diameter, a 16" head diameter, and a 36" length. The square of the difference of diameters is 64, three times the square of the sum of the diameters is 4800, and the sum of these times the length is 175104. That multiplied by 0.06566 gives 11497.329 which, divided by 231, equals 49.77 gallons. Pike's formula for the same barrel would give a gallonage of 50.95. However, using Haswell's 0.52 vs. Pike's 0.55 in Pike's formula gives a gallonage of 49.76 and shows that to be the primary mathematical difference between the two. Another comparison of the two formulas for the first type barrel where Haswell and Pike agree on a 0.7 factoring gave gallonages of 63.32 and 63.62. Although somewhat more deviant than for the fourth type barrel, this is still close enough to make one wonder about the rationale for the extra complexity of the Haswell formulas.

One of Haswell's formulas for any type barrel uses a

A "TAKE-DOWN" EXTENSION STICK,
or
SOMEBODY GOOFED!

Philip Stanley

In 1947 the Stanley Works resumed offering their No. 510 extension stick, a popular measuring tool which had been last listed previously in 1943.

Prior to the second world war they had manufactured a whole series of extension sticks, The Nos. 240, 360 (until 1932), 480, 510 (until 1936), H510 (so-called from the folding hook located on one end), and 612. Except for size, all were basically alike (see Fig. 1), consisting of two brass-tipped equal length maple sticks sliding against one another and held in alignment by a pair of cast brass guide fixtures. From the closed position the front stick could be slid out to the right almost its full length, and be clamped anywhere in that range by a thumb screw in the right hand guide fixture. The sticks were graduated in such a way that the markings on the back section at the left end of the front section read directly the overall length of the pair at that setting.

---

diameter reading taken midway between the bung and the head in addition to those two diameters. This is assumed to provide an equivalancy to the factoring figure. The other formula, which is indicated to be less accurate, uses only the bung and head diameters. Haswell also provides formulas for determining the contents of partly filled barrals (ullage) by using the wet inch readings on a dip-stick dropped through either the bung or, in a standing barrel, the head.

It should be noted that none of the Haswell or Pike formulas would be completely accurate regarding such stave shaving variations as are cited by Kilby. If anyone wants to know the absolutely true capacity of a given barrel, one has to carefully fill it with another container of an accurately known capacity - which is probably the way it was done before someone dreamed up the idea of using a barrel gauging rod in the first place. □

REFERENCES:

(a) Stanley, Philip E. *Boxwood & Ivory; Stanley Traditional Rules, 1855-1975*. Westborough, Mass.: The Stanley Publishing Co., 1984.

(b) Kilby, Kenneth. *The Cooper and His Trade*.

(c) Haswell, Charles H. *Mensuration and Practical Geometry*. 1858.

(d) James J. Hicks Catalog, London (late 1800's)

(e) Misc. Encyclopedias and Mathematical Books.

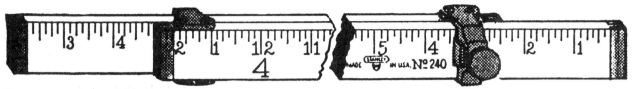
FIGURE 1: Stanley Extension Stick, Pre-World War II

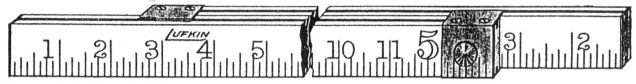
FIGURE 2: Lufkin Extension Stick

(The extension sticks were only a few of the traditional rules discontinued before and during the war to reduce costs and make room for increased war production. Some reappeared in the catalogue in 1947 and 1948, but many others were gone for good. The other size extension sticks, for instance, the Nos. 240, 480, and 612, were never offered again).

The return of the No. 510 to the catalogue in 1947 did not mean, however, that the Company had resumed production of their original stick. Apparently it would have been too expensive to restart production, or perhaps the Stanley type of stick, with cast brass fittings, was now too costly to make and sell at a profit. At any rate, instead of making their own extension stick, Stanley purchased Lufkin No. 7165 extension sticks, specially marked "STANLEY" and "No. 510", and offered those instead.

Lufkin extension sticks differed in a number of details from the prewar Stanley sticks (see Fig. 2). The guide fixtures were steel, not brass, and instead of encircling the other stick, ran in narrow milled slots in its edge. The

WHY 12?

William N. Gilliland*

Once a unit of lineal measure is established, there are two arithmetic processes that inevitably follow: multiplication, for lengths of more than one unit, and subdivision, for lengths of less than one. The former requires only simple multiplication; the latter requires a system. Perhaps the most common systems are the binary (2), the decimal (10), and the duodecimal (12). Among more than 100 "feet" in use before the 1800s, almost almost all consisted of 12 "inches", each subdivided into 12 "lines" A notable exception to this was the Amsterdam "foot" of 11 "inches".

The widespread early adoption of 12 as a basis for subdivision was undoubtedly based on its easy divisibility and its small magnitude. Of all numbers from one through 50, only one, 48, is evenly divisible by nine numbers in addition to one; only one other, 36, is similarly divisible by eight numbers; only three others, 24, 30, and 40, by seven; only eight others, 12, 18, 20, 28, 32, 44, 45, and 50, by five; and only one other, 16, by four.

12 is the smallest number significantly and easily divided eveny into fractions. These are halves, thirds, quarters, sixths, and twelfths, and are apparently simple enough to be competently handled arithmetically by a generally uneducated population.

Also, twelfths of inches may well have been near the ultimate limit of physical demarcation until the invention of sophisticated dividing engines in the late 18th century and except when the binary system is used to subdivide relatively large units. The number 16, also easily divisible, was probably well within the general population's arithmetic competency. Certainly many early cloth measures (e.g.: ells) usually several feet long were commonly divided by the binary system into 16ths and even 32nds. Fractions derived from numbers higher than 16 probably strained arithmetic and demarcating abilities.

Both 12 and 16 have some distinct advantages for subdivision. Twelve, when divided by its five whole number divisors, produces five convenient fractions, halves, thirds, quarters, sixths, and twelfths. 16, when divided by its four whole number divisors produces only four fractions: halves, quarters, eights, and sixteenths. 16, on the other hand, is better suited to binary subdivision (repeated division by 2); if factored this way, 16 produces halves, quarters, eights, and sixteenths, while 12 will produce only halves and quarters.

The "foot" was almost always divided into 12 "inches", many of which were also subdivided into 12ths. For example – the French *Pied du Roi* (32.484 cm) established by Charlemagne, consists of 12 *pouces* (inches) each in turn divided into 12 *lignes*. Later, with improved techniques, many inches were divided by the binary system into halves, quarters, eighths, sixteenths, and even 32nds or 64ths.

The Chinese used the decimal system of subdivision long ago. Today, those of us who prefer the decimal point and the metric system and hate fractions are still wondering – "Why 12?". □

* *Editor's note:*
William N. Gilliland died on 11 June 1988

front stick slid to the left, instead of the right as the instrument was extended. Also, the Lufkin style of figures, particularly the large ones marking the foot points, was different from the style used by Stanley.

At this point, human error crept into the equation. Stanley, apparently not wishing to have it known that they were remarketing a competitors' product, instructed the artist preparing the illustrations for the 1947 catalogue to show the Lufkin-made stick in such a way that it would resemble the earlier Stanley product. In compliance, the artist drew what is basically a Lufkin extension stick (this is apparent from the style of the foot figures, and from the fact that the rule depicted extends to the left instead of to the right), but modified it by omitting the Lufkin steel guide fixtures and drawing the traditional cast brass Stanley ones in their place (see Fig. 3).

Unfortunately, This effort to make the picture look as much like the earlier Stanley rule as possible backfired. As examination of Fig. 3 will show, the stick, as drawn, is totally unusable; any attempt to extend it more than 2 or 3 inches beyond the position shown would cause it to fall apart!

The artist should have reversed the position of the two guide fixtures in the drawing, putting the front one on the left, to compensate for the fact that the Lufkin stick extended to the left, instead of to the right. He neglected to do so, however, and whoever was checking his work didn't catch it. Perhaps he was just not mechanically inclined, and couldn't visualize how the stick operated; maybe it was a Monday or something. For whatever reason, this flawed drawing made it into the catalogue, and Stanley ended up offering an "extension stick that couldn't be used"!

(It is worth noting that Stanley continued to use this same illustration in their catalogue for the next 10 years, until the No. 510 was last offered in 1958. Someone must have noticed the error in that time, but either they didn't want to spend the money to fix it, or [more likely] somebody didn't want to admit they'd made a mistake!) □

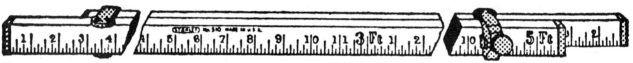

FIGURE 3: "Stanley" Extension Stick Catalogue Illustration, 1947-1958

"FULL AND COMPLETE INSTRUCTIONS..."

Philip Stanley

During the 19th century, many books of instructions were published explaining the use of the carpenters' sliding rule and the engineers' rule. Most were written and published commercially, either included in mathematics texts, or in connection with some particular trade or craft such as gauging or surveying. Others, however, were published and sold by major manufacturers of these rules in order to encourage their purchase and use. Their customers were mostly carpenters and artisans, capable of little more than arithmatic and a little trigonometry, and needed instruction before attempting to use anything as sophisticated as the slide rule.

These books are interesting, avoiding as much as possible any discussion of the theory of logarithms or the slide rule, and concentrating primarily on rote procedures which the user could memorize. Proportions and the "Rule of Three" were emphasised, and most examples were drawn from civil engineering, mining, and textile manufacturing (the major fields of industrial endeavour prior to 1850).

At least three makers, Chapin, Rabone, and Stanley, are known to have offered such instructions. Roberts has reprinted the Rabone instructions (Refs. 1 & 2), and a copy of the Chapin instructions (Ref. 3) is in the library of the Connecticut Historical Society. Until now, however, no example of the Stanley instructions, offered from 1860 to 1902, has been found.

It seems now that a copy of this elusive book has finally surfaced at a Richard Crane listed auction in 1988, appearing in the list simply as a "Book on slide rules". The buyer didn't recognize it for what it was; he happened to notice the Stanley name stamped with a round rubber stamp on the front flyleaf (see Fig. 1), and bought it under the impression that it was a book which had at one time been in the Company library.

FIGURE 1: Stanley name stamped on front flyleaf

When he got home, however, he examined his find more closely, comparing it with the description in the old Stanley catalogues, and has now concluded (and we agree — Ed.) that his purchase is the first known example of the elusive Stanley instructions.

The book he purchased is entitled *Utility of the Slide Rule*, by Arnold Jillson, of Providence, R.I. It is a small book, only 3.7" X 4.7", bound in dark cloth, with the name printed on the front in gold letters (see Fig. 2). It has 202 pages, of which 113 deal with the use of the carpenters' sliding rule, and 80 with the engineers' rule. It was published in 1874 in Hartford. Conn., by Case, Lockwood & Brainard.

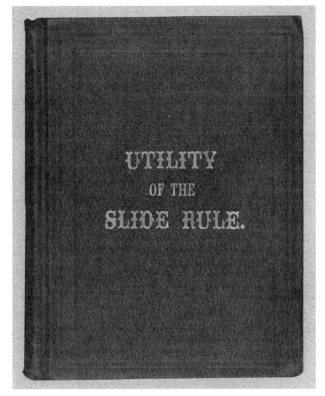

FIGURE 2: Instructions for carpenters'/engineers' sliding rules

The Stanley description in their catalogues reads as follows:

"We have an improved Treatise on the Gunter's Slide and Engineers' Rules, showing their utility, and containing full and complete instructions, enabling Mechanics to make their own calculations. It is also particularly adapted to the use of persons having charge of cotton or woolen machinery, Surveyors, and others. 200 pages bound in cloth. Price $1.00 net."

Several correspondences support the contention that this book is the actual Stanley instructions: The number of pages matches almost exactly. The book covers both the carpenters' sliding rule and the engineers' rule, unusual in this type of book, but specifically mentioned in the description in the Stanley catalogue. Finally, the publisher, Case, Lockwood & Brainard, was the same firm which was printing the Stanley catalogues at that time.

The most telling evidence, however, is a small 1/3 page advertisement for Stanley 2 foot 2 fold carpenters' sliding rules tipped (pasted) into the book inside the back flyleaf (see fig. 3).

Such an advertisement would never have been pasted into a book in the Stanley library, but would have been a perfectly natural adjunct to a publication intended for customers. Its presence, coupled with the match between the book and the description, establishes the identity of this book with near certainty.

No. 12, Rule, Two Fold, 24 inch,
1½ inch wide, with Gunter's Slide. 8ths, 10ths and 16ths of inches, 100ths of a foot, Drafting and Octagonal Scales. PRICE $1.00.
Sold by all Hardware Dealers.
☞ Sent by Mail, post-paid, on receipt of the price. ☜
STANLEY RULE & LEVEL COMPANY,
NEW BRITAIN, CONN.

FIGURE 3: Advertisement attached to back flyleaf

It is apparant, then, that Stanley never actually published their own instructions. They simply purchased Jillson's book from Case, Lockwood & Brainard, stamped their name inside the front cover and pasted an advertisment into the back, and shipped them to any customer ordering a copy of their "Treatise".

One wonders how many more copies of the Stanley instructions are gathering dust on the shelves of some used book dealer, passed over by collectors because it didn't say "Stanley" on the cover or on the title page and wasn't in the trade catalogue section. How often do we take the time to check inside the front and back covers as well as the title page, and what do we miss in the process? □

REFERENCES:

(1) —. *The Carpenter's Slide Rule, Its History and Use.* 3rd ed. Birmingham, England: John Rabone & Sons, 1880. (reprinted by the Ken Roberts Publishing Co.)

(2) —. *Instructions for the Use of the Practical Engineers' & Mechanics' Improved Slide Rule, as Arranged by J. Routledge, Engineer.* Birmingham, England: John Rabone and Son, Ca. 1900. (reprinted by the Ken Roberts Publishing Co.)

(3) —. *Instructions for the Engineer's Sliding Rule, with Examples of its Applications.* Pine Meadow, Conn.: Hermon Chapin, 1858

(4) Jillson, Arnold. *A Treatise on Instrumental Arithmetic: or, Utility of the Slide Rule.* 3.7" X 4.7", Cloth bound, viii + 194 pp. Hartford, Conn.: Case, Lockwood & Brainard, 1874

THE NEXT ISSUE ...

Vol. II No. 1 will feature the article by the late William Gilliland, SOME MEASURES OF HISTORY, which appeared this year in the *Bulletin of the Scientific Instrument Society*. This article on foriegn measures identifies and lists the lengths of literally hundreds of standards of linear measure which were in use in Europe prior to the Metric System. This will be an invaluable aid to the members in identifying European measuring instruments as to their geographic and historic origins.

V
Materials, Construction, and the Rulemaking Process

The folding rules made by American rule makers during the second half of the nineteenth century are wonderful examples of traditional form tools as manufactured by the evolving factories of that era. They used the same selected types of materials as when made by hand, and were constructed to the traditional patterns, but were made in lots of from 50 to 100 rules at a time. The level of mechanization employed primarily replaced human power with machine power, but did not significantly alter the fundamental sequence or nature of the operations performed.

MATERIALS

Folding rules were made of three types of materials: wood, ivory, and metal. Wood was the most common, with ivory reserved for premium-quality rules, and metal only where its special properties were essential. The trim on the rules (the hinges, tips, etc.) was always metal, usually brass, but sometimes German silver, and occasionally steel. Hinge pins and the small pins joining the various parts together were either brass or steel.

Wood

A number of different woods were used for rules. The most common was boxwood, but sometimes, for reasons of cost or physical characteristics, maple or hickory was used instead. Occasionally fancy woods, such as rosewood, satinwood, or mahogany, were used, but this was rare. The preferred woods were light in color, and very fine grained. Coarse grained woods, such as hickory, were harder to graduate, as the lines and figures had to be cut deeper in order to be adequately visible. Dark woods such as mahogany and rosewood presented even more severe problems. An incised line filled with black paint or wax (the traditional method for marking wood rules) is almost invisible on a dark surface. Using a white filler is not a great improvement either; the contrast is not nearly as great and becomes even less as the rule becomes dirty with use and wear.

(Turkey) Boxwood

(Turkey) Boxwood (*buxus balearica*) is the traditional, premium material for rule manufacture. It is a hard, heavy wood, with a creamy yellow color, and is among the finest textured of commercial woods, being almost grainless. It is an excellent wood for turning (such as chess pieces) and detailed carving (such as wood engraving). When properly seasoned, boxwood is dimensionally stable and insensitive to ambient humidity, and it is this property, coupled with its ability to be clearly and visibly marked with fine graduations which make it so desirable as a rule material.

Turkey box is a small tree (up to 36 feet high) which occurs locally from southern Europe through Turkey to Iran, the majority growing in Turkey and southern Russia. The logs of the tree are small, with an average diameter of from six to seven inches, and will sometimes have an irregular grain pattern. Until the turn of the century the primary source of boxwood was Turkey, which was the shipping point for not only its own timber, but also that originating in Russia. By 1900, however, the timber had become increasingly scarce and expensive, due to exhaustion of mature trees in areas accessible to transportation. It thus became necessary for American rule makers to find a substitute, and Maracaibo Boxwood, or Zapatera, a Venezuelan timber which had recently become available, proved to be nearly as satisfactory and was adopted. The use of the European variety was never resumed. Improved transportation made new supplies available, but political and economic turmoil in the countries of origin had raised its cost to the point where the supply from South America was cheaper, and also more reliable.

Maracaibo Boxwood (Zapatera)

Maracaibo Boxwood (*gonypiospermum praecox*), also known as Zapatera, closely resembles boxwood, having the same fine, even texture and an only slightly lighter color. It has a straight grain, and, like boxwood, almost no figure. It is about 20 percent lighter than boxwood, and not quite so hard, and is thus marginally easier to work (although it splits more easily).

Zapatera is a small tree, similar in size to the Turkey box, which grows in the southern Caribbean. It grows in Cuba and in the Dominican Republic, but in large, commercially useful quantities only in Colombia and Venezuela. This wood has an advantage over boxwood in that it is available in slightly larger sizes (logs from this tree have a typical diameter of from 6 to 10 inches), and is less likely to have irregular grain.

Hard (Rock) Maple

Hard Maple (*acer saccharum*), sometimes known as Rock Maple, is a hard, dense wood, very pale, and usually straight-grained (although examples with "birds-eye" or "fiddle-back" figure are not uncommon). When straight-grained it will accept a smooth finish, and when thoroughly dry is reasonably stable. Hard maple has exceptional wear resistance (hence its use for bowling alleys, shoe lasts, etc.). The wood has a fine, even texture, and will easily accept fine graduations.

Hard maple is widely distributed in the North American continent, being common throughout Canada and the northern and central parts of the United States. The tree is medium to large in size, and its wood is available commercially in good-sized logs. Due to its ready availability, hard maple is a relatively inexpensive wood.

Hickory

Hickory (*carya spp.*) is a tough, heavy, coarse-grained wood produced by four species of North American trees. The sapwood is almost white, the heartwood a reddish brown. The paramount quality of hickory is its high strength and impact resistance; it is frequently used for the handles of impact tools, etc. The coarse, ring-porous texture of the wood makes it difficult to mark with fine graduations, but it was still sometimes used for rules and sticks likely to be subjected to rough usage. Hickory is very dense, and seasons only very slowly, with much shrinkage.

Hickory is a medium to large size tree, indigenous to southern Canada and the eastern and southern United States. If properly grown, the logs are of good size, and the grain straight.

Mahogany

Mahogany (*swietenia mahogani* and *swietenia macrophylla*) is one of the world's great cabinet woods, extensively used for fine furniture ever since the middle of the eighteenth century. Medium to dark red-brown in color, it exhibits a fine, even grain, and occasionally a spectacular figure. The Cuban mahogany (*S. mahogani*), native to the Caribbean islands, was a darker and heavier wood than the Honduras or American mahogany (*S. macrophylla*) found on the Central and South American mainland. Mahogany is easily dried and when properly seasoned is dimensionally stable and resistant to decay and insects. This, coupled with its ability to be easily worked to a high finish and to be carved with fine detail, has made mahogany probably the most widely used premium wood. In addition to fine furniture manufacture, mahogany is used in boat-building, pattern-making, and wood-engraving.

The mahogany tree is quite large, often attaining 150 feet in height, approximately half of which is trunk bare of branches. This trunk can be as large as 6 to 8 feet in diameter, so large as to require reduction to timber at the site before shipment to the ocean for transport overseas. It was probably the Honduras mahogany that was used for rulers by Stanley in 1862-1865; by that date the Cuban variety was very scarce, and would have been much too expensive for use in commercial articles.

Rosewood

Rosewood (*dalbergia spp.*) is a tropical wood of spectacular appearance, frequently used for cabinet work, as well as other purposes. A number of varieties exist (Indian, Brazilian, Honduras, etc.), but all have similar characteristics. The heartwood is purplish brown, with dark, almost black markings creating an exceptional figure. The sapwood is much paler, and is frequently discarded. The wood is heavy, and possessed of natural oils which enable it to slide easily on smooth surfaces (hence its occasional use for plane stocks and soles). Rosewood turns and carves fairly easily, and is frequently used for knife and tool handles (until the 1950s it was regularly used for the handles and knobs of metal planes, and for the stocks of squares, bevel gauges, etc.). Rosewood has the disadvantage of being difficult to glue, due to its oiliness, but in spite of this is frequently used for veneering because of its appearance.

The rosewoods are small to medium-sized trees, widely distributed in the tropical regions of the world. The various species are found in commercial quantities in India, Madagascar, and Brazil, as well as Honduras and other Central American countries.

(West Indian) Satinwood (Yellowwood)

West Indian Satinwood (*zanthoxylum flavum*) is a fine cabinet wood, one of a large number of different woods of African and Caribbean origin which have been loosely grouped under the name Yellowwood. Found in Puerto Rico, Santo Domingo, Jamaica, and in some of the smaller Caribbean islands, it had been a popular furniture wood in England 100 years prior to the popularity of mahogany in Georgian times, and was still occasionally used as late as the end of the nineteenth century.

The wood is pale yellow to cream in color when cut, subsequently turning slightly darker with age. It has a fine texture, and is found with both straight and irregular grain, the latter producing an exceptional figure. Satinwood is a dense and somewhat oily wood, and gives off an odor similar to that of coconut oil when freshly worked.

The species can develop as either a shrub or a tree, depending on local conditions; when the latter, heights of 40 feet are common, and logs can be up to 20-24 inches in diameter.

IVORY

Ivory is the dentine-based living substance which forms the tusks of elephants and certain other mammals. It is a slightly creamy, almost white material, very dense, with fine close pores containing a gelatinous solution which renders it easy to work and polish. It is tougher than bone, but not so tough as horn, and will break under sufficient strain or shock. Almost all ivory comes from elephants, the other mammals having much smaller tusks, and most of that from the African elephant, whose tusks are much larger than the Indian and Ceylonese species. Some ivory also originates as fossil ivory, being the tusks of mammoths found frozen in northern Siberia. At the peak of the ivory trade (1880-1890) 750 tons of African ivory a year were entering the world markets, about 60 percent through London, and the remainder through Bombay, Amsterdam, New York, and other cities. The properties of ivory which rendered it valuable (aside from its scarcity) were its hardness, its creamy whiteness, and its ability to be finely carved and polished. Until the invention of celluloid and other plastics, it was much used for billiard balls, cutlery handles, the faces of piano keys, and decorative pieces. Its hardness and brittleness required special techniques to work it, and it would gradually yellow with age, especially if kept away from the light. Ivory shrinks while drying, and must be seasoned like wood. The so-called "bleaching" process described by Stephens (Ref. 1, p. 376) was actually to ensure that the ivory was fully seasoned before making it into rules; sunlight will not bleach ivory, only slow down the rate at which it yellows. Even after seasoning it is not as stable with varying humidity as boxwood; Strelinger, (Ref. 2) describes a two foot ivory rule which had shrunk a full $1/8$th of an inch.

Prior to the Civil War, ivory was received from the importer in the form of whole tusks (Ref. 3), to be sawn into slabs and resawn into rule blanks by the rulemaker. By the turn of the century the initially "slabbing" was done by the wholesaler prior to delivery (Ref. 1, p. 376), and all that remained to the maker was to reduce the slabs to blanks.

METAL

Three types of metal were usually employed in the manufacture of rules. Brass was the most common. German (nickel) silver was usually used in place of brass on ivory rules. Steel, of course, was used where high strength or wear resistance was required (for machinists' rules, for instance, and for joint pivot pins, assembly pins, etc.).

Brass

Brass is an orange to yellow alloy of copper and zinc (sometimes with the addition of small amounts of other metals) which is almost ideal for the manufacture of the joint and trim parts of folding rules. Brass is not as cheap as steel, but is still relatively inexpensive, and it has a variety of other properties which are highly desirable. It has a fairly low melting point, and can readily be cast. It is ductile enough to be easily rolled and drawn, and it machines freely. Brass will accept a high polish, which can be preserved by subsequent lacquering, and when so polished presents an extremely attractive appearance.

The wear resistance of brass is not exceptional (hence the number of well-used folding rules encountered with very loose joints), but its corrosion resistance is fairly good, so that, while it will tarnish, it does not have the extreme tendency to rust encountered in steel.

The brass alloy used for the joints and trim of wood and ivory rules was a so-called "high" brass, containing about 62 percent copper and 38 percent zinc, and was particularly suited to cold-rolling, stamping, and machining. Rules made from sheet brass without any wood, such as Blacksmith's rules, were made from a harder "spring" brass, containing 66 percent copper, 32 percent zinc, and 2 percent tin, which was more difficult to work, but was more elastic, and less subject to bending.

German (Nickel) Silver

German silver is a white alloy of copper, zinc, and nickel which was frequently used for the trimmings on premium-quality rules, due to its fine silver-like appearance, and its improved durability. This alloy originated in China, where its composition is said to have been known since time immemorial, but was only introduced to the English-speaking world in 1830, when a sample was brought to England from Germany (hence its name). For many years the alloy was very widely used for electroplated tableware and for scientific instruments, and only recently has it been largely replaced in these applications by other materials.

German silver (also known as nickel silver, Chinese white silver, Packfong, Electrum, etc.) is basically a modification of the alloy for brass, containing about 60 percent copper, 20 percent zinc, and 20 percent nickel. It has all of the workability characteristics of brass, and can be cast, rolled, and drawn. At the same time, it possesses the properties of being hard, tough, and not easily corroded. When polished it very closely resembles real silver, being only slightly more gray than that metal, and tarnishing to a light yellow, instead of an orange-black.

German silver was used by most rulemakers to trim ivory rules, just as brass was used with the wood ones. There were some exceptions, of course; some makers offered ivory rules with a choice of brass or German silver trim, but generally speaking the rule was brass for wood, German silver for ivory. Even when the trim was of German silver, the rolls in the joints were still of brass; apparently German silver was not available in the thick wire needed for the automatic roll-making machinery, and rule manufacturers did not feel it was necessary to have it specially made up.

Steel

Steel is, of course, the alloy of iron with a small amount of carbon (less than two percent) and was the mainstay of the industrial revolution. With its hardness and formability, this metal would have been more extensively used in rule making but for several disadvantages. Compared to brass it is too hard to scratch and stamp with markings; if it had been used for the joint plates and tips, the scales could not have been extended to the end of the rule legs. Additionally, this same hardness would have interfered with the scraping and filing operations which followed rule assembly (see THE RULEMAKING PROCESS, below).

Equally a problem is steel's tendency to rust. Rust could probably be cleaned off exposed surfaces, but rust forming on inner surfaces would tend to discolor the boxwood with black stains. Prior to the Civil War, American rulemakers followed the then English practice of using steel for the tips of their rules. By about 1850 to 1860, however, they had abandoned steel for brass, and did not resume the use of steel for tips until after World War II, when it was sometimes employed in the manufacture of cheap rules and yardsticks.

The hinge pins of the joints were steel, of course; steel running on brass provides a smooth, low-friction joint. Steel was also used for the assembly pins used to fasten on edge binding and bits and to pin the buried plates of round and middle plate joints. It was never used to fasten plates on the flat surface of the rule; the ends of the steel pins would have been visible in contrast to the brass of the plates, and spoiled the plates' appearance.

CONSTRUCTION

Almost all rules were made using the same structural elements, varying only in their dimensions and in the number of "sticks" (pieces of wood or ivory) linked together by metal joints to form the rule. A rule made of two sticks with a single joint was known as a two fold rule, one with four sticks and three joints as a four fold rule, etc. Rules were classified in catalogs by their length, number of folds, and width when folded (e.g., two foot, four fold, narrow, or one foot, four fold, broad).

THE MAIN (RULE) JOINT

The joint of a two fold rule, and the middle joint on four and six fold rules consisted of a circular metal disk attached to one of the two legs (the "head" stick), pinned between two thinner disks attached to the other (the "hollow" stick). The combined thickness of these three disks was the same as the thickness of the leg, and they were so arranged as to cause the two legs, when folded, to lie edge to edge. This joint was variously named the "rule," "center," "head," or "main" joint.

Three types of main joints were used on the vast majority of folding rules, all based on the same structure, but differing in their appearance, and in the method of attaching the disks to the legs. These were the round joint, the square joint, and the arch joint. In the round joint, the center disk was attached to a plate which was embedded in a slot in the end of one leg, and held in place by steel pins driven through the leg and the plate, while the two outside disks were attached to a similar plate embedded in the other leg. When assembled, only the circular disks were visible (hence the name "round" joint). This joint was the cheapest, weakest, and least decorative of the three types of main joint.

In the square joint, two rectangular plates (hence the name of the joint) were used to attach the disks to each leg, one on each outside surface of the leg. These plates were usually "let in," so that they were flush with the surface of the leg, and were held in place by brass rivets driven through them and the wood in between. This joint, while more expensive than the round joint,

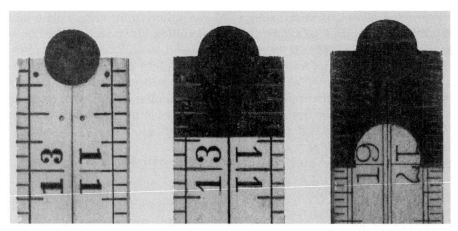

Figure 5-1: Round, square, and arch joints

was also much stronger, and presented a much nicer appearance.

A more elaborate form of this type of joint was the arch joint, dating from about 1806. In this joint, as in the square joint, two plates were used to fasten the disks to each leg, but they were longer, and instead of being rectangular they were cut out in a cove shape on their inside edges. This joint was probably slightly stronger than the square joint, due to the extra length of the plates, but its primary advantage lay in its showy appearance. Due to the extra hand work involved in cutting the more elaborate rabbets for these plates and fitting them, the arch joint was even more expensive than the square joint, and was invariably the mark of a high quality rule.

In 1912 the Stanley Rule & Level Company made a change in the shape of the plates in their arch joint rules to reduce the amount of hand work involved in fitting them. The sharp corner where the curve met the straight end of the plate could not be cut by machinery, but had to be done by hand. To eliminate this expensive step in the framing process, the contour of the end of the plate was changed to a continuous curve, a sort of modified ogee, with no corner. The rabbet for this shape could be cut entirely by machine, significantly reducing manufacturing cost.

The Chapin-Stephens Company, the only other major American rulemaker at that time, opted not to follow Stanley's example, but continued to use the traditional shape. The Lufkin Rule Co., on the other hand, followed Stanley's lead when they began making boxwood rules in about 1924, and used this modified shape for their arch joints also.

On extra-thin rules the main joint plates were not "let in" flush with the surface; it would have weakened the wood at the joint too much. Instead, on these rules the plates were mounted on the surface of the wood.

Occasionally joint designs differing from these three basic patterns will be encountered. Stephens, and subsequently Chapin-Stephens, offered their extra-thin two fold rule with fancy fleur-de-lys shaped plates. J. & G. H. Walker, the New York makers, offered a main joint in which the plate on each leg had a full arch shape, instead of a half arch. One maker in Middletown, Connecticut produced a "V" or "diagonal" joint, similar to a square joint, but with the edges of the plates beveled so as to make the outside edges of the plates longer than the inside edges. English makers such as Rabone and Preston, with their "Grecian," "Vulcan," and "New" joints, also produced rules with unusual joint plates.

Most, but not all, multi-section rules had main joints of this type. There were exceptions in certain cases, however. English blacksmiths' rules used a round friction joint; American blacksmiths' rules used a locking spring joint of the same kind used on Stanley Zig-Zag rules. Ship carpenters' bevels had their tongues set into riveted friction joints. Extension sticks had a sliding, instead of a rotating, joint.

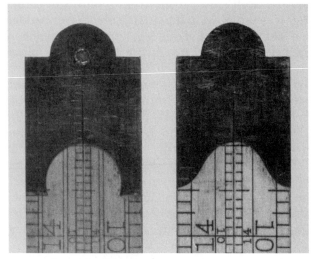

Figure 5-2: Old (left) and new (right) arch joint shapes

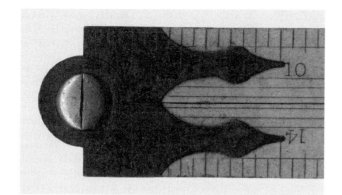

Figure 5-3: Fleur-de-lys shaped joint plates on a Chapin-Stephens rule

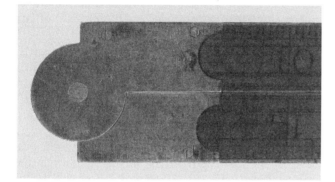

Figure 5-4: Unusual arch-shaped joint plates on a J. & G. H. Walker rule

THE MIDDLE (KNUCKLE) JOINT

The other joints in four fold and six fold rules, the joints which connected the head and hollow sticks to the "tip" sticks, were similar in construction to the main joint, but had their axis of rotation parallel to the surface of the legs. As a result of this difference, the legs united by this joint, when folded, laid surface to surface instead of edge to edge.

This joint was known variously as the "middle" or "knuckle" joint, or sometimes as the "plates" of the rule, with particular reference to the details of its construction (middle plates, edge plates, etc.). Because the middle joints lay parallel to the surface of the leg, the joint had to be correspondingly longer (from ¼ inch, on some one foot, four fold rules, to 1½ inches, on the No. 31½ two fold shrinkage rule. The circular disks which formed the actual pivot occupied only a small part of this distance. In order to space the disks apart, and to fill up the remaining space between them and the end of the rule, small free-turning brass rolls or collets were added to the joint during assembly.

Two types of middle joints were used in rules, depending on the quality of the rule and the strength required in the joint, known respectively as "middle plates" and "edge plates."

Middle plates was the simpler form of middle joint, similar in construction to the round joint form of main joint. In middle plates the four plates, two to a leg, which formed the pivot were set in slots in the end of the leg, and held in place by steel pins driven entirely through the leg and the encased plates. Like the round joint, this was the weaker and less attractive of the two styles of plates, but was frequently used in the cheaper rules due to its lower cost.

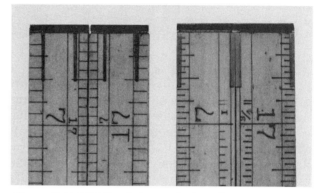

Figure 5-5: Middle plates and edge plates

In most rules employing middle plates each joint consisted of two pairs of plates, but there were exceptions; sometimes only a single pair of plates was used, giving a joint which was even less expensive, but also correspondingly weaker. The pattern of application of this minimal joint is not clear, but seems to have been confined to the period 1900 through 1920, and only to the lower quality one foot, four fold rules.

Better quality rules used edge plates, the other form of middle joint, in which the plates, instead of being set into slots cut in the end of the leg, were set flush with its outside edge, in rebates equal in depth to their thickness. This was a much stronger form of middle joint, for two reasons. First, the two sets of plates were further apart, giving a much better resistance to twisting. Second, the plates were fastened to the leg with through rivets instead of pins (having the metal on the outside made this possible). Edge plates were also a better-appearing type of joint, presenting more brass to the eye, and transmitting a feeling of quality.

In certain rules the edge plate joint was made even stronger, by adding intermediate plates set into the end of the leg in the same manner as middle plates. If only a single pair of additional plates was added, this was called "double-plated" edge plates; if two pairs, then "triple-plated." Double- or triple-plated edge plates were used most frequently on the longer rules, such as the four foot carriagemakers' rules, where the extra length made possible more severe wracking of the joint, and on broad, premium quality rules.

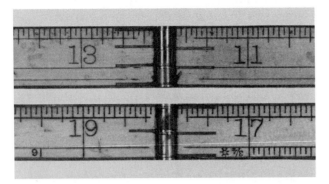

Figure 5-6: Double- and triple-plated edge plates

It is characteristic of English rule making practice to use more plates in a middle joint than would be used in an equivalent American rule. Where a Stanley No. 69 rule would have two plates in each middle joint, the Rabone No. 2320, its English equivalent, had four. This was generally the case, with English rules having almost twice the number of middle joint plates as American. This is particularly noticeable in English ivory rules, with two foot, four fold examples sometimes having as many as eight or ten plates in each middle joint. This plethora of joint plates in English rules is one of the two best indicators (the other being the direction of graduation) to consider when determining if a rule is of English or American origin.

The Double Arch Joint

A variation of rule construction was the so-called "double arch" joint, offered on a few four fold rules. In this scheme, the main joint was an arch joint, and arch-shaped plates were set into the outside surface of the wood of all four sticks at the middle joints to reinforce them. The inside surfaces of the sticks would be left unchanged. The middle joints could be either edge plates or middle plates.

It is not clear just how effective these extra plates were in strengthening the middle joints. In the author's experience, the typical failure mechanism for these joints was to break either when twisted or when accidentally forced past the fully open position. In either case these plates, being on only the outside surface of the rule, could not have been of much use. They must have been primarily for decoration, and not because of any extra strength they could provide.

The double arch joint was the showiest of the four main joint designs, and was only used on the very finest rules. The extra brass or German silver presented a very fine appearance, and lent a weight to the rule which made it feel more substantial, and of better quality.

Two minor variations of the double arch joint have been observed, and should be mentioned. One of these is the E. A. Stearns & Co. No. 34 two foot, four fold rule with what could be called a "double square joint." This rule had a square main joint, and square plates set into the outside surface of all four sticks at the middle joints. Stearns described this rule as having a square joint and "back" plates.

The other variation are the trio of two foot, two fold rules with what resembled a double arch joint offered by H. M. Chapin prior to the Civil War. This resemblance resulted from the tips

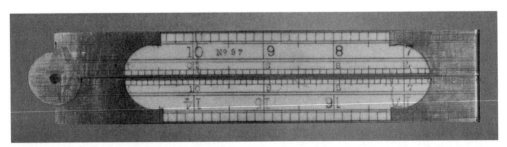

Figure 5-7: Rule with double arch joint

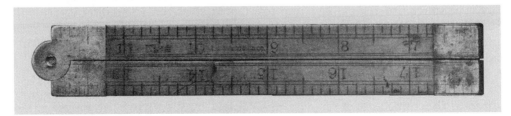

Figure 5-8: Stearns No. 34 four fold rule with "back plates"

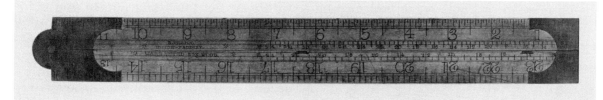

Figure 5-9: Chapin No. 9 two fold rule with "arch tips"

being given an arch shape similar to the arch shape of the main joint plates. These rules, the Chapin Nos. 4, 9 and 14, were described as having arch joints and "arch tips."

BINDING

Rules were sometimes edged with thin metal strips to strengthen them, and to protect the edges of the rule from wear, a construction feature called "binding." The metal used was either brass or German silver, whichever was the appropriate trim metal for that rule, and was held in place by steel pins driven through it into the body of the rule. Two forms of binding were offered, half binding and full binding.

On half bound rules only the outside edges were bound, the edges which were exposed when the rule was fully folded. These were the edges which would suffer wear while in the pocket or tool chest, and needed the most protection.

On full bound rules, all edges of the rule, both inside and out, were edged with metal. This was the most elaborate form of binding, offering the most protection, but also having the highest cost.

Full bound two fold, four fold, and six fold rules were offered by various makers, and half bound four fold and two fold rules as well. It is interesting to note that Stanley was the only maker known to offer half bound, two fold rules, and then only for four years. One foot sticks bound on only one edge tended to warp and twist, and were usually avoided.

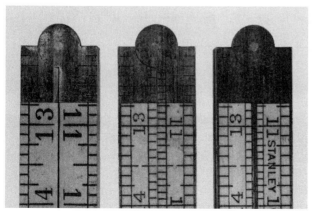

Figure 5-10: Unbound, half bound, and full bound rules

Bound four fold rules always had the edge plate type of middle joint, and, in such cases, the binding plates also served as the plates of the joint. This was true whether the rule was half bound or full bound.

Binding had other disadvantages in addition to cost. A half or full bound rule was significantly heavier than its unbound equivalent, enough so that one could feel the difference. Binding a rule prevented the use of the edge for an additional scale; bound carpenters' sliding rules, for example, do not have the 100ths of a foot scale on their outside edge, a common feature of the unbound ones. Finally, in the case of ivory rules, binding frequently caused the stick to break. As mentioned earlier, ivory shrinks markedly with reduced humidity, and binding would resist this shrinkage; the brittle ivory would often crack across when subjected to this tension.

ALIGNMENT (CLOSING) PINS AND BITS

A common problem with two fold and four fold rules was the bending and breaking of the joints which could occur if the joint was bent past the fully open position, or the leg was inadvertently bent sideways when shut. Not much could be done to prevent the first type of abuse, but the second could be largely prevented by alignment pins. These alignment or "closing pins," as some makers referred to them, were perpendicular steel pins set in the inside surface or edge of the rule which mated with corresponding holes in the opposite surface when the rule was closed. They prevented the legs containing the pin and socket from sliding across one another when shut, and thus prevented damage to the joint connecting them.

The pin and hole protecting the main joint were located on the inside edge of the rule near the tips of the legs (if a two fold rule), or near the middle joints (if a four fold rule). High quality two foot, two fold rules would sometimes have two sets of pins and holes, the second set being located midway between the joint and the tips.

Four fold rules also had alignment pins to protect the middle joints. These were located on the inside surfaces of the rule near the tips and the main joint, one pin and hole set to protect each joint.

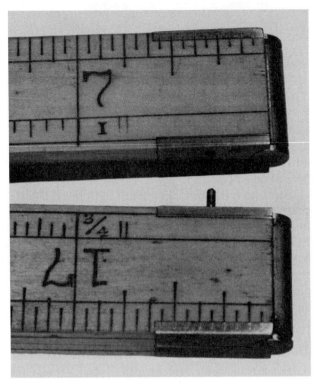

Figure 5-11: Alignment pins on a four fold rule

Some carpenters must have felt that the alignment pins were in the way, by preventing them from laying the edge or surface of the rule flush on the work when measuring or marking. A number of examples have been found in which the surface pins, and sometimes the edge pins as well, have been carefully removed. Apparently they were more willing to put up with the possibility of damaging the rule joints than with having the pins in the way.

The alignment holes in the edge of the rule were rather delicate, and it is common to find examples where the pin has been pulled sideways from the hole, tearing the wood and damaging the edge. Bound rules, and unbound four fold rules with edge plates did not have this problem; on such rules the hole would naturally fall on a part of the edge protected by the binding or plate and thus the pin could not tear out. Unbound and half bound two fold rules, and unbound four fold rules with middle plates had no such protection, and were often damaged.

On a few boxwood rules of these last three types it was sometimes the practice to install "bits" to protect the edge from this pin damage. These bits, substitutes for the binding or joint plates which protected the more elaborate rules, were rectangular brass plates, perhaps 1/4 to 1/2 inch long, set flush and pinned into the edge of the rule and drilled to receive the alignment pin(s).

The selection of which rules should have bits seems to have been somewhat arbitrary, and seemed to vary from maker to maker

TIPS

The tips of the rules were U-shaped pieces of metal which wrapped around the two ends of the rule and were set flush with its surface. Early rules, that is, rules made prior to about 1850, followed English practice and had steel tips. Later, beginning at about the time of the Civil War, the tips were changed to be the same metal which formed the binding and joint plates of the rule; this was usually brass for wood rules, and German silver for ivory ones.

Three types of tips can be found on long, non-folding measuring sticks: U-shaped brass tips similar to those used on the folding rules; a two piece sandwich design which set and riveted a separate plate into each of the two opposing surfaces, but left the end of the stick exposed; and a flat plate riveted to the end of the stick leaving the surface of the stick exposed. Judging from examples, the second and third types were the earlier forms, and became less common after 1900.

Figure 5-12: U-shaped and "sandwich" tips on yard sticks

SLIDES AND CALIPERS

A number of rules were equipped with "slides" set into the surface of one of their legs. These slides were made of thin brass, with a hat-shaped cross section, and ran in an undercut groove in the wood or ivory of the rule leg; thus they could slide back and forth freely in the groove, but could not be removed except by sliding them out.

Two types of slides were made, ordinary and caliper.

Ordinary slides had no protrusion at either end, and could slide freely in either direction as far as the groove in the leg extended. The simplest form of ordinary slide was the so-called "plain" or extension slide. This was a slide graduated in common inches, and was used to extend the length of the rule when taking inside measurements, or to measure the depth of narrow holes. These slides were sometimes put on two foot rules, both two fold and four fold, and occasionally on special purpose rules, such as those used by hatters.

The other form of ordinary slide was the Gunter's slide, a primitive type of slide rule. Gunter's slides were standard on both the carpenters' sliding rule and the engineers' rule, in both the two fold and the four fold configurations. The slide, which was normally made of the trim metal appropriate to that type of rule, would have both edges of its surface graduated with logarithmic scales, and both edges of the groove in which it ran were so graduated as well. By moving the slide back and forth in the rule leg one could perform all ordinary computations dealing with multiplication, division, and ratios. On some rules with Gunter's slides, the back of the slide was graduated in common inches, thus allowing it to function as an extension slide as well.

A few elaborate rules were equipped with more than one slide (see MULTIPLE SLIDES, in Chapter VII). If the slides were separated, running in different legs of the rule, this created no special problems. If the slides were in the same leg, running side by side, then a special structure was required (see figure opposite). One slide had to be twice as thick as usual, and run in a deeper track in the rule; the other would be the usual thickness, and run with one edge housed in a groove cut in the side of the first slide. The legs of the rule had to be made thicker to accommodate this structure.

Caliper slides were used for measuring the outside dimensions of small objects (typically up to two to three inches). These slides had an L-shaped or T-shaped head fastened across one end which overhung the end of the rule leg in which the slide was housed. The length of the slide was graduated in inches, beginning at the jaw; by placing the object between the jaw and the tip of the leg and bringing the jaw down against it, its dimension could be read on the graduations on the slide.

On some small (six inch, two fold) caliper rules the leg carrying the slide would be covered all around with metal to strengthen it. The metal

Figure 5-13: Adjacent-slide structure

used was appropriate to the rule material, brass for a wood rule, and German silver for an ivory one. This was referred to in the catalogs as having a "cased" slide. This type of construction indicates manufacture some time before about 1880.

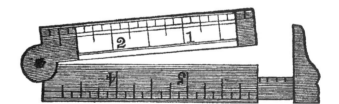

Figure 5-14: Caliper rule with "cased" slide

Until about 1935 the standard caliper slide arrangement was left hand, that is, with the caliper set in the leg of the rule and graduated in such a way that when measuring across the top of an object with the caliper jaw pointing down, the caliper would be held in the left hand and the object in the right. This arrangement was the customary arrangement in the United States, in contrast to England and continental Europe, where right hand calipers were the norm.

In 1934 Stanley, by then one of the two remaining rulemakers in America (the other was Lufkin) changed from left hand calipers standard to right hand calipers standard. Lufkin did not

148 MATERIALS, CONSTRUCTION, AND THE RULEMAKING PROCESS

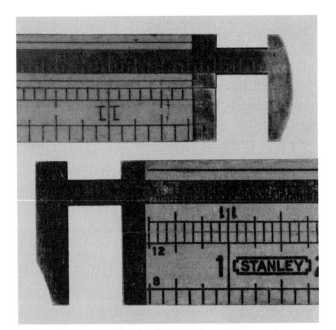

Figure 5-15: Top: Left-hand caliper slide; bottom: Right-hand caliper slide.

follow Stanley's lead, but continued to graduate their calipers for left hand use.

THE RULEMAKING PROCESS

SOURCES OF INFORMATION

Prior to about 1850, rule making was a craft, conducted mostly in small five to ten man shops, and the processes and techniques were passed from master and journeyman directly to apprentice. The large rulemaking firms which introduced mechanization after that date usually made their own specialized machines, and chose to protect their methods by secrecy. As a result, relatively little published material exists describing the process of making folding rules.

A few articles have been found, however. One such, describing rulemaking at Stanley, was published in the Stanley newsletter *Tool Talks* (Ref. 4). Another, describing rulemaking at Stephens & Co., was published in the book *Great Industries of the United States* (Ref. 5). Roberts has reprinted two articles on rulemaking, one relating to Stephens & Co. (Ref. 1, p. 176) and the other to Rabone & Sons in England (Ref. 6) (both included in the articles in Chapter III of this book). *Tomlinson's Cyclopedia* provides a good description of the process of hand graduation (Ref. 7), and an oblique reference to the method of hand-shaping joint plates by filing (Ref. 8). A crude picture of the setup for hand graduation can be found in Sturmy's *The Mariner's Magazine* (Ref. 9). (The fact that some of these references describe rulemaking in England and others depict rulemaking in America does not create any conflict; the basic steps in the process were common to all makers, the differences among them being primarily in the degree of mechanization and in minor manufacturing details.)

THE RAW MATERIALS

The process of manufacturing a rule actually began several years prior to its final assembly, with the acquisition and preparation for use of the boxwood and ivory.

The boxwood was initially received in the form of logs or rough-cut balks at a seasoning and storage yard in or near the manufactory. If in the form of whole logs, they would be split into either halves or quarters (depending on diameter) prior to seasoning. The timber was then stacked under roof in open-air drying sheds.

The ivory was acquired either in the form of whole tusks or in precut blocks from wholesalers who imported it from London or Zanzibar (the centers of the ivory trade in the 19th century). Whole tusks would be cut into blocks after receipt, and then set aside in that form to season. Elephant ivory was the type used, the tusks of the other mammals being too small to yield the six inch lengths needed for two foot rules.

When the boxwood was fully seasoned (a matter of several years) the logs were then ready to be made into sticks. As the first steps in this process they were "blocked up," that is, cut into blocks of the proper length, and then sawn and resawn lengthwise into rough sticks slightly wider and thicker than the desired final size. Great care was taken during these sawing operations to ensure that the cuts were made parallel to the grain of the wood. The rough sticks were then "dressed off," milled straight and smooth on all four surfaces, and sized to their final width and thickness. This was done by different means at various times, at first by the traditional method of hand planing with special high angle planes, then later by methods more adapted to large volume manufacture such as drum sanding and milling machines.

The ivory was also allowed to season before use, first in block form, as it was received from the dealer, and then again later after it had been sawn into sticks. It was particularly important to season the ivory completely; it is characteristic of ivory that it can shrink one-half of one percent or more as it dries. Any rule made from incompletely seasoned ivory would shrink to be too short and/or warp, and if bound, would probably crack from the strain.

When fully seasoned, the ivory pieces would also be "blocked up" by sawing and resawing.

This was done on special fast-running, fine toothed saws, fed very slowly to prevent scorching and cracking.

The second period of drying the ivory came next. The sawn sticks would be washed, doused with hydrogen peroxide, clamped in a press to prevent warping, and placed in a "bleach house." These bleach houses were long, narrow frame structures, triangular in section, whose roofs and walls were covered with glass. The hydrogen peroxide and the exposure to sunlight would bleach the ivory sticks from the brownish-yellow of the raw tusk to the desired milky white. Bleaching could take as long as four or five months. (Ref. 10). At Stephens & Co. the bleach house was a glassed-in cupola on top of the factory buildings.

The metal to form the various parts of the rules was received in roll form: flat brass sheet for the various plates, slides, and trim pieces, large diameter brass wire for the rolls in the joints, and smaller diameter steel and brass wire for the joint and assembly pins.

Manufacturing and Assembly

There were three distinct steps in rule making: joint making, framing, and finishing. Joint making was concerned with the shaping of the different metal parts for the joints and putting together the joint assemblies ready for use. Framing was the process of fitting the assembled joints and trim pieces to the sticks to form a complete rule. Finishing dealt with applying the graduations, figures, and other markings, and applying a protective finish to the surface.

Joint Making

In the early part of the 19th century, joint plates were formed by cutting out and then filing to final shape by hand. This was accomplished by clamping a group of rough plates between a pair of hardened steel templates, and then filing down to the template edge (Ref. 8). Later, with advances in mechanization, the metal parts for the joints were cut and shaped on a press using formed punches. The hinge plates, tips, binding strips, and bit plates were die cut from sheet brass, and the tips folded and squared in a press. The rolls for joints were turned and drilled on automatic lathes (screw machines). The steel hinge pins, and the steel and brass assembly pins were cut from wire on still other machines.

After being cut to shape the edges of these metal parts had to be "scraped" to eliminate any burrs left by the presses and lathes, and bring them to their final shape. Any holes required were made at this time, and the large holes which held the pivot pins were countersunk with either a five-sided hole (Stanley) or an oblong hole (other makers) to match the shape of the head of the joint pin upon assembly (keying the pin to the plate in this fashion guaranteed that the pin would be locked to the outside plates of the joint, and rotate only with respect to the inside plates).

After the parts were prepared, they were "driven up," that is, assembled into finished joints. The various plates and rolls would be placed in a jig which would hold and align them, and then clamped while the pin was inserted and then headed on the other side.

At the same time that the joints were being driven up, the heads of any calipers would be cross-slotted and pinned to their associated slides, after which the heads would be shaped against grinding wheels. The heel and toe of the head and the far end of the slide were left a few thousandths oversize during this operation; this excess would be removed later in the process of fitting the slide to the rule.

Figure 5-16: Caliper head pinned to slide

Joints for wood rules would be assembled from all brass components. Those for ivory rules would use mostly German silver parts, except for the rolls which spaced the plates apart; for some reason these were always brass irrespective of the type of trim.

Framing

The first step in framing was to cut the dressed off sticks to exact length, and then to cut the various recesses in their surfaces and ends for the trim and type of joint they would receive. Quarter circular cuts were made for the rolls which spaced the joint plates apart (on both ends of the head and hollow sticks, on only one end of the tip sticks). If the rule was to have a round joint or middle plates, slots were cut to accommodate the joint plates; if a square, arch, or double arch joint, or edge plates, shallow rabbets would be cut in the stick surface to house them. The tips (and

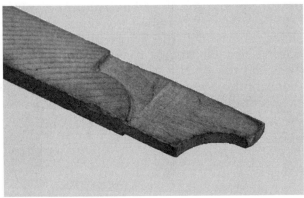

Figure 5-17: Rule parts ready for assembly. Top: main and middle joints, driven up. Bottom: stick rabbeted for arch main joint.

bits, if installed) would also require rabbets or dadoes. The slot for a slide or caliper would not be cut at this time, however; this was postponed until after the tips were installed, so that both could be grooved in one operation.

Assembly (called "framing") was next. The trim (the tips, bits, and binding) was attached to the separate sticks first, riveted to the surface by pins driven into predrilled holes in the wood or ivory, and then any slide or caliper slot would be cut. After this the joints were attached. The first joint fitted was the main joint, connecting the head and hollow sticks. After it had been pinned/riveted in place, then the two middle joints were fitted, attaching the two tip sticks. At this point, the length of the rule was adjusted to its exact final value. This was done by sanding the tips; the combined length of the sticks and the thickness of the brass tips was initially made such that the newly assembled rule came out slightly oversize; removing a few thousandths brought it down to exactly the correct length (the resultant thinning of the tip plates where they wrap around the end of the stick is usually visible if looked for).

After the length of the rule had been corrected, the caliper slides were fitted, the ends of the slides being sanded down to an exact fit. Then the various trim and hinge plates and the ends of the caliper head were brought flush with the surface of the sticks. Being (by design) a little thicker than the rabbets prepared for them, they protruded

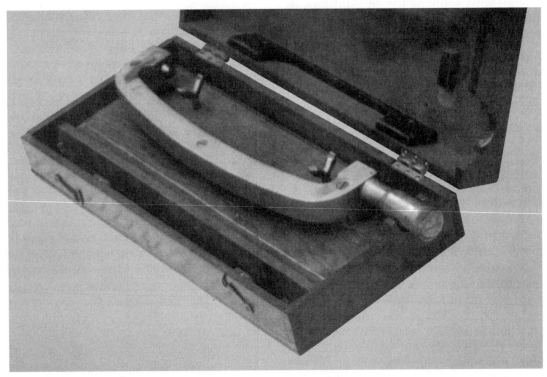

Figure 5-18: Special bench micrometer used by Coffin & Leighton, Syracuse NY (1885-1901) to check the length of one foot steel rules during manufacture. The rule rested on the Y-shaped supports while the measurement was made.

slightly above the surface of the sticks, and had to be scraped/sanded flush.

Figure 5-19: Rule main joint ready for scraping

Before about 1850 this was all done by hand, using special planes (see "Rulemaker's Planes," in Chapter III). By about 1870 special machinery had been developed to scrape the outside face and edge of the rule; the inside face and edge, where the joint protruded above the surface, required a combination of machine and hand work.

One machine for scraping up and finishing the surfaces of rules (and try squares & levels) was invented by George Storer, of New Britain, Connecticut, and patented by him on June 19, 1860 (U.S. Patent No. 28,818, entitled "Machine for Finishing Works in Wood or Metal"). This was a large machine, similar in arrangement to a vertical belt sander, but with a belt consisting of linked scraper plates instead of sandpaper. Storer assigned the patent for this machine 50 percent to the fledgling Stanley Rule & Level Company, where (presumably) it was used in the finishing of framed rules.

The final step in framing was to drill for and install the closing pin(s). They would have interfered with the scraping/filing process, so they were not put on until the rule was otherwise ready for graduation. Only the edge pin(s) were put in at this time. The surface pins had to wait until even later, after the graduating process was complete; they would have prevented the rule from lying flat when graduating the outside, and blocked the dividing knife when scribing the inside.

After the closing pins were installed the rule was ready to be turned over to the finishers to be graduated and figured.

Finishing and Graduation

Finishing was an area in rule making where it was considered suitable (and economical) to hire women. No great physical strength was required, only a good eye and a steady hand, patience and concentration (and a willingness to work for low wages).

The first step was to seal the surface of the blank rule by applying a coat of clear finish. The basic graduation process (until the introduction of printed graduations and figures after 1900) was to scratch or stamp fine lines into the surface of the rule and then fill them with a black paste to make them visible. By sealing the rule surface prior to this operation the paste was prevented from staining it when being rubbed into the lines.

Initially the graduation lines were scribed by hand, in the manner described in Rabone (Ref. 6) and Tomlinson (Ref. 7) (see Figure 5-21):

> "Straight scales and rules are usually divided by placing the article to be divided and the original pattern side by side, then passing a straight-edge with a shoulder fixed at right angles to serve as a guide along the original, and pausing at each division; then a corresponding line is made on the copy by the dividing knife."

This was a more rapid process than might be imagined. A skilled worker could do an entire scale in a very short time (Rabone states that the one foot Gunter's scale of the carpenters' sliding rule, comprising 540 lines, could be cut by hand in ten minutes).

The introduction of factory methods to rule making included the mechanization of graduation as well. The first known machine to graduate rules was invented by Lemuel Hedge, a gifted machinist and inventor from Windsor, Vermont, and patented by him on June 20, 1827 (U.S.

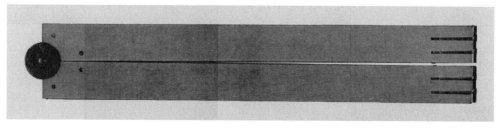

Figure 5-20: A framed four fold rule ready for graduation

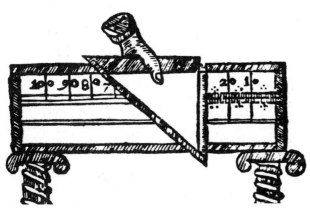

Figure 5-21: Hand graduation in the 17th century

Name&Date Patent No. X4799, entitled "Engine for Dividing Scales, &c."). This machine, which could graduate only one rule at a time, was initially sold to S. M. Clark, a rulemaker in Brattleboro, Vermont, whose firm ultimately became the E. A. Stearns Co. Hedge developed an improved version of this machine in about 1836, and for a number of years sold machines to other rulemakers in New England. Ultimately Hedge set up as a rulemaker in his own right, making rules in New York City under the name of Hedge & Co. The Hedge machine could also figure the rules (that is, stamp the figures into the wood after the graduation lines were in place). It was not as fast as later graduating machines, but performed very well; one was still being used to graduate ivory rules at Stanley as late as 1923 (Ref. 11).

Another graduating machine, operating on a different principle, was invented by Samuel C. Hubbard, of Middletown, Connecticut, and patented by him on June 16, 1857 (U.S. Patent No. 17,606, entitled "Machine for Graduating Lineal Measures"). This machine rolled the graduations into the rule's surface in a single pass under a heavy steel drum with the chisels mounted on the surface, instead of stamping them one at a time as the Hedge engine did. Hubbard's patent was assigned to his brother Charles, also of Middletown, one of the partners in the rulemaking firm of Hubbard & Johnson.

Other early inventors who developed graduating machines were Joseph R. Brown, of Providence, Rhode Island, and Samuel Darling, of Bangor, Maine, both makers of machinists' scales.

Toward the end of the 19th century larger graduating machines were developed where 50 or 100 rules at a time could be clamped to a large table, each below a separate dividing knife; these ganged knives were controlled by a complicated mechanism which moved them in unison to simultaneously scribe each line in succession on all the rules at once. Some of these machines used a lead screw to advance the knives each time and a cam to control alternating line length; others controlled them via a pantograph-type linkage which traced a large scale master pattern.

Slides and calipers were graduated separately from the rules to which they were fitted. At first this was done by scribing, in the same way that the wood was marked, but by about 1900 methods had been developed to stamp the graduations into the metal with a rolling die, at a great savings in time. Prior to separating the body and slide to graduate them, they were marked, so that each slide could be reunited later with the body it had originally been fitted to. This was done with a number stamped on the back of the slide and an identical number stamped on the inside of the groove. Each rule/slide in a lot would have a different number, thus making identification simple. (It is interesting to observe that these numbers all seem to fall in the range 1 to 50; this may indicate that the usual lot size for graduation was 50, but this is only a surmise.)

At some point at about this time in the process the embellishment lines (if any) were scribed on the rule. When hand graduation was performed, the lines would be put on before the graduation, as a guide to the workman scribing the graduations. Once graduating machines came into common use, this order of operations was no longer obligatory, and the embellishment lines would be

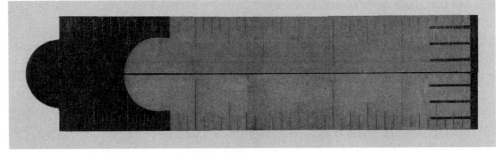

Figure 5-22: A four fold rule graduated with 8ths, ready to be graduated next with 16ths.

put on either before or after graduations, at the option of the maker. (The Hubbard engine scribed the embellishment line at the same time the graduation was being performed.)

After the graduations had been scribed, the figures were then stamped into the rule. Special stamps were used, with a finer edge than those for steel, which would cut the wood fibers instead of crushing them. Ivory, being so hard and brittle, was steamed to soften it temporarily for stamping. Like most other phases of rule making, figuring was eventually automated, with stamping and figuring machines being developed for this task.

The final step in marking the rule was to apply the rule number and the inside and/or outside trademarks. This was also done by stamping, with special stamps each of which could mark one or more lines of text at a time. Sometimes the maker's name and trademark would be omitted, so that a finished rule could be sold to a wholesaler or retailer who wished to market rules under their own name, or so that it could be transferred to another maker as part of the process whereby the rulemakers' association adjusted production volumes to sales (Ref. 1).

Finally, the lines created by this scribing and stamping were then filled with a black paste. Boiled linseed oil mixed with lampblack was used at the Chapin-Stephens Co.; other makers may have used this also, or some other type of paint or wax instead. It was applied liberally with a brush, and then the surplus wiped off with a soft cloth; the residue left in the graduations and figures rendered them both visible and durable.

After the blacking had dried, a second, and possibly a third coat of clear finish was applied to the rule. This protected the surface and the markings from damage and wear, and gave the rule a smooth, attractive finish. The practice of polishing the final coat, to make it extra smooth, was apparently discontinued about the time of the First World War; rules made during the 1920s and later have nowhere near the fine surface finish seen in the earlier product.

The last step prior to final inspection and shipment was the installation of the inside closing pins and their mating holes. This final operation, as mentioned earlier, had been postponed until after graduation. Once they were in place, the rule was complete and ready for use.

REFERENCES:

1. Kenneth D. Roberts, P.E., *Wood Planes in 19th Century America; Vol. II: Planemaking by the Chapins at Union Factory, 1826-1929* (Fitzwilliam NH: Ken Roberts Publishing Co., 1983).
2. —. *A Book of Tools, Machinery and Supplies* (catalog) (Detroit, Michigan: Charles A. Strelinger & Co., 1895).
3. —. "Bigger Yet," *The Vermont Phoenix*, Brattleboro, Vermont, January 26, 1866.
4. —. "I Supervise the Making of Boxwood Rules," *Tool Talks*, no. 11 (New Britain, Connecticut: The Stanley Tools Division of the Stanley Works, 1938).
5. —. "Rules," *Great Industries of the United States* (Hartford, Connecticut: J. B. Burr & Hyde, 1873).
6. John Rabone, Jr., "Measuring Rules," *Birmingham and the Midland Hardware District, the Resources, Products, and Industrial History*, ed. Samuel Timmins (London: 1866).
7. Tomlinson, "Graduation," *Cyclopedia of Useful Arts and Manufactures* (London: 1858).
8. Tomlinson, "Filing," *Cyclopedia of Useful Arts and Manufactures* (London: 1858).
9. Samuel Sturmy, *The Mariner's Magazine* (London: 1669).
10. Matthew Roth, "Comstock, Cheney and Company Factories (1847)," *Connecticut - An Inventory of Historic Engineering and Industrial Sites* (Connecticut: Society for Industrial Archeology, 1981).
11. Guy Hubbard, "The Development of Machine Tools in New England, Part IV: The Ornamental Period of Machine Design," *American Machinist*, 59 (August 30, 1923), 9.

VI
Graduations and Markings

The primary purpose of measuring instruments is to act as vehicles for linear scales marked on their surface. These simple scales, divided into inches/fractions and/or centimeters/millimeters allow us to quantify the distance between two points by placing the edge of a rule in contact with them.

These scales are only the beginning, however. A wide variety of applications existed (and still exist) that required some form of calculation or manipulation to be performed on a measured value, and many different special scales have been devised to automatically apply these calculations as the measurement is made. Sometimes it was even desirable to provide the user with entire tables of data to be used in manipulating the measured value.

In this chapter are described and explained most of the common schemes for graduating and marking rules with these scales and tables. The listing is fairly complete, but like everything else associated with measuring instruments, new examples are always turning up.

EMBELLISHMENT LINES

Embellishment lines are the longitudinal lines, parallel to the rule edge, marked on all but the least expensive rules. These lines were marked on both the edge of the rule and the face, each pattern of rule having its own combination of lines.

Face embellishment lines were marked either singly or in pairs, depending on the style of the particular rule, and were usually placed between the figures and the edge, crossing the graduation lines. Some rules also had additional lines on the other side of the figures as well.

The original purpose of face embellishment lines was to aid the early rule makers during graduation. When marking a rule by hand it was very helpful to the workman scribing the graduations to have a line marked on the rule surface to indicate the correct length of each graduation. The 16ths graduations could be started a little short of this line, the 8ths graduations just at it, and the 4ths graduations slightly beyond it, etc. Even after the advent of machine graduation of rules in the 1800s, the face embellishment lines were retained on all but the very cheapest rules. Just as they had guided the workers in hand marking the graduations correctly, so they also made it easier for the user to read them, distinguishing the fine and coarse divisions from one another. Additionally, they improved the appearance of a rule, "framing" the graduations and figures, and giving them an attractive finished look.

The simplest scheme of face embellishment line was a single line crossing the graduations even with the tips of the 8ths graduations. More elaborate schemes (reserved for premium rules) could have as many as four embellishment lines per scale, two in the region of the graduations, and two more, purely decorative, very close together on the other side of the figures, at the point where the inch graduation lines ended. Some of the early rules even had as many as seven, including a line at the outer edge of the graduations.

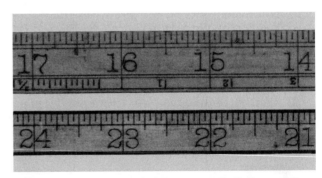

Figure 6-1: Simple and elaborate embellishment line schemes

Edge embellishment lines were much simpler, consisting of a single line or pair of lines scribed the length of the edge, dividing it into either two or three equal width areas. They were purely decorative, and were omitted if the edge was bound, or if it was graduated with a 100ths of a foot scale.

GRADUATIONS

Rules were graduated with a wide variety of different scales, the particular combination on any rule being determined by its intended use and whether it was an ordinary or a quality product. Broadly speaking, three classes of graduations were marked on the rules: linear, 1:1 (full size) graduations; linear graduations which were not 1:1; and nonlinear graduations.

LINEAR, 1:1 GRADUATIONS

The most common graduations were those which formed "true measure" or "1:1" scales, scales which were intended to measure linear distances and indicate the result in commonly accepted units of uniform size such as inches, millimeters, yards, etc. The vast majority of rules had scales of this class. Scales of inches subdivided into 8ths, 10ths, 12ths, and 16ths were all common. Scales finer than this are rare. The only rules where the wood was marked with scales finer than 16ths of an inch were rules with metric scales (1 millimeter equals approximately 1/25 of an inch) and the No. 36/036 combination rule, where the body and blade both had scales of inches/24ths.

Graduations would be either left-to-right or right-to-left; that is, when the rule is unfolded and held horizontally with the figures upright, the numbering begins either at the left end, left-to-right, or the right end, right-to-left. During most of the 19th century and until the end of traditional rule production in the 20th century it was the American practice to figure carpenters' rules right-to-left. English practice during this same period was to figure rules left-to-right. This difference is one of the two best indicators (the other is the number of plates in the joint) of whether a rule is of English or American origin.

English, or Common, Graduations

8ths & 16ths of An Inch

Although these graduations usually went together, this was not universally the case. Sometimes a scale would be graduated in 8ths only.

Each line in the 8ths/16ths scale was a different length, as a function of the fineness of the division it represented. The 16ths markings were the shortest, not quite reaching the embellishment line. The 8ths markings were longer, ending at the line, and the 4ths markings even longer, extending past it (to the second embellishment line, if such was present). The half inch markings in their turn were correspondingly longer still, and the inch marks, of course, extended all the way past the figures (to the embellishment lines beyond them, if any).

The 8ths and 16ths scales were always marked adjacent to the outside edge (when folded) of two fold, four fold, and six fold rules. This allowed these commonly referenced scales to be used free of interference from the alignment pins located on the inside edge. On four fold rules the usual arrangement was to mark the 8ths scale on the outside surface (when folded), and the 16ths scale on the inside surface. Presumably this was done so that when the 16ths scale was used, the rule could be laid flat with this scale uppermost without similar interference from the inside surface alignment pins. This arrangement also had the advantage of protecting the 16ths scale from dirt and pocket wear when the rule was closed.

10ths & 12ths of An Inch

It was not possible to mark the 10ths and 12ths scales with the same hierarchical arrangement that had been applied to the 8ths and 16ths scales, with the 16ths lines shorter than the 8ths lines, and the 8ths lines shorter than the 4ths lines, etc. Only the whole inch graduations, and those subdivisions which coincided with 4ths of inches were emphasized (5/10ths on the 10ths of an inch scale, and 3/12ths, 6/12ths, and 9/12ths on the 12ths of an inch scale). All other subdivisions were of the same length, reaching exactly to the embellishment line.

As far as can be determined from catalogs and other printed matter, up until the 1870s, only a few rules were available with the 10ths of an inch scale, and when such a scale was a feature it was marked on the edge of the rule. Subsequent to

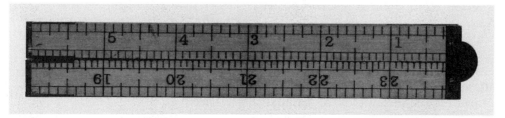

Figure 6-2: Four fold rule with 8ths, 10ths, and 12ths of inches graduations

THE ORIGINS OF THE LEFT-TO-RIGHT VERSUS RIGHT-TO-LEFT DIFFERENCE IN GRADUATING RULES

Careful study and research has convinced Don and Anne Wing of Marion, Massachusetts that America and England both started out graduating their rules to read from right to left, and that English rule makers changed over to graduating their rules from left to right in the late 1820's while the Americans continued the original direction. There are exceptions to this scenario, but this appears to be generally the case.

Examination of a large number of examples and the study of numerous editions of 18th century instruction books explaining the use of Coggeshall's sliding rule has clearly established that in the 18th century and before, the conventional method of marking the inches on a folding rule in England was to figure it from right to left. This appears to be consistent for all English rules, including Gunter scales and sectors, up to the mid-1820s. In fact, this tradition was so strong that the three fold two foot carpenters' rules (two nine inch wood legs with a six inch brass extension folding out of one leg) had to be made with their logarithmic slide placed in the bottom limb to make all the features work with the right-to-left numbering. (It is interesting to note that the rules illustrated in Stone's translation of Bion (1758) are numbered from left to right. Perhaps this is due to Stone's engraver directly copying the plates in Bion without realizing that French and English rule making differed in this respect).

A small number of English rulemakers emigrated to America in the early part of the 19th century and began practicing their trade in Boston, Connecticut, and New York. Not surprisingly, the first manufactured rules made by these makers and the American workmen they trained were distinctly English in construction and layout, and were graduated from right to left in the traditional English style. In subsequent years, even though the American makers made more or less significant changes in rule construction and patterns, the practice of graduating scales from right to left continued unchanged.

Shortly after the mid-1820s, for reasons which are not completely clear, the English switched and started figuring their rules left-to-right. This is made very apparent by examination of hundreds of examples of English rules made by various datable makers. This change was made quite suddenly, which is surprising given the strong English craft tradition with its resistance to change. There must have been some strong impetus to force this switch. It could not have been just a casual idea of one of the larger manufacturers such as Rabone; there were so many small makers that some major external cause must have intervened.

It is possible that this change in England was related to the 1824 Act of Parliament (actually finalized in 1826) which established the Imperial system of weights and measures. Most English rules which had a scale for calculating volume measurements in the old style of Wine Gallons and Ale Gallons are numbered with right-to-left inch readings, while those using the new Imperial Gallon scales are numbered from left to right. Perhaps something in the text of this act required rule makers to lay out their rules so they matched the way people read, from left to right. It is suggestive that the printed English tape measure, first patented in 1845, consistently reads in this direction also.

It was not as difficult for the English rule makers to execute this change to left-to-right graduation as one might think. In the late 1820s most English commercial rule making was centered in Birmingham, and was just beginning to change from a cottage industry to factory production; machinery for graduation and figuring was still in the future, and the work was done almost exclusively by hand.

American makers did not emulate this change in the direction of graduations. There was no statutory requirement to do so, and rule making in this country was well enough established so that makers would be reluctant to make gratuitous changes. A second reason for not changing may have been the Hedge graduating engine, patented in 1827, which was just coming into use at that time. This machine was set up to mark rules from right to left, and it would have been very difficult and expensive to redesign or replace it so as to graduate in the other direction.

The author wishes to thank Don and Anne Wing, of Marion, Massachusetts for providing the information upon which this discussion is based.

that time, however, this scale became a more common feature of many more rules, and, equally of interest, from then on was usually placed on the surface of the rule. This change was probably the result of the introduction around this time of new graduating machines capable of marking 10ths scales. These machines could typically only mark graduations on the surface of the rule (hence the change in scale location) but reduced the cost of applying this scale to the point where it could be much more widely offered.

A similar situation prevailed at this period with regard to the 12ths of an inch scale. By about 1880 12ths scales, which had previously been relatively rare, began to be offered on a great many rules. This was probably due to the same cause as the wider availability of 10ths scales. Graduating machines were being introduced which could mark this scale, or, more likely, the machines introduced a few years previously had been modified to mark 12ths scales as well.

24ths of An Inch

The 24ths of an inch scale on the No. 36/036 Stephens-patent combination rule was one of the two exceptions to what was apparently an industry-wide policy not to graduate the wooden body of a rule with any scale finer than 16ths of an inch. Lines spaced closer than 1/24th of an inch were easily obscured as the rule became soiled with use, and, also, the wood fibers between adjacent lines would be so short that they would tend to separate from the surface with only slight wear. The 24ths scale, however, was a recognized feature of the No. 36 combination rule invented by L. C. Stephens in 1857. Stephens had graduated the blade of his rule and the body in 24ths so that the rule could be used to calculate brace measure; 24ths, instead of 12ths, being necessitated by the fact that the legs and blade of the rule were six inches, instead of a foot, long.

Only every third line in this scale (the lines coinciding with the 8ths of an inch mark) was emphasized, all the others ending at the first embellishment line; the 8ths lines extended slightly beyond it, the 4ths lines even further, and the half inch and inch lines all the way beyond the figures to the second embellishment line.

100ths of a Foot

This scale was not generally offered, as its applications were rather specialized (it was useful in engineering calculations and measurements dealing with distances expressed entirely in feet and decimals of feet (e.g., 2.16 feet). The Gunter's slides of the engineer's and carpenters' sliding rules operated in decimal notation, and wherever possible would have a 100ths of a foot scale on their edge. In ordinary carpentry, where measurements were usually given in feet and inches, it was not applicable, and would not have justified the extra cost of marking it.

This scale, when present, was always marked on the outside edge of the rule (presumably by hand). It had no embellishment lines, but simply consisted of lines halfway across the edge at each 100th of a foot, with the 20ths lines slightly longer, and the 10ths lines going completely past the figures. It was figured in 100ths, with a figure (10, 20, 30, etc.) each tenth of a foot. It was not marked on rules which were bound, as the binding completely covered the edge of the stick.

Fractions of a Yard

Fabric for sewing, drapery, and upholstery purposes has always been priced and sold by the yard, and the traditional measuring instrument for this purpose was a three foot stick marked on one side with a scale of fractions of a yard. All makers manufactured a selection of "yard sticks" graduated in this fashion. The usual arrangement was to have one surface of the rule/stick devoted solely to this scale, with markings at the 1/32, 1/16, 1/8, 1/4, 3/8, 1/2, and 3/4 yard points, and another surface divided into inches/8ths. This system was not uniformly adhered to, however, and variations, such as the addition of 5/8 and 7/8 yard markings, are not uncommon. The graduations extended completely across the stick, which was usually devoid of embellishment lines. The figures were in the form of proper fractions (1/4, 3/8, etc.), either superimposed on these lines, or placed next to them. On early yardsticks the 1/32, 1/16, and 1/8 yard markings were sometimes labeled as 1/2 NAIL, 1 NAIL, and 2 NAILS (a nail was an early measure of length equal to 2 1/4 inches).

Before about 1925 these yardstick scales were usually graduated from right to left; after that date left to right seems to have been the more common.

The yard graduations on the folding yardsticks such as the Stanley No. 66 and the Chapin No. 84 differed in two ways from those on solid yard

Figure 6-3: Nail and fraction markings on an early Belcher Bros. yardstick

sticks. On these rules the scale was graduated in both directions, with markings and figures running from both left to right and from right to left, and two equally spaced embellishment lines were added. The right to left graduations were above the upper line and the left to right graduations were below the lower line; the space between them was used for decorative asterisks emphasizing the graduation lines. Both sets of figures used the graduation lines as the separators for their fractions.

Another variation of the fractions of a yard scale was to graduate the first quarter of the scale into nine inches, and the remainder into 8ths of a yard. This is the scheme which was used on the Stanley and Lufkin counter measures. By the time metal counter measures were in general use, the nail had been replaced by inches for expressing small fractions of a yard.

32nds of An Inch

Although graduating the wood parts of rules with markings finer than 16ths of an inch was not usually done, this was not a limitation when graduating steel machinists' rules or brass caliper and extension slides. Some of these rules had 32nds graduations standard on their slides, others had them as an option. Ultimately they became standard on all slides.

The system of accenting these finer markings in this scale was a simple extension of that used for 8ths and 16ths graduations, with the 32nds markings slightly shorter than the 16ths.

Metric Graduations

This is the well-known system of measurement which was invented by the French Academy in about 1805, shortly after the revolution, and spread across Europe by the conquests of Napoleon.

Based on a fundamental unit called the meter (equal to 39.37 inches), subdivided into 100ths (centimeters) or 1000ths (millimeters), it had become the only legal standard in most European countries by the middle 1800s, was in general use in their various colonies by about 1900, and today is almost universally used outside the English-speaking world.

Metric scales were usually graduated in millimeters, with the centimeter lines (every 10th line) made about twice as long, and the ½ centimeter lines (every fifth line, in between) about 1½ times as long. The 10 centimeter lines were frequently extended all the way across the rule. Every centimeter line was marked with figures representing the distance from the beginning of the scale in centimeters (11, 12, 13, etc.).

Foreign Measures

Occasionally rules graduated in measurements other than inches/fractions or centimeters/millimeters will be encountered, frequently marked with the name of a country or region, such as Burgos or Rhineland. These rules are graduated in the measures used in the indicated locale prior to the introduction of the metric system. Such rules were made for sale to users who had not yet converted to the metric system at the time of their manufacture. These rules are usually graduated into inches and 12ths of an inch, with the inches usually identified as "thumbs," and the 12ths as "lines" in the local language (in France, for instance, the words *pouces* and *lignes* would be used. This practice of dividing a "thumb" into 12 "lines" is a carryover from the Roman system of measurement which was prevalent throughout Europe prior to the Dark Ages.

A more complete discussion of foreign non-metric measures can be found in Chapter VIII. Examples of rules of this type are shown in the section FOREIGN MEASURES OR COMPARISON RULES, in the next chapter.

Other 1:1 Scales

1:1 scales other than inches/fractions or centimeters/millimeters were sometimes placed on rules. These were usually special purpose scales used in a particular trade, and were divided into the traditional units of that trade. The following are the most commonly encountered examples:

Printers' Measure

"Printers' Measure" is not a single scale, but rather the set of scales, one per size (called "font") of type, which a printer uses in laying out pages and composing. It is much easier for the printer to do all his measuring in units of type size rather than in abstract units like inches or millimeters.

Type is sized in units called "points," where one point is equal to .013837 inches (approximately 1/72 of an inch). Prior to the 20th century each commonly used font had its own name, and the scale for that font would be labeled with that font's name. The names and sizes of the most frequently used fonts are listed on the next page.

Another scale that was frequently encountered in typesetting and composition is Agate. An Agate is 1/14 of a "column" inch (slightly more than five points), and is the unit used by printers to measure the length of non-text copy such as display advertising and illustrations. A two column advertisement, four inches high, for instance, would be described as being "2 columns by 56 Agate lines."

Font	Height
PEARL	5 points (= 0.070")
RUBY	5½ points (= 0.078")
NONPARIEL	6 points (= 0.083")
MINION	7 points (= 0.100")
BREVIER	8 points (= 0.111")
BOURGEOIS	9 points (= 0.125")
LONG PRIMER	10 points (= 0.141")
SMALL PICA	11 points (= 0.156")
PICA	12 points (= 0.167")
ENGLISH	14 points (= 0.200")
GREAT PRIMER	18 points (= 0.250")

Figure 6-4: Commonly used sizes of type

The usual method of graduating printers' scales was to make a mark for each line of that size type (e.g., the marks on the Pica scale would be twelve points apart, those on the Minion scale seven points, etc.). Every fifth graduation line would be accented by extending it past the single embellishment line, and would be figured with the number of lines it indicated (5, 10, 15, etc.) The intermediate lines would all end at the embellishment line.

Watch Glass Size

Scales of watch glass size were established by watch manufacturers to classify the diameters of the crystals for the different sizes of watches which they made. They would be used by a watch repairman to measure an existing crystal when selecting a replacement, or when ordering a new one from the factory.

The fundamental unit of watch glass diameter was the Geneva (Switzerland) Line, a unit equal to .1334 inches. Sizes were expressed in lines/16ths; thus a glass 2.426 inches in diameter would be described as being an "18³/₁₆ glass".

As marked on a watch glass gauge this scale was diagonally divided. It consisted of 16 vertical lines, at right angles to the step which located the other edge of the watch glass, said lines being numbered 0 through 15. Crossing these lines were a set of lines spaced exactly 0.1334 inches apart, numbered 5 through 27; these lines were not parallel to the step, but were slightly diagonal, each one beginning exactly opposite the far end of the previous line (see figure 6-5). By resting the edge of the glass on the step, and shifting it back and forth until it exactly touched one of the diagonal lines, the size could be read. The diagonal line it just touched was the integer portion of the size; the vertical line passing through the point of contact was the fractional portion.

40ths of An Inch (Button Size)

Fortieths of an inch were the traditional units used to measure and size buttons and buttonholes when sewing and tailoring, and makers accordingly placed a 40ths of an inch scale on the caliper slide of their button gauges for this purpose.

This scale had all graduations the same length, except that every 10th line (coinciding with the ¼ inch points), was made extra long. These accented lines were figured with the number of 40ths from the beginning of the scale (30, 40, 50, etc.). This scale was never augmented with embellishment lines.

Horse Measure

It is customary to measure the height of a horse at the shoulder, using units called "hands." A hand is equal to four inches; a height would be expressed as "15 hands," "16¼ hands," etc.

The only place where hands are encountered is on horse measures (see Horse Measures in the next chapter), special instruments designed for the express purpose of measuring horses. On these measures the scales have major graduations four inches apart, numbered sequentially, subdivided into unnumbered individual inches.

LINEAR GRADUATIONS, NOT 1:1

A second group of scales graduated on rules were the linear scales which were graduated in units other than full scale. This group included, for example, such scales as those put on patternmakers' shrinkage rules, where all the "inches" were slightly oversize to compensate for the shrinkage of the casting during cooling; architects' scales, which were graduated into scale feet and inches to compensate for the reduced size of

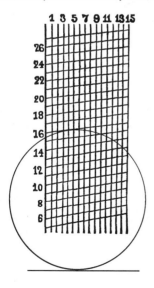

Figure 6-5: Geneva scale for a watch glass gauge

the drawing as compared to the size of the actual object; chain scales, used by surveyors and engineers; and the circumference scales which appeared on tinners' rules, used to calculate the amount of tin plate needed to make a particular size container.

These were really types of computing scales, each including in its graduation interval some factor which represented the difference between its nominal spacing and the size of the object which it represented. Without these scales, it would have been necessary to take a dimension in true inches, and then apply the needed factor arithmetically, a much slower and more error-prone process.

Shrinkage Graduations

Shrinkage graduations are a special form of inch scale used by patternmakers in making the patterns for foundry use. A shrinkage scale has all the appearance of an ordinary inch scale, but differs in that it is between $1/4$ percent and $1 1/2$ percent oversize. By using this oversize scale the patternmaker can compensate for certain shrinkages (hence the name) which occur during the foundry process. Shrinkage scales were graduated in 8ths, 10ths, 12ths, and 16ths, and at different times, on different shrinkage rules, were sometimes accompanied by embellishment lines. Accenting of major divisions by lengthening them was exactly as on ordinary inch scales.

If a rule with the desired shrinkage scale was not available, a patternmaker would sometimes make his own rule marked with that special scale. This was easily done, hand graduating the special scale onto the rule as shown in the illustration below.

Additional information on shrinkage graduations and shrinkage rules can be found in the section SHRINKAGE RULES, in the next chapter.

Architects' Scales

Architects' drafting scales are special scales used to make measurements on scale drawings, drawings where the dimensions of the drawing are some fixed percentage of the dimensions of the object being depicted. An example of this would be a drawing with a scale factor of $1/2$ inch equals 1 foot, that is, where each $1/2$ inch on the drawing represents 1 foot on the subject of the drawing. Scale drawings are always used in architectural work, and are common in mechanical design and engineering work, where the object drawn is either too large or too small to be conveniently represented by a 1:1 (full scale) drawing.

Architects' scales were typically divided into intervals equal to scale "feet" ($1/2$ inch intervals in the case of the scale factor mentioned above), and then have one or more of the "feet" at the end of the scale subdivided into scale "inches" ($1/24$ inch intervals in this case), with the lines representing 3, 6, and 9 "inches" made longer to emphasize them. This type of scale, where only one or two of the scale feet is fully divided, is called "open divided."

Whether building furniture or houses, carpenters would frequently have to deal with scale drawings, and it was convenient to have architects' scales in commonly-used scale factors marked on their rules. The exact details of the scales on any particular rule would vary as a function of factors like the rule pattern, its size, or the preference of the maker.

Two foot, two fold rules, for instance, usually had four drafting scales, each nine inches long, with scale factors of $1/4$, $1/2$, $3/4$, and 1 inch per foot. On rules with slides these scales were placed on the same side of the rule as the slide, on the other leg, and were graduated from left to right, beginning near the main joint. On rules without slides they were placed so that two were on each

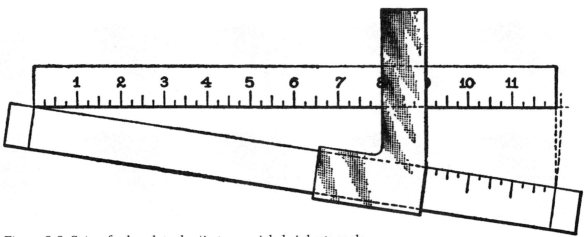

Figure 6-6: Setup for hand graduating a special shrinkage scale.

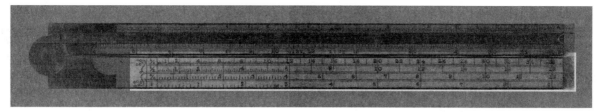

Figure 6-7: Drafting scales on a two fold rule

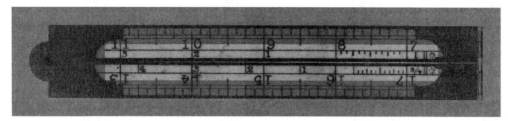

Figure 6-8: Drafting scales on a four fold rule

leg, adjacent to the inner edge, and were graduated from right to left, beginning near the tips. The most common scheme of graduation was to divide the first three inches of each scale into scale feet and inches, and the remaining six inches into scale feet only. This is only generally true, however and many variations exist.

Two foot, four fold rules also had four scales with the same scale factors, but they were shorter (four inches), and arranged slightly differently. On these rules the drafting scales were located on the inside surface of the rule, one scale per stick, adjacent to the inside edge. These scales were graduated from left to right, and only one scale foot (the leftmost one) was divided into scale inches. Also, in accordance with contemporary practice, the zero point of each scale was not at the end, but one scale foot to the right of it, at the right hand end of the foot which was divided into inches; the feet were read to the right of this zero point, the inches to the left, thus making slightly easier the process of converting the ordinary form of scale dimension (X feet, Y inches) into a measurement. As in the case of two fold rules this arrangement was only the most common, and many variations exist.

The two foot, four fold rules with inside beveled edges (such as the Stanley No. 53½) were graduated identically to the ordinary four fold rules, except for the scale factors of the four drafting scales. On these rules, the scale factors of the four scales were ⅛, ¼, ⅜, and ½ inch per foot, instead of ¼, ½, ¾, and 1 inch per foot. On each scale the leftmost scale foot was divided into 12 scale inches (except for the ⅛ inch per foot scale, where it was divided into six intervals of two scale inches each).

Chain Scales

The scales preferred by engineers and surveyors differed slightly from those used by architects. In the scales used by these two professions all of the major scale divisions would be fully subdivided, the type of subdivisions depending on the intended use of the scale. Scales divided in this fashion are known as "chain" scales.

Engineers preferred scales with the major divisions being either scale "feet" or scale "inches," with the former subdivided into scale inches and the latter into either fractions (8ths, 16ths, etc.) of an "inch" or decimals (10ths, 50ths, etc.) of an "inch." Surveyors preferred scales with the major divisions being scale "chains," subdivided either into "poles" (¼ of a chain) or 10ths of a chain.

The Circumference Scale

The circumference scale was a characteristic feature of the tinners' rule, the specialized steel rule designed for use by tinsmiths, coppersmiths and other sheet metal workers. This scale was graduated like a common inch scale, except the "inches" were only $1/\pi$ (equal to 0.3183) inches in length. Thus, on a typical three foot tinners' rule, the circumference scale would be about 113 scale "inches" long.

This scale was used by the tinsmith to convert from diameter directly to circumference without recourse to multiplication. The diameter of the circular object to be mated with could be measured using this scale, with the resulting reading giving the length in common inches of the workpiece to be bent into a circle to fit that object.

The Plumbers' Scale

The plumbers' scale was designed to simplify the process of determining the length of pipes running at a 45° angle. When it was necessary to "jog" a run of pipe when plumbing a building it was common, particularly in heating systems, to introduce the jog as a short length of pipe running at 45°. (This had mostly to do with the steam and water flow requirements of "common-return" radiator systems, the details of which do not concern us here.) This scale was graduated like a common inch scale, except the "inches" were only $1/\sqrt{2}$ (equal to 0.7071) common inches in length. Using this scale to measure either the vertical or the horizontal component of the jog would automatically yield the length of the 45° section of pipe required to accomplish it.

Although not common, these scales have been observed on both four fold and zig-zag rules, labeled specifically as plumbers' scales.

Octagonal (Four Square) Scales

This is a pair of scales, the E & M (Edge and Middle) scales, designed to allow a carpenter to determine how much to cut off each corner of a square to make it an octagon, or how long to make a diagonal brace when framing a building. The Rabone instructions for the carpenters' sliding rule describe these scales as being used by mast and spar makers, but they were probably more often employed to solve problems in the building trades.

These scales were only marked on two foot, two fold rules. They were both the same length, but had different graduations. The E scale was graduated from 0 to 24; the M from 0 to 32. Both were used to find the point at which the bevels must begin in order to turn a square into an octagon, but used different points on the square as a reference. The E scale located that point with reference to the edge (corner) of the square; the M scale with reference to the middle (the center point of face). The length of the side of the square would be found on either the E or the M scale, and opposite that point on the common inch scale would be the distance from the desired reference point to the point at which the bevel must begin.

These scales were always placed one on each leg on one side of the rule, and were both graduated from the end of the leg toward the joint. They were not placed at either edge of the leg, but located on the surface between the two ordinary scales, and used one of their two embellishment lines to define their apparent edge. On these scales the minor graduation, which only extended to the second embellishment line, represented ½ inch; the foot markings extended slightly beyond the line, and the two foot markings all the way past the figures with which they were marked.

For a more detailed description of the E & M scales, see the article in *Mensuration*, Vol. I, No. 1, reprinted in Chapter IV.

Board Scales

Board scales were computing scales, calculated to convert the length and width of a one inch board into its equivalent area in board feet (a board foot is one square foot of wood one inch thick). Each scale would be calculated and graduated for a specific length board (12 feet, 14 feet, etc.), such that opposite the width of any board of that length would be a figure representing its equivalent in board feet. On the scale for 14 foot boards the mark representing 7 board feet, for example, would be 6 inches from the origin of the scale (6 inches divided by 12, times 14 feet, equals 7 board feet), the mark representing 8 board feet 6.86 inches from the origin, etc.

These scales would be marked on a board stick or rule, from one to four to each surface, depending on the stick's cross sectional shape. Flat sticks would have three on each face (for a total of six); square sticks would have four per face (for a total of sixteen); octagonal would have either one or two, depending on the thickness of the stick (giving either eight or sixteen). When multiple scales were marked on a single face, they would be separated by embellishment lines. Board scales were graduated at each whole board foot only (1, 2, 3, etc.), with a line running from edge or embellishment line to edge or embellishment line, and always ran from left to right, beginning at the head or hook of the stick.

Additional information on board scales and board sticks can be found in the next chapter, in the section BOARD MEASURE, BOARD STICKS, AND BOARD TABLES.

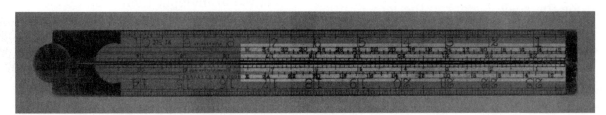

Figure 6-9: Octagonal (4 square) lines on a two foot, two fold rule

Nonlinear Graduations

The most complex scales put on rules were the nonlinear scales, scales where the interval between graduations of equal difference were not uniform, but varied along the length of the scale. These were computing scales of an advanced type, involving functions more complex than simple multiplication by a constant, etc. Some of these scales, like the Doyle's log scale and those on the Gunter's slide, were based on mathematical formulas. Others, like Scribner's log scale, were derived from measured data, and did not represent any specific mathematical expression.

These scales were almost invariably cut by hand, using brass or steel patterns in the manner described in Chapter V. It is possible that methods to do this by machine were developed after the turn of the century; machines capable of cutting nonlinear scales, based on the principle of the pantograph, with ganged graduating knives controlled by a stylus guided across a template, were well known by that time. No evidence has been found to suggest that such machines had been applied to rulemaking, however.

Log Scales

A log scale is a specialized scale designed to simplify the problem of estimating how much sawn lumber can be obtained from a given log. Sticks calibrated with these scales are employed by lumbermen and sawyers for selecting trees for cutting and as a basis for payment at the mill, and are widely used, even today.

The basis for a log scale is a log rule, a table or formula which has been developed to relate log length and diameter to the number of board feet of lumber which can be sawn from such a log. A large number of such rules have been published over the years, to deal with different types of trees and varying sawmill methods (Ref. 1).

Log scales can be readily distinguished from board scales. The graduations on log scales are uniformly spaced one inch apart, with each graduation line figured with the nearest integral number of board feet which it represents. On board scales, the graduations are also linearly spaced, but with a different spacing for each scale (1 inch apart for the 12 foot scale, approximately 1.091 inches apart for the 11 foot scale, etc.), and are marked such that the figures on adjacent lines differ only by one. A more complete description of log rules can be found in the section LOG STICKS, LOG CANES, AND LOG CALIPERS in Chapter VII.

Gauging and Wantage Scales

Gauging and wantage scales are specialized computing scales designed to simplify the work of the gauger, the individual responsible for measuring the capacity of liquid containers, and the quantity of liquid therein, if less than full.

The gauging stick is used for capacity measurement. It is designed to measure the linear distance from the filling hole in the side of a cask to the corner opposite, and, via its gauging scale, to read the capacity of the cask directly in gallons. It is a simplistic form of measurement, depending for its accuracy largely on the fact that casks tend to have proportional dimensions whatever their capacity (that is, the head diameter is always the same fraction of the cask length, etc.). A gauging stick less than 3 feet long could measure the capacity of casks up to 120 gallons; one about 3½ feet long was sufficient for casks up to 180 gallons.

Measuring rods graduated with gauging scales were useful for quick measurements where great precision was not required, and were widely used for many years.

The wantage scale was another computing scale, used to convert the linear distance from the bung hole to the liquid level in a cask to the number of gallons of liquid needed or "wanted" (hence the name) to fill the cask. These scales were barrel-specific, each size barrel requiring a different scale, and usually would be marked on wantage rods in groups, like the multiple scales on board sticks.

For more information on gauging and wantage rods, see GAUGING AND WANTAGE RODS, in the next chapter.

Degree and Pitch Scales

The No. 36/036 Stephens patent combination rule was marked on its blade with scales for both degrees and pitch. Both are measures of angular distances, but are expressed in different terms. Degrees is the familiar system of angular measurement we all learn in school; pitch is the framers' and millwrights' expression for slope, in which the angle relative to the horizontal is expressed as the ratio of rise (change in height in a given horizontal distance) to run (the horizontal distance). Workers in these trades did much of their angular measurement using the two legs of their framing square, and it was easier and more natural for them to think in terms of rise and run, and of pitch as the ratio of the two. Thus, where a mathematician would refer to a slope of 30 degrees, a house carpenter would speak of a pitch of nearly 7 inches per foot.

The tongue of the combination rule was marked with pitch and degree scales, one scale on each side. Although these are linear functions, the geometry of the rule required that these scales be nonlinear, being compressed as the measured angle increased. The degree scale was marked at each degree, with every fifth degree accented by extending its graduation line past the single embellishment line. The ten degree lines were made longer still, and every fifth degree was marked with a figure (5, 10, 15, etc.).

The pitch scale was graduated in 8ths of an inch in inches of rise, with only the half and whole inch lines extending past the embellishment line, and only the latter marked with figures. Since the length of each leg of the No. 036 (the run) was six inches, the resulting scale would read in inches per one-half foot. It was probably more common, however, for the user to mentally double that number so that it was expressed in inches per foot, the more common way (along with feet per foot) of stating pitch.

The No. 036 had a second degree scale marked on the inside edge of the leg containing the spirit level. This scale was designed for use in setting the angle of the tongue relative to a line normal to the leg containing it, as in when the tongue functioned as a protractor or bevel gauge. This scale was very simple, consisting of a series of short (1/8 inch) transverse lines in the middle of the brass edge surface, devoid of embellishment lines, and with degree figures adjacent to every fifth mark. The location of this scale, and the geometry of its use was such that it could be linearly graduated, but still indicate angular degrees correctly.

The "Gunter's" Slide

"Gunter's Slide" was the name applied during the eighteenth and nineteenth centuries to the logarithmic slide rule, after Edmund Gunter (1581-1626), the English mathematician. Gunter, who had a knack for finding practical applications for the mathematical discoveries of his day (he was the inventor of the Gunter's Chain, used in surveying, and two navigating instruments, the Gunter's Quadrant, for taking observations, and the Gunter's Scale, for performing computations), was popularly considered to be the inventor of the slide rule, although his actual contribution was only the invention of the logarithmic scale. The true inventor, the man who conceived of two or more logarithmic scales which could be shifted (slid) relative to one another, was William Oughtred (1574-1660), one of Gunter's contemporaries.

The two forms in which the slide rule was applied to carpenters' rules were the "Soho" rule, popularized by Watt and Boulton at their Soho works in about 1810, and "Coggeshall's" rule, developed in about 1677 by Henry Coggeshall for use in timber measurement. The Soho rule had been developed at Boulton and Watt to deal with the engineering problems associated with steam engine design, and it was in this form that it was placed on the engineers' (sliding) rule. Coggeshall's rule was particularly developed to deal with problems of timber mensuration, and it was this form that was placed on the carpenters' (sliding) rule.

The two slide rules were similar, having four scales, the A and D scales on the body, and the B and C scales on the slide. The A, B, and C scales were what today is referred to as two-cycle scales, non-folded. That is, each scale began at 1 on the left, progressed through 9, to 1 or 10 in the middle, and then continued on, repeating the cycle, to 10 on the right. Two copies of this scale (the B and C scales) were required on the slide to compensate for the lack of a cursor (the sliding vertical line which is found on all modern slide rules).

The D scale was where the two forms of slide rule differed. On both rules it was a single-cycle scale, but was marked differently on one type than the other. On the Soho rule it was non-folded; that is, it began with 1 at the left, and progressed through 2, 3, ... to 5 in the middle, and then through ... 8, 9, to 10 on the right. On Coggeshall's rule it was "folded" at the value 4; that is, it began with 4 at the left, progressed through 5, ... 8, 9, to 1 or 10 in the middle, and then continued on through 2 and 3 to 4 again on the right. Coggeshall's D scale was also labeled GIRT LINE near its midpoint, and had either two or three particular values marked: the value 12, marked with those numerals, the value 1715, marked with the letters WG, and (on early examples) the value 1895, marked with the letters AG. The first of the three, the number of inches in a foot, is valuable when using the rule for timber measure; the other two, the number of cubic inches in a wine gallon (171.5) and the number of cubic inches in an ale

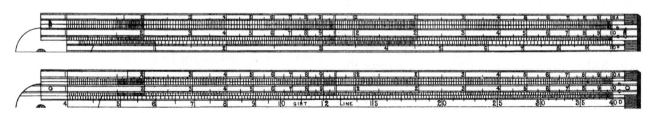

Figure 6-10: "Soho rule" (top) and "Coggeshall rule" (bottom) form of slide rule

gallon (189.5), were valuable when using the rule for cask gauging calculations.

Additional information on the two forms of Gunter's slide will be found in the sections THE CARPENTERS' (SLIDING) RULE and THE ENGINEERS' (SLIDING) RULE, in the next chapter.

Trigonometric Scales

This is a group of eight scales developed in the 17th century for use in solving problems in plane and spherical trigonometry.

Each scale graphically represents a different trigonometric function, and was marked accordingly: the Line of Numbers/Line of Equal Parts, labeled LEA; the Line of Rhumbs, labeled RUM; the Meridian Line (from Mercator's chart); labeled M*L; the Line of Chords, labeled CHO; the Line of Sines, labeled SIN; the Line of Tangents, labeled TAN; the Line of Secants, labeled SEC; and the line of Sines of Tangents, labeled S*T.

Calculations were performed using these scales graphically, using a pair of dividers and a common rule. The graphical sine of an angle, for instance, could be determined by setting the dividers to the distance represented by that angle on the SIN scale, and then transferring that distance to the LEA scale, where the value could be read off.

The method of developing these scales can be seen in the accompanying drawing (opposite). A circle is drawn whose radius is equal to the length of the Line of Equal Parts, and its periphery is divided into 360°. Then, using a compass and straight edge the other lines are drawn either as radii, tangents, or chords, and divided in the manner shown. For more information on these scales, their development, and their method of use, see Bion (Ref. 15), Book I, Chapter 6.

When placed on a plain scale or Gunter's scale, a dozen tiny brass plugs, each with a small dimple punched in its face, were set flush into the surfaces at the origins of the various scales, and at other key values in the graduations. The purpose of these plugs was to provide a good location for the tip of a pair of dividers when setting them or reading their setting. They also served to protect the surface of the wood from damage due to repeated piercing by those tips.

TABLES

In some cases data was put on rules in the form of tables, with the vertical and horizontal lines scribed, and different values stamped into the resulting boxes, one figure at a time. The most elaborate of these were the tables of gauge points marked on the Nos. 6 and 16 engineers' (sliding) rules, but there were other tables for such pur-

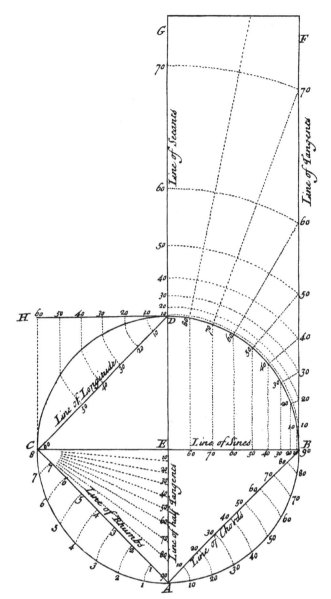

Figure 6-11: Diagram used to develop the trigonometric scales

poses as board measure calculations and hat size conversion as well.

Data tables were expensive to apply to rules, involving as they did a great deal of slow, one figure at a time hand work. The last rules with data tables were the four fold rules with board measure tables, which were discontinued by 1919.

Board Tables

Board Measure Tables were simply that: tables of board foot values for boards, having as their two axis board length in feet and board width in inches. These tables, usually called just "board tables," had a column for each board length, and a row for each board width; the entry in each box was the number of board feet and 12ths of board feet (board "inches") in a one inch board of that length/width. A rule would be marked with two

WIDTH IN INCHES

LENGTH IN FEET	1	2	3	4	5	6	7	8	9	10	11	12
6	·6	1·	1·6	2·	2·6	3·	3·6	4·	4·6	5·	5·6	6·
7	·7	1·2	1·9	2·4	2·11	3·6	4·1	4·8	5·3	5·10	6·5	7·
8	·8	1·4	2·	2·8	3·4	4·	4·8	5·4	6·	6·8	7·4	8·
9	·9	1·6	2·3	3·	3·9	4·6	5·3	6·	6·9	7·6	8·3	9·
10	·10	1·8	2·6	3·4	4·2	5·	5·10	6·8	7·6	8·4	9·2	10·
11	·11	1·10	2·9	3·8	4·7	5·6	6·5	7·4	8·3	9·2	10·1	11·
12	1·	2·	3·	4·	5·	6·	7·	8·	9·	10·	11·	12·

WIDTH IN INCHES

LENGTH IN FEET	1	2	3	4	5	6	7	8	9	10	11	12	13	14	15	16	17	18	19
13	1·1	2·2	3·3	4·4	5·5	6·6	7·7	8·8	9·9	10·10	11·11	13·	14·1	15·2	16·3	17·4	18·5	19·6	20·7
14	1·2	2·4	3·6	4·8	5·10	7·	8·2	9·4	10·6	11·8	12·10	14·	15·2	16·4	17·6	18·8	19·10	21·	22·1
15	1·3	2·6	3·9	5·	6·3	7·6	8·9	10·	11·3	12·6	13·9	15·	16·3	17·6	18·9	20·	21·3	22·6	23·9
16	1·4	2·8	4·	5·4	6·8	8·	9·4	10·8	12·	13·4	14·8	16·	17·4	18·8	20·	21·4	22·8	24·	25·4
17	1·5	2·10	4·3	5·8	7·1	8·6	9·11	11·4	12·9	14·2	15·7	17·	18·5	19·10	21·3	22·8	24·1	25·6	26·11
18	1·6	3·	4·6	6·	7·6	9·	10·6	12·	13·6	15·	16·6	18·	19·6	21·	22·6	24·	25·6	27·	28·6
19	1·7	3·2	4·9	6·4	7·11	9·6	11·1	12·8	14·3	15·10	17·5	19·	20·7	22·2	23·9	25·4	26·11	28·6	30·1

Figure 6-12: Small and large board measure tables

such tables, arranged side by side: a small table for boards 1 to 12 feet long and 6 to 12 inches wide, and, to its right, a larger table for boards 1 to 19 feet long and 13 to 19 inches wide.

Additional information on board tables can be found in the section BOARD MEASURE, BOARD STICKS, AND BOARD TABLES, in the next chapter.

Weights and Measures Tables

These are the tables sometimes put on rules used in the metal trades. The most common tables are those giving the weight per foot of square iron bars, round iron bars, and flat iron stock, of various sizes. Because these tables all dealt with iron shapes, the rules they appeared on were usually referred to as "ironmongers' rules" (see the section IRONMONGERS' RULES, in the next chapter). A typical table of this sort is shown in Fig. 6-13 opposite.

These tables are almost exclusively a feature of English rules. The only American maker known to have put such tables on one of its rules was the Stanley Rule & Level Co., and then only one time. From 1860 to 1863 this company offered a six fold rule which was marked with no less than 18 different tables of weights and measures for various shapes and materials, the tables being so densely crowded together that there was literally no space on the rule for the maker's name (see the article: "The Stanley No. 58½ Rule," reprinted in Chapter III).

FLAT IRON PER FOOT BY ¼ IN THICK		
IN	LB	DEC
1	–	83
1⅛	–	94
1¼	1	04
1⅜	1	15
1½	1	25
1¾	1	46
2	1	67
2¼	1	88
2½	2	08
2¾	2	29
3	2	50
3½	2	92
4	3	38

Figure 6-13: Typical wieghts & measures table

Wantage Tables

These are tables which allow any ordinary measuring instrument with an inch scale to determine the number of gallons of liquid needed to fill a cask (how many gallons it "wants" of being full). They are two-column tables, with the figures in the left hand column being various distances in inches from the bung hole down to the liquid level, and the corresponding figures in the right hand column showing the number of gallons required to completely fill the cask. The distances are ordinarily given as integral values of inches; the gallon values are expressed to tenths of gallons in decimal form.

Each type of cask requires a different table, for the same reason that different scales are required on a wantage rod. The function being represented is determined by the capacity and shape of the cask. The table for a barrel would not work for a hogshead, nor for a pipe, etc. Probably a dozen or more different tables are known to have been used at one time or another.

To use the table, the person doing the measuring would lower his measuring instrument vertically into the bunghole until the tip just touched the surface of the liquid. He would then note the inch mark opposite the inside of the stave at the bunghole, and then look at the appropriate table and find the gallon value opposite that inch figure. For distances other than even inches he would extrapolate as required.

Hat Size Table

This table had three columns; the first, SIZE, for the nominal hat size, the second, HEAD, for the circumference of the head, and the third, HAT, for the circumference of the hat (outside the band).

This table is constructed according to the formulas:

HEAD = (3 x SIZE) + $5/8$
HAT = (3 x SIZE) + $15/8$

Properly speaking, the correct formulas for these relationships should have had the HEAD and HAT sizes be related to the SIZE measurement by the factor π (3.1416), but this multiplication/division would have been time-consuming and require pencil and paper. Undoubtedly hatters had early on decided that the formulas given were accurate enough for all practical purposes, and could be worked out quickly in the head.

These tables were used by hatters to compare the measurement of the customer's head with the proper size hat to fit it. More information about this table, and about the hatters rule on which it was marked, can be found in the section HATTERS' RULES, in the next chapter.

Engineers' Tables

These were the elaborate tables of gauge points devised around 1811 by Josiah Routledge for his so-called "engineers' (sliding) rule."

The most popular of a number of similar "engineers'" rules introduced during the early nineteenth century, Routledge's rule consisted of a two foot, two fold rule marked with his tables and equipped with a Gunter's slide (a simple form of slide rule). Consisting of 168 data values in five categories, the tables provided the core of the material needed to solve many of the relatively primitive engineering problems of the time. The tables consisted of five separate tables:

Conversion factors relating the volumes of various geometric solids;

Conversion factors relating various volumes of different materials to their weight in pounds;

The areas of various regular polygons;

A table of numbers relating the diameter, area, and circumference of a circle to the dimensions of an inscribed or circumscribed square;

A table of values for computing the required cylinder diameter for a steam engine driving a pump of known diameter.

Many rule makers manufactured rules of Routledge's pattern until after the turn of the century. These tables are probably the most elaborate markings made on any American folding rule (with the possible exception of the Stanley No. 58½ six fold rule with weights of metals tables). Occupying a space only ¾ inch wide by 10½ inches long, they were framed by a grid of vertical and horizontal embellishment lines, and stamped totally by hand, one digit at a time. It is probable that only the most skilled and painstaking workers were used for these rules, and that at best each worker could only produce a few each day.

SIZE	HEAD	HAT
5⅞	18¼	19¼
6	18⅝	19⅝
6⅛	19	20
6¼	19⅜	20⅜
6⅜	19¾	20¾
6½	20⅛	21⅛
6⅝	20½	21½
6¾	20⅞	22⅞
6⅞	21¼	22¼
7	21⅝	23⅝
7⅛	22	23
7¼	22⅜	23⅜
7⅜	22¾	24¾
7½	23⅛	24⅛
7⅝	23½	24½
7¾	23⅞	25⅞
7⅞	24¼	25¼
8	24⅝	25⅝

Figure 6-14: Hat size table

It has been found that these tables were notoriously prone to error, with an average of probably one misstamped figure on every rule. The apparent cause for this was probably a combination of worker error (the figures are very small, and the slightest inattention can cause mistakes) and the tendency of rulemakers to copy such tables off another maker's rule, mistakes and all.

Additional information on Routledge's rule and the engineers' tables can be found in the articles by the Author and others reprinted earlier in this book, and in the section THE ENGINEERS' (SLIDING) RULE, in the next chapter.

Load and Hundredweight Tables

Early two fold carpenters' rules were sometimes marked with tables for extending the unit cost of some material to the cost per load, to simplify the calculations required of an artisan. These tables had four columns, the first giving the unit cost of the material in pence (D) and farthings (¼ pence). (These tables appear only on English rules, and deal in exclusively in English currency.) The other three columns give the cost of a load of that material in pounds (L), shillings (S), and pence (D).

Two typical tables of this type are shown opposite. The one on the left extends the cost per pound to the cost per hundredweight (an English term meaning 112 pounds, a common unit of measure). The one on the right extends the cost per cubic foot/pound to the cost for 50 cubic feet/pounds, another common method of quantifying materials.

As mentioned above, these tables are only found on English rules, and indicate a date prior to the middle of the 19th century.

D	L	S	D
1	0	9	4
¼	0	11	8
½	0	14	0
¾	0	16	4
2	0	18	8
	1	1	0
	1	3	4
	1	5	8
3	1	8	0
	1	10	4
	1	12	8
	1	15	0
4	1	17	4
	1	19	8
	2	2	0
	2	4	4
5	2	6	8
	2	9	0
	2	11	4
	2	13	8
6	2	16	0
	2	18	4
&c.			

D	L	S	D
2	0	8	4
¼	0	9	4
½	0	10	5
¾	0	11	5
3	0	12	6
	0	13	6
	0	14	7
	0	15	7
4	0	16	8
	0	17	8
	0	18	9
	0	19	9
5	1	0	10
	1	1	10
	1	2	11
	1	3	11
6	1	5	0
	1	6	0
	1	7	1
	1	8	1
7	1	9	2
	1	10	2
&c.			

Figure 6-15: Two typical tables for extending cost.

REFERENCES:

1. Frank Freese, *A Collection of Log Rules (General Technical Report No. FPL 1)* (Madison, Wisc.: US Department of Agriculture Forest Service, 1974).

VII
Special Rule Types and Uses

Over the years rule makers and users have invented and developed a large number of special types and patterns of rules for particular applications or uses. For some types of work, a specific size would be found to be particularly appropriate; for other jobs special graduations could solve or simplify calculations; still other jobs required a particular material or mechanical configuration.

This section describes some of the more frequently encountered types and patterns of rules intended for particular jobs and applications. It is, admittedly, far from complete. Rule types not frequently encountered have been omitted, as have applications so exotic as to be of minimal interest to the average collector. If any reader feels that some rule type has been wrongly omitted, I can only reply that the selection of types and examples to be included/omitted seemed appropriate to me at the time, but that I would be happy to discuss the subject further.

ADVERTISING & THIRD PARTY RULES

An advertising rule is a rule marked with a company or product name or advertising message, intended to be sold or, more usually, given away, to promote business or good will. Such a rule could be a simple standard product, marked with the advertiser's name or desired message, or a special pattern of rule, with scales or features not usually offered. By their very nature advertising rules were a special order product, different for each customer.

Third party rules (so called because they were marked with the seller's name instead of the maker's) were rules made for sale by a wholesale or retail firm which did not have rulemaking capability, but still wanted to include a selection of rules as part of the product line bearing its name. This firm would contract with an established maker to produce rules either marked with its (the seller's) name, or unmarked, so that the seller could later affix a paper label bearing its name.

Most major rulemakers made advertising rules at one time or another, the most notable being Stanley, Chapin-Stephens, and Lufkin. Stanley and Lufkin even went so far as to advertise in their catalogs that they were willing to make advertising rules upon demand. Occasionally a maker would produce an advertising rule promoting its own product line, usually as a giveaway item.

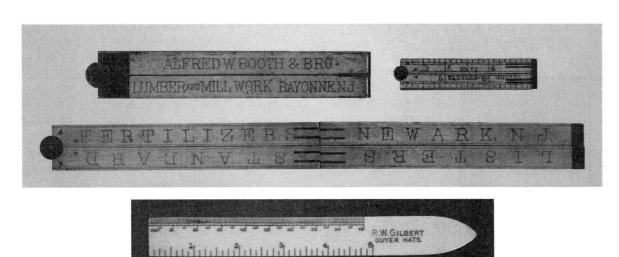

Figure 7-1: Rules advertising others' products

Figure 7-2: Rules advertising rulemakers' own products

DEARBORN'S "ANGLET"

The earliest U.S. patent for a rule was that issued for the "Anglet," a combination rule and protractor invented by Benjamin Dearborn, of Boston, Massachusetts and patented by him on April 29, 1808 (U.S. Name&Date Patent #X0866, entitled "Instrument Called the Anglet"). This instrument was a two fold rule with the two legs joined by a sliding arm that indicated the numerical angle to which the rule was opened (Fig. 7-3, opposite).

This sliding arm, which formed the key element of Dearborn's invention, was joined to one leg by a fixed pivot about halfway down the inside of that leg, and attached to the other leg via a sliding pivot that moved along the inside of that second leg. This sliding pivot carried a cursor line which ran along a scale marked on the surface of the rule along the inside edge of that leg, and the indicated value on this scale was the opening of the rule in degrees.

This method devised by Dearborn for indicating the angle of opening of his "Anglet" is interesting, but has two problems. First, the geometry involved is such that it was only accurate for angles in the middle of the 0-180° range; for very small angles, where the rule was nearly shut, or large angles, where the rule is almost fully open, the motion of the index was so small as to yield no meaningful reading on the scale. Second, the scale marked on the rule was nonlinear and (given the technology of the time) had to be marked by hand from a master pattern. This method of graduation is inherently less accurate than the circular graduation used on ordinary protractors. Dearborn recognized the first of these problems, as can be inferred from the details of the patent. The drawing shows a scale marked only for angles in the range from 20° to 110°, and the patent text (Fig. 7-4, page 172) refers to a "small arch of a circle [at the joint] to ascertain the distance or number of degrees at which the instrument is opened, before the index will note the degrees with accuracy."

It is not quite clear if Dearborn had any specific application in mind for his "Anglet." He suggested several possibilities in the text of the patent: as a protractor in drafting or carpentry, as a graphometer in land surveying (with a compass in one of the legs and a pair of sights on each leg), as an inclinometer (with a spirit level in one leg). He did not indicate any of these as its primary purpose, however, and it may well be that he was

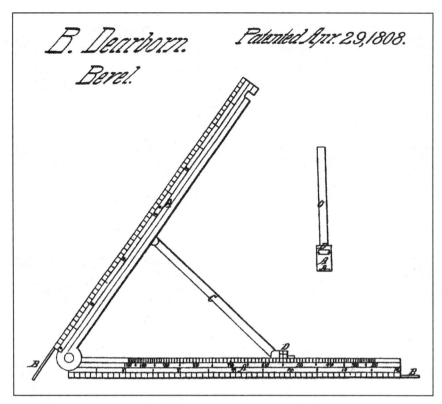

Figure 7-3: Patent drawing of Dearborn's "Anglet"

BENJAMIN DEARBORN

Benjamin Dearborn (1754-1838) was a printer, publisher, educator, inventor, and manfacturer, originally from Portsmouth, New Hampshire, but later from Boston, Massachusetts. In his youth he (presumably) attended Major Hale's Latin grammer school in Portsmouth, the best local education available, and subsequently attended Harvard College. At the outbreak of the Revolution in 1776 he left the loyalist newspaper, the *New Hampshire Gazette*, and founded his own newspaper, the *Freeman's Journal*, in Exeter, New Hampshire. He operated the Journal for six months and then sold it to Daniel Fowle, owner of the *New Hampshire Gazette*.

In 1780, after a brief career dealing (not too successfully) in "English, West Indies and other goods," he opened a "Writing School for Misses." This was an early effort at providing equal education for girls; he taught not only the basics of the "three R's," but also spelling and grammar. Later he added instruction in music, French, and needlework. Like many educators of his time, Dearborn recognized the deficiencies of most imported textbooks, and wrote and published several of his own in arithmetic, grammar, and music. In 1791, after an unsuccessful attempt to expand his school into a full academy, he moved it to Boston, operating it there until 1799. In 1799 he turned it over to a new master while he devoted his time to inventing and manufacturing.

From his youth Dearborn was a prolific inventor. He had received two patents prior to the revolution from the New Hampshire colonial government, for a waterwheel and for scales for weighing. Between 1799 and 1819, he received five more patents from the new national government. These patents were for a steelyard, a candlestick, the "Anglet," a mason's hoist for bricks and mortar, and a balance (scale). He also made a number of other inventions which were never patented, among which were: a "Machine for Drawing in Perspective," a "Perpetual Diary," a "Table for Computing Interest," and a "Method of Printing in Color from Wood Blocks."

The patented invention for which he is best known, and which was the basis for his success as a manufacturer, was the steelyard (weighing device) which bears his name. He began the manufacture of these balances in 1799 in Taunton, Massachusetts. Subsequently, in 1801, he transferred production to his own works in Boston on Theatre Street. Manufacture continued at this location until Dearborn's death in 1838. This "vibrating steelyard," as he described it, was made in a wide range of sizes, from "Ounces to Tons," and was so widely used that examples still surface occasionally at antique shows and flea markets today.

> *Benjamin Dearborn of Massachusetts*
>
> # 𝕷ETTERS 𝕻ATENT
>
> *Dated April 29, 1808*
>
> *The Schedule referred to in these Letters Patent and making part of the same containing a description in the words of the said Benjamin Dearborn himself of his instrument called the Anglet.*
>
> *This instrument consists of the following parts, viz: 1st, Two bars or legs united by a joint in such a manner as to open and shut like a jointed rule. 2nd, an index made to slide on or in one of the bars or legs before mentioned. 3rd, A Brace the two ends of which turn on two pins, one end being thereby being attached to one of the legs, and the other to the index. To the foregoing the following subordinate parts are added or omitted as may be expedient: 1st, A ruler or rules sliding in one or both of the legs above mentioned. 2nd, A Spirit Level, Water Level, or Plumb Line. 3rd, a small arch of a circle to ascertain the distance or number of degrees at which the instrument is opened, before the index will note the degrees with accuracy. 4th, A magnetic needle and proper sights for the purpose of surveying land, &c. On that leg where the index moves a scale is laid down on one or both sides, showing by means of the index at what angle the instrument is open at any degree from the lowest to the highest marked thereon. On other parts of the instrument scales of inches and such other scales may be added as shall be required for mathematical calculations or measurements in any way in which the sector, plain scale, or sliding rule, or common rule can be applied. This instrument may be made of Ivory, Bone, Metal or Wood, or any of these materials may be united therein; and it may be of any size required to suit the purpose for which it is wanted. The largest which has yet been made is of five feet radius, opening to a ten feet pole, the smallest six inches radius, opening to a twelve inch rule. In addition to the purposes above described for which the Anglet may be used, it may also be graduated with a scale of equal parts on the inward edge of one leg corresponding with a similar scale on the other leg and a sliding square or right angle applied thereto will ascertain the latitude and departure of a Ships course. But a different instrument has long been known with a sliding square applied to the same purpose. Therefore the application of the square in this way is not considered as being secured by the exclusive right of this Patent.*
>
> *Witnesses: William Thornton*
> *Hez^b Rogers*   *Benjamin Dearborn*

Figure 7-4: The text of Dearborn's patent for the "Anglet"

more concerned with the mechanism than with an application.

Dearborn also described several additions to his rule which were not related to its patented feature, the ability to measure angles. He postulated the addition of various scales such as the plotting lines from a plain scale or a pair of sector lines, or the incorporation of slides with scales for calculation (two examples of these last are shown in the patent drawing). These were just speculative proposals, however, and did not form any part of the basic patent.

No examples of Dearborn's "Anglet" are known to exist. In the patent Dearborn states that several have been made, from a very large ten foot version to a diminutive one foot one, but these were probably prototypes. In all likelihood, given the problems with measuring very small and very large angles mentioned above, the "Anglet" was never made commercially, and the prototypes have been lost or destroyed over the years.

It is only by chance that Dearborn's patent for the "Anglet" is available for study. The U.S. Patent Office suffered a catastrophic fire in 1836, and all records, including the Patent Office copies of all patents issued up to that time, were destroyed. After the fire efforts were made to reconstruct the Patent Office files. All ten thousand patentees were contacted at their post offices of record and requested to lend their inventor's copy of their patent(s) to the Patent Office so that duplicate copies could be made and placed in the Patent Office files. Despite all efforts, however, only about one-quarter of the patentees complied with these requests, and their patents are the "restored" patents whose text and drawings can be found in the early Patent Office files (now kept in the U.S. National Archives). Since Dearborn was still alive during this restoration process, he was able to comply with the Patent Office's request. As a result his is one of the 2,500 or so restored patents, making the drawings and text available for reproduction here.

BENCH RULES

For working at the bench, when compactness and transportability were not important, a folding rule was unnecessary, and could be replaced by a stronger and cheaper nonfolding one, a so-called "bench" rule. Such rules were wider, from 1 1/4 to 1 1/2 inches, and often thicker, up to 1/4 inch, than the ordinary folding rule, and were typically one or two feet in length.

Most major rulemakers offered bench rules as part of their standard line. Three types were common: one foot bench rules, two foot bench rules, and two foot bench rules marked with board measure tables. The material was usually maple, although some makers also offered them in boxwood, and at least one two foot bench rule was made of satinwood.

Bench rules were usually graduated on all four edges in 8ths and 16ths of an inch. The graduations ran from left to right on the bottom edge, and from right to left on the top edge.

Some one foot desk or school rules are similar in appearance to one foot bench rules; the only way to tell one type from another is to look at the maker's description in the catalog. Sometimes, as in the case of the Stanley Nos. 34 1/4 and 34 1/2 rules, a rule would be listed as a "school" rule for several years, and then suddenly be changed a "bench" rule thereafter.

Rules longer than two feet in length were not usually considered bench rules, but were instead listed in catalogs as other types of rules, such as glaziers' rules, saddlers' rules, or forwarding sticks.

BEVELS, SHIP CARPENTERS'

The bevel gauge is a widely used woodworkers' tool, consisting (in its most common form) of a thin tongue or blade set into a thick stock, with the joint between the two adjustable so that the tongue can be set and held at any desired angle to the stock. The purpose of the bevel gauge is to transfer an angle from a drawing or protractor, or from some other piece, to a workpiece for either marking or testing purposes.

Wooden ships are possibly the most extreme example of a structure in which no two pieces join at a right angle. The planks form compound angles as they follow the hull shape; the frames are curved, and beveled to fit the planking; the deck is crowned from side to side, and has a hollow curve from bow to stern. Because of this the ship carpenter used the bevel gauge more than any other woodworker.

The bevel used by the ship carpenter was somewhat different from the sliding T bevel used by ordinary carpenters. In its most common form, it was both longer (12 inches) and slimmer (5/8 of an inch) than the sliding T bevel. It was made of boxwood, and was usually graduated in 8ths and 16ths of an inch along its length, allowing it to double as a rule. It usually had two tongues,

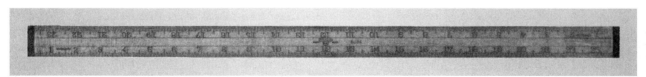

Figure 7-5: Stanley #34 bench rule

Figure 7-6: Double tongue ship carpenters' bevel

174 SPECIAL RULE TYPES AND USES

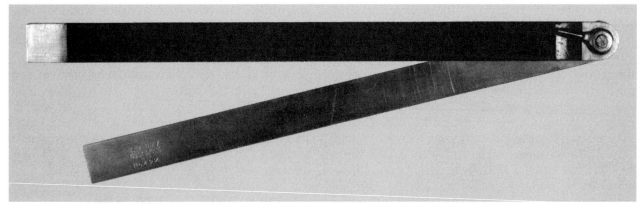

Figure 7-7: 14" rosewood & brass ship carpenters' bevel

instead of only one, of unequal lengths, attached to the stock at opposite ends. These tongues were attached with pivoting friction joints, instead of the screw- or lever-clamped joint used on the sliding T bevel.

Variations existed in the pattern, of course. Sometimes the stock was of rosewood, instead of boxwood (in which case the graduations were always omitted); sometimes the rule was longer than 12 inches; sometimes it had a single long tongue instead of two shorter ones.

Tools made of exotic woods such as rosewood, live oak, etc. are more often found in the kits of ship carpenters than any other group of woodworkers. There was a tendency among these craftsmen to make their own tools, due to the specialized work for which they were used. Since most of them worked near seaports, they had access to foreign woods unavailable to the inland cabinetmaker. It is not surprising, therefore, that makers should offer premium bevels of rosewood to appeal to ship carpenters whose other tools were extra fancy. An example of such a premium tool is shown above.

BLACKSMITHS' RULES

The rule used by a blacksmith had to be made of metal; the hot iron would have quickly charred or destroyed one made of wood. Most were made of brass, being easy to keep bright and tarnish-free, but steel was sometimes used instead, for its durability.

The most common pattern was a two foot brass rule jointed in the middle. There were also non-folding versions, however, and versions with a hook on one end to make it easier to measure across wide stock.

Two types of joint were used on folding blacksmiths' rules. The earlier rules used a simple friction joint. Later rules (after about 1900) were generally jointed with a spring joint similar to

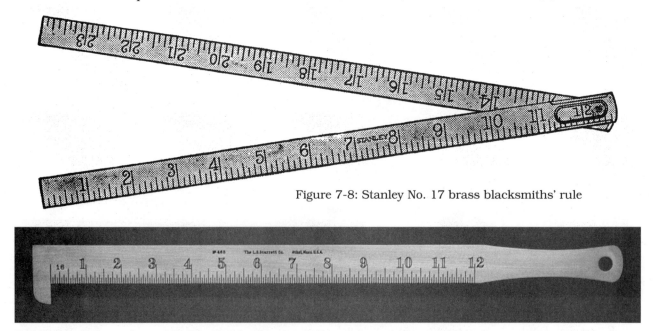

Figure 7-8: Stanley No. 17 brass blacksmiths' rule

Figure 7-9: Starrett brass blacksmiths' hook rule

that used on the Zig-Zag rules which came into common use about that time.

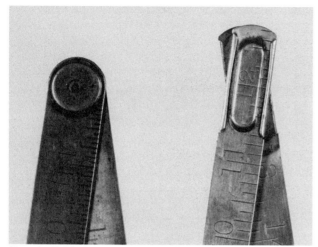

Figure 7-10: Friction joint (left) and spring joint (right)

"BLINDMAN'S" RULES

Prior to 1900, the graduations and figures on rules were always put on by the same method: by scribing or stamping them into the wood, and then filling the resulting incisions with black paint or wax. The resulting markings were thin, clear, and as durable as the surface of the wood itself. The only disadvantage of this method was that the markings were not as dark and heavy as some would have; they were hard to read if the light was dim, or if the user had impaired vision.

In 1900, partly as a response to this problem, and also partly as a result of improvements in ink and printing technology, two rulemakers began to manufacture and sell rules whose markings were printed on, instead of being incised into, the surface of the wood, and thus could be made extra wide and extra dark. Additionally, the figures were made almost twice as large as on ordinary rules, improving their readability even further. These rules were always broad (one inch wide) to provide room for the extra large figures.

Stanley produced two such rules, the two foot, four fold No. 7 and the three foot, four fold No. 8. Lufkin produced one, identical to the Stanley No. 7. Chapin-Stephens produced one, also similar to the Stanley No. 7, but differing in one respect: the graduation lines were printed in black, but the figures were printed in red. Chapin-Stephens referred to this as their "Nearsight" rule.

An unusual feature of these rules was that they were invariably graduated from left to right, in contrast to the right to left graduations typical of American manufacture during the period when they were offered (1900-1950).

BOARD MEASURE, BOARD STICKS, AND BOARD TABLES

BOARD MEASURE

A common method of measuring gross sawn lumber quantities is to reduce the volumes of the various pieces in a lot to the equivalent number of square feet of lumber one inch thick, add up the results, and to refer to this sum as the number of "board feet" in the lot. Wood is frequently priced and sold by the board foot, and material estimates for jobs are often prepared in these units.

The process of measuring lumber and computing the number of board feet is not a difficult one, but it is time consuming, and subject to computational error. It is not surprising, therefore, that rulemakers should manufacture special rules with features that would aid the woodworker in this task. Two types of such rules were offered: "board sticks," graduated with special scales called "board scales," and ordinary rules imprinted with "board measure tables."

BOARD STICKS

Board sticks were sticks, usually two to four feet in length, often equipped with a head hook, and graduated with a set of computing scales for measuring sawn lumber and directly expressing the result in board feet. Each scale would be designed for directly reading the content of a board of a specific length. A typical stick would have anywhere from 6 to 16 of these scales, depending on its shape or complexity, for that number of board lengths, with each scale marked at its origin as to what length board it applied to.

Figure 7-11: Lufkin No. 7 "Blindman's" rule

176 SPECIAL RULE TYPES AND USES

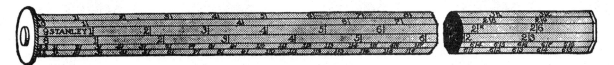

Figure 7-12: Stanley three foot octagonal board stick

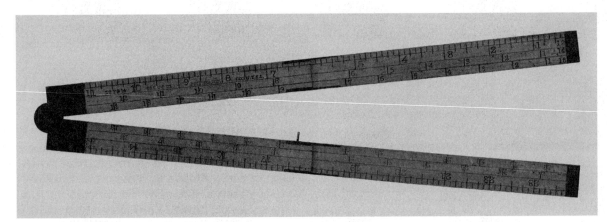

Figure 7-13: Chapin-Stephens four fold rule with board scales

A board stick was used as follows: Knowing the length of a board, the stick would be laid across it, with the scale for a board of that length uppermost, and the end of the stick just even with the far edge (this is where the head hook was useful); the number of board feet in that board could then be read off the scale without the need for a table lookup.

Board sticks were made in a number of different configurations and materials. They could be square, octagonal, or flat in cross section, with several scales marked on each face. Sticks intended for use in a sawmill or lumberyard would be made of a tough wood like hickory or maple, be three or four feet long, and almost always had a steel or brass hook on the end to locate the far edge of the board when measuring or to slide and turn the lumber when going through a whole stack. Sticks intended for use in the shop were shorter, usually made of boxwood or maple, and frequently had the end hook omitted. Some makers, specifically Stephens, Chapin-Stephens, and Belcher Brothers also offered ordinary two foot four fold rules marked with board scales.

The board scales took up all the space on the stick, leaving no room for ordinary 8ths and 16ths of an inch scales (although the markings on the scale for 12 foot boards, being one inch apart, could be used as a substitute), and thus these were special purpose tools intended only for lumber measurement.

Probably the most elaborate form of board stick is the example shown opposite, made by J. Watts of Charlestown, Massachusetts, which includes a tally mechanism for carrying a running total of board feet as boards are measured. This stick is three feet long, 1/2 inch by 2 1/4 inches in cross section, brass bound on the edges, and equipped with a substantial brass head for hooking on the far edge of the board. Two sets of three sliding brass tally markers run in slots in the brass binding on the two edges of the stick (two sets of markers allow two independent tallies to be kept). In each set one marker keeps track of single board feet on a scale of 1 to 99 marked on the surface of the stick, one keeps track of hundreds of board feet on a scale of 100 to 900, and the third keeps track of thousands of board feet on a scale of 1,000 to 10,000. Each tally marker has a spring latch which must be depressed before it can be moved, to prevent it from being moved inadvertently resulting in the loss of the current total.

BOARD MEASURE TABLES

Board measure tables were simply that: tables of board foot values for boards, having as their two axes board length and board width. These tables, usually called just "board tables," had a column for each board length in feet, and a row for each board width in inches; the entry in each box was the number of board feet and 12ths of a board foot (board "inches") in a one inch board of that length/width. A rule would be marked with two such tables, arranged side by side: a small table for boards 1 to 12 feet long and 6 to 12 inches wide, and, to its right, a larger table for boards 1 to 19 feet long and 13 to 19 inches wide.

Three different patterns of rule were offered by various makers with these board measure tables:

SPECIAL RULE TYPES AND USES 177

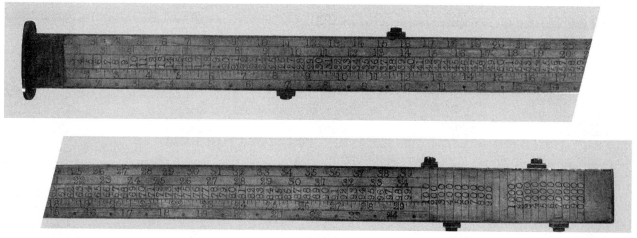

Figure 7-14: J. Watts board stick with two sets of tally markers

Figure 7-15: No. 35 bench rule with board tables

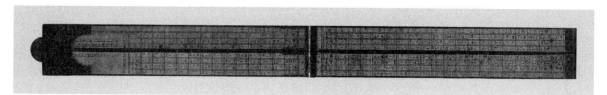

Figure 7-16: Hubbard Hardware No. 22 two fold rule with board tables

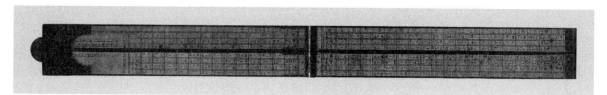

Figure 7-17: Stearns No. 22 four fold rule with board tables

bench rules, two foot two fold rules, and two foot four fold rules.

When placed on folding rules these tables were marked half on each leg (on the inside surface in the case of four fold rules), and thus could only be used when the main joint was closed. Furthermore, although they were stamped in extremely small figures, they were still physically quite large, and were only placed on broad two foot four fold rules.

The tables were used by rounding the measured board length to the nearest even foot and the width to the nearest even inch, and then looking up the value in the table at that column and row. The result was a number of board feet and "board inches," separated by a decimal point. Extrapolation could be used to improve accuracy slightly, but probably most users did not bother.

BOARD & LOG CANES

It is not known when or where the idea of a three foot measuring stick shaped like a walking cane with a brass head and ferrule originated. It was a standard pattern, offered by most of the major makers, and could easily date from the 18th century or earlier.

The combination of cane and measuring stick is an interesting one. It conjures up the image of an elderly sawyer or timber buyer, clambering over piles of sawn lumber with the aid of his cane, and then laying it across a log or board and calling out or recording the indicated figure.

These canes were designed to stand up to rough usage. They were invariably made of extremely tough hickory, and were tipped with a brass ferrule to reduce wear. The round brass

Figure 7-18: Stanley No. 48 board cane

head served a double duty, providing a good grip when used as a cane, and acting as a shallow hook to aid in placing the head of the cane at the board edge when measuring. The cane itself was octagonal, with a single scale on each of its eight faces, and was tapered from 3/4 inch thick at the head to about 1/2 inch at the ferrule.

BUTTON GAUGES

The button gauge is a short sliding caliper rule used by tailors and seamstresses to measure button diameter, so as to be able to size the associated button hole. Typically about three inches in length, these devices were provided with a sliding caliper calibrated in 40ths of an inch, the commonly accepted scale for button size. They were made in both boxwood and ivory, to suit both the taste and the pocketbook of prospective buyers.

Button gauges were made by several American makers during the mid-19th century. Stanley offered three: the boxwood No. 210 in 1862, and the boxwood No. 23 and ivory No. 24 in 1867. Chapin also offered the boxwood No. 49 in its pre-1856 product line, and Lufkin offered the boxwood No. 024B. Both Belcher Bros. and Kerby also offered button gauges.

Several examples have been observed of the Stanley No. 136 four inch caliper rule with scales for use as a button gauge. These are advertising pieces, however, marked with the name of a manufacturer of overall buttons, and were never part of the standard line.

CALCULATOR RULE

One unusual rule which is sometimes encountered is the "Perfection Self-Adding Ruler," a 15 inch desk rule which has a simple stylus-type adder on its top surface for use by an accountant or bookkeeper in, for example, totaling up columns of numbers.

This adder operated as follows: a longitudinal slot in the top surface gave access to an endless loop of ribbon running the length of the rule, mounted on two internal pulleys. The slot had numbers marked next to it; the integers from 1 to 45 on one side, and the integers from 46 to 90 on the other. The ribbon had 100 equally-spaced holes in it, marked with the integers from 0 to 99. A number was added to the total by placing the tip of a stylus or pencil in a hole next to that number and then sliding it to the end of the slot. A window beyond the end of the slot displayed the total by allowing the user to see one of the 100 numbers on the ribbon. A small nib mounted on the surface of the ribbon at one edge would cause a small auxiliary dial near the window to increment each time the total on the ribbon passed zero, thus keeping track of hundreds.

This calculating rule was invented by Robert McClelland, of Williamsville, Illinois, and patented by him on January 8, 1895 (U.S. Patent No. 532,241, entitled "Computing-Machine"). The maker is unknown.

CALIPER RULES

A caliper rule is a rule equipped with a caliper slide, a T-shaped metal slide which allows the rule to easily measure material thickness or other outside dimensions of small objects (typically up to two or three inches). This slide consists of a metal strip with a hat-shaped cross section, running in a matching slot running the length of the

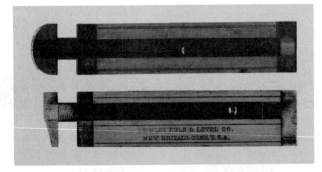

Figure 7-19: Belcher Bros. (top) and Stanley (bottom) button gauges

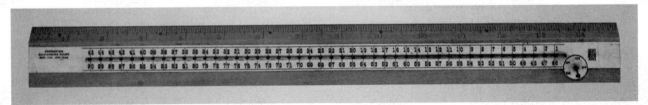

Figure 7-20: McClelland's calculating rule

rule leg. On its end the strip has fastened a right-angle metal head or jaw, giving it an L-shaped or T-shaped configuration, where the head/jaw overhangs the end of the rule leg in which the slide is housed. The length of the slide is graduated beginning at the jaw. By placing the object to be measured between the jaw and the tip of the leg and bringing the jaw down against it, its dimension can be read from the graduations on the slide.

The most common patterns of rules to be equipped with a caliper slide were the small ones: six inch two fold rules, one foot two fold rules, and one foot four fold rules. Other configurations were also offered with calipers, but less frequently. A few two foot four fold rules were offered with this feature; Lufkin offered a whole series of short nonfolding caliper rules. These were the exceptions, however, and not frequently encountered.

Since the ability of a sliding caliper to measure the diameter of a circle is limited by the depth of the jaw, folding caliper rules were configured such that the jaw extended fully across both legs of the rule when half open, thus doubling its capacity.

On some small (six inch, two fold) caliper rules the leg carrying the slide would be covered all around with metal to strengthen it. The metal used was appropriate to the rule material, brass for a wood rule, and German silver for an ivory one. This was referred to in the catalogs as having a "cased" slide. This type of construction indicates manufacture sometime before about 1880.

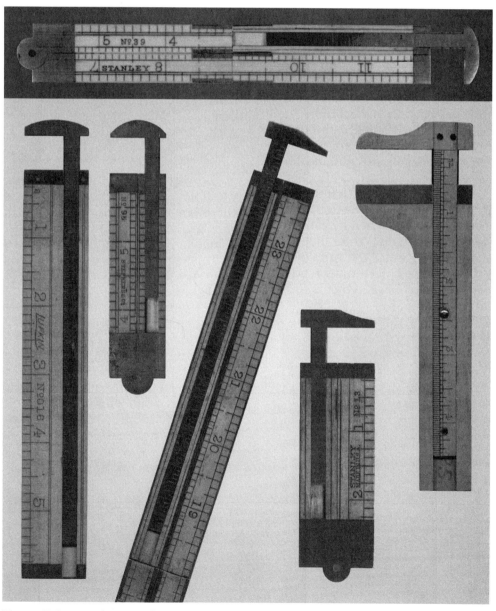

Figure 7-21: A selection of caliper rules

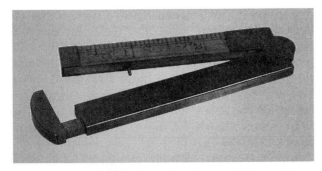

Figure 7-22: Chapin #71 caliper rule with "cased" slide

In England small caliper rules were often called "Ironmongers' Rules." When so identified, they were frequently marked with tables of data relating to standard types and shapes of steel, or with tables allowing the easy extrapolation from the cost per pound to the cost of a hundredweight.

In 1932 Stanley introduced a pair of four and six inch calipers, the Nos. 136 and 136½, capable of measuring inside dimensions and the diameters of holes. This was made possible by a feature which, while common in machinists' hand tools, had not hitherto been applied to boxwood caliper rules. On these two rules the tips of the caliper jaws were milled on the outside into a curved shape of an exact diameter. On the body of the rule an index line was marked that same distance from the end of the leg, the edge used for reading the scale on the slide when measuring an outside dimension. By using this index line instead, the distance between the outside surfaces of the jaw tips could be accurately read, and the rule be used for inside measurements as well.

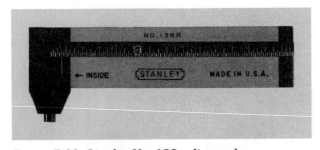

Figure 7-23: Stanley No. 136 caliper rule

CARPENTERS' (SLIDING) RULES

The carpenters' sliding rule is a two foot, two fold rule similar in design to the ordinary two foot, two fold carpenters' rule, but with the usual graduations on the back of one leg replaced by a simple logarithmic slide rule, with the slide set right into the leg, and those on the back of the other leg with a set of drafting scales.

The slide rule, consisted of a brass slide marked with two logarithmic scales, the B and C scales, set into the upper leg of the rule, said leg having two more logarithmic scales, the A and D scales, marked on the wood surface adjacent to the slide. Three of these scales (the A, B, and C scales) were the same, being two cycles in length, beginning at the left at 1, progressing through 2, 3, etc. to 1 at the middle, and then progressing through 2, 3, etc. again to 1 at the right. The D scale was only one cycle in length, and was "folded" at 4 (that is, it began at 4 on the left, progressed through 5, 6, etc., to 1 near the middle, and thence through 2, 3, etc., to 4 again on the right). This D scale was also labeled GIRT LINE near its midpoint, and had either two or three particular values marked: the value 12, marked with those numerals, the value 1715, marked with the letters WG, and (on early examples) the value 1895, marked with the letters AG. The first of the three, the number of inches in a foot, is valuable when using the rule for timber measure; the other two, the number of cubic inches in a wine gallon (171.5) and an ale gallon (189.5), were valuable when using the rule for cask gauging.

This particular arrangement of scales in a slide rule was the invention of Henry Coggeshall, an Englishman, in the mid-17th century, for use in timber measurement, and was described by him in a pamphlet published in 1677 (Ref. 1).

At that time the usual method of calculating the amount of sawn timber which could be gotten from a log was to take ¼ of the circumference ("girth") of the log, square it, and then multiply the result by the length of the log. This "quarter-girth rule" was not particularly accurate, but the dimensions were easy to measure, and the calculation within the capabilities of the elementary arithmetic then in common use. This was actually

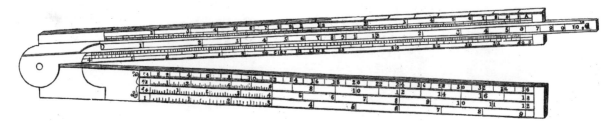

Figure 7-24: Stanley No. 12 carpenters' sliding rule

a crude early form of log rule (see LOG STICKS, LOG CANES AND LOG CALIPERS, later in this chapter), one which assumed a wastage of approximately 20% during reduction to square.

Coggeshall arranged the C and D scales to facilitate the calculation of log content according to the quarter-girth rule. By making the C scale (where the length of the log was entered) two cycles in length, and the D scale (where the girth of the log was entered) only one cycle this would automatically perform the squaring of the girth. Further, by "folding" the D scale at 4, this automatically divided the girth by four, eliminating another step in the calculation. Since the D scale was the scale on which the girth of the log was entered, Cogggeshall called it the "GIRT LINE," and all makers of these rules followed this nomenclature.

Not all carpenters' sliding rules were two fold rules. Four fold rules were sometimes offered, but their disadvantages made them uncommon. On a four fold rule the slide was only six inches long, and the logarithmic scales were compressed accordingly. This reduced the accuracy of the calculations which could be performed with the slide. Also, a four fold carpenters' sliding rule had a tendency to damage the track the slide ran in if not handled carefully; if the middle joint was opened while the slide was extended in the direction of that joint, the end of the adjacent head stick would press against the underside of the slide and forcibly pry it from the groove!

The set of drafting scales occupied the entire surface of the lower leg. There were four of them, each nine inches long, graduated from left to right (that is, beginning near the joint and ending near the tip of the leg), for scale factors of ¼, ½, ¾, and 1 inch per foot. On each scale the first three inches was divided into scale feet and inches, and the remaining six inches into scale feet only.

The front face of this rule was graduated in the usual 8ths, 10ths, and 16ths of an inch, and was marked with the so-called octagonal or "four-square" lines. Also, if the rule was not bound, the outside edge was graduated in 100ths of a foot. This last was a very useful feature, as slide rule calculations deal in decimal quantities, and this 100ths scale allowed dimensions to be taken in decimal form.

It is not known when the carpenters' sliding rule was developed. It must have been after 1677, when Coggeshall invented his pattern of slide rule. It has been suggested that this rule formed the basis for the various engineers' sliding rules, as developed by Routledge, Carrett, Hawthorne, and others, and thus must be earlier than 1811, but this is only a supposition, and needs more research to prove or disprove it. Whatever its origin, the carpenters' (sliding) rule was a highly standardized pattern by the middle of the 19th century, produced by a large number of makers, and widely used.

Carpenters' sliding rules were made by all of the major American rulemakers except Lufkin. These rules had ceased being manufactured by about 1915; since Lufkin only began rulemaking in 1924 they were never part of its product line.

CARRIAGEMAKERS' RULES

Extra strong four foot four fold rules were frequently called carriagemakers' rules, from their wide use by carriage and wagon makers. The size of the work, and the cluttered condition of the workspace made it imperative that these measuring rules be both longer than usual, and have double plated joints and brass binding to prevent them from breaking if stepped on.

COMBINATION RULES

An interesting class of rules are the so-called "combination rules," rules which could perform functions in addition to simple measurement. Some of the functions which could be performed

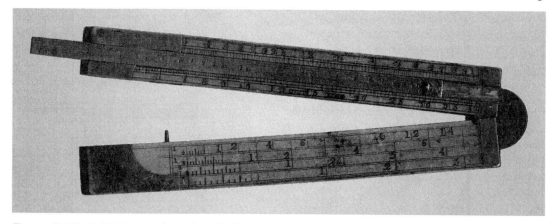

Figure 7-25: J. Watts two foot, four fold carpenters' sliding rule

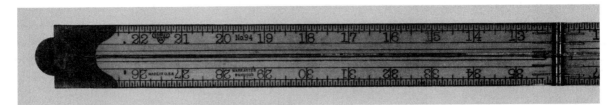

Figure 7-26: Stanley No. 94 carriagemaker's rule

by different combination rules were: measuring and laying out angles, plumbing and leveling, drafting, and calculation. The examples described here are representative of what was available.

LUFKIN THREE-FOLD RULES WITH LEVELS

This group of rules represents probably the simplest form of combination rule. They were three-fold rules, with a level vial mounted in the edge of the center stick. The joints allowed the sticks to fold flat against one another, displaying the level vial even when the rule was folded. Thus the unopened rule could be used as one would use a simple spirit level.

This type of combination rule was invented by Fred Buck, of Saginaw, Michigan, and patented by him on December 18, 1914 (U.S. Patent No. 1,120,443, entitled "Level-Rule"), the patent being assigned at the time of issue to the Lufkin Rule Co., also of Saginaw.

These rules were made in two lengths: one foot or two foot, and with two styles of trim: either unbound or half bound (on the bottom edge). In addition to the four possible resulting configurations there was a fifth, produced by adding a tiny plumb vial on the face of the two foot, half bound rule.

STEPHENS #36 COMBINATION RULE

The Stephens patent combination rule is probably the most universal measuring tool ever made in America. In form, it was a simple one foot, two fold rule with a 4½ inch swinging blade on the inside of one leg, and a spirit level on the outside of the other. A small slot and clamping screw in the tip of the second leg allowed the blade to be clamped to this leg at any point, thus making it possible to lock the rule partially open at any angle between 0 and 45 degrees.

It could be used as a rule, protractor, plumb, level, square, bevel gauge, or inclinometer. The features which made the No. 036 so versatile and useful were not its construction details so much as its graduations. The blade was marked with

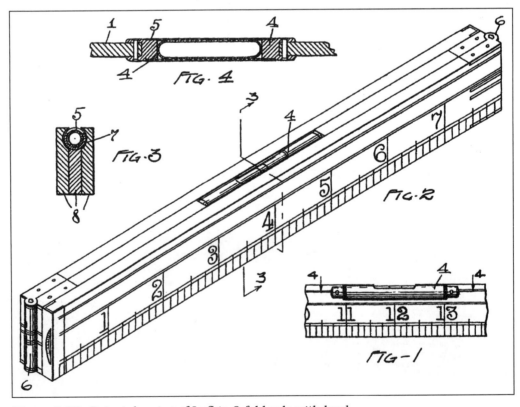

Figure 7-27: Patent drawing of Lufkin 3-fold rule with level

Figure 7-28: Stanley No. 036 combination rule

scales of both degrees and pitch to the ½ foot, to facilitate its use as an inclinometer. The body, in addition to the usual graduations of 10ths, 12ths, and 16ths of an inch, was also marked with a 24ths scale, two drafting scales, and a second degree scale.

By proper use of the various features and scales of the No. 036, it was possible to use the rule in any of the following ways:

As a 1 foot, 2 fold rule;

As a plumb or level;

As a try square or bevel gauge;

To measure slope in degrees or pitch to the foot;

To scale braces (as in framing buildings);

As a drafting rule (on scale drawings);

To lay out angles;

To measure the height of inaccessible objects.

This rule was invented by Lorenzo Stephens, of Pine Meadow, Connecticut, founder of the rule-making firm of L. C. Stephens & Co., and was patented by him on Jan. 12, 1858 (U.S. Patent No. 19,105, entitled "Carpenters' Rule"). It was offered in boxwood & brass (Stephens No. 36) and in ivory & German silver (Stephens No. 38) (A few unnumbered examples have been found made of ebony & German silver, probably custom or special order pieces). This rule was for many years the leader of the extensive Stephens line of rules, and was sold at a premium price.

In 1901, L. C. Stephens & Co., by then known as D. H. Stephens & Co., was acquired by the competing rule and plane making firm of H. Chapin's Son & Co. The resulting combination, renamed The Chapin-Stephens Company, continued to offer most of the Stephens line of rules, including the No. 36 (renumbered as the No. 036). It is not known whether the ivory and German silver No. 38 was still in production at this time; it was not offered in the 1914 Chapin-

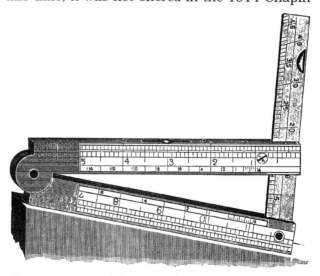

Figure 7-29: No. 36 as an inclinometer

Stephens catalog, but could have been discontinued only a few years before.

In 1927, Chapin-Stephens, by then severely pressed financially by the shrinking market for wood planes and traditional folding rules, was reorganized. Two years later, in 1929, it finally ceased production, selling all rights to its line of rules and all stock in process to the Stanley Works.

Stanley continued to offer the 036 combination rule, referring to the rule in its large dealers' catalogs as a "Miners' Combination Rule;" and in their smaller No. 34 catalogs simply as a "Combination Rule." The No. 036 was part of its product line for more than ten years, only being discontinued sometime after 1941. It is not known whether the No. 036 rules offered between 1929 and 1941 were made by Stanley, or were partially finished Chapin-Stephens stock marked with the Stanley trademark. The relative scarcity of examples marked Stanley, especially in view of their relatively recent date, would tend to support the latter as more likely.

Lufkin Nos. 863L/873L Two Foot, Four Fold Rules

These two rules were relatively simple combination rules, each consisting of a two foot, four fold rule with a protractor scale on the main joint and a level vial mounted on the inside of one leg. The only difference between the two was that the No. 863L was unbound, while the No. 873L was half bound. They were graduated in 8ths and 16ths of an inch, and marked with a set of four drafting scales for ratios of ¼ inch, ⅓ inch, ½ inch, and 1 inch to the foot.

In order to make the protractor scale more precise and accurate, it was not marked on the joint plate itself, but on a larger ($1^{5}/_{16}$" diameter) disk which was pinned to the joint plate. This larger graduated edge allowed the degree scale to be divided into five degree intervals.

In addition to simple measuring this rule could also be used to lay out angles, plumb and level surfaces, and measure pitch.

This design and combination of features was not original to Lufkin. Both Rabone and Preston in England had begun offering such rules before the turn of the century. Lufkin apparently recognized the value of the design and included this pattern in its product line when it began making boxwood rules in 1924.

Acme Two Foot, Two Fold Rule and T Square

This Acme rule is an unusual combination rule in that it was developed to serve both as a common rule and as a drafting T square and protractor. This flexibility was achieved by adding a single middle joint to one leg at a point four inches from the main joint. When this middle joint was open, the rule functioned normally (Figure 7-31, below); when the leg was folded over and the main joint half open the folded leg crossed the main joint, effectively forming a head for a drafting T square whose blade was the other leg. (Figure 7-32, opposite)

When the rule was folded to form a T square, a small angle scale near the 21 inch mark, divided in five degree increments, was visible near the main joint and allowed the "blade" to be set at any desired angle to the "head," thus extending the

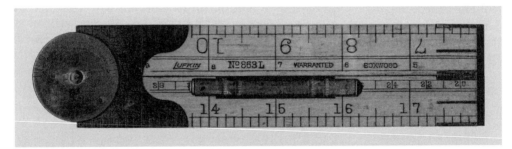

Figure 7-30: Lufkin No. 863L combination rule

Figure 7-31: Acme two fold rule unfolded to allow use as a rule

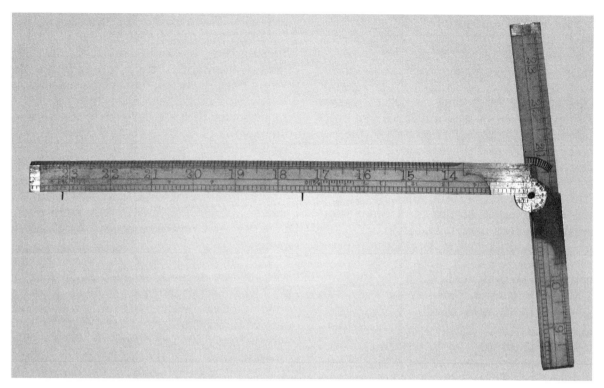

Figure 7-32: Acme two fold rule folded for use as a T square

rule's functionality to that of a protracting T square (albeit a primitive one).

This rule was invented by William H. Jones, of Montreal, Quebec, Canada, and was patented by him on December 21, 1886 (U.S. Patent No. 354,865, entitled "Pocket-Rule"), with three-fourths of the patent being assigned to George Alexander Gray, also of Quebec. Early examples are marked with the name of a Canadian maker, the Canada Rule & Level Co., of Montreal, but the patent must have been later sold to the the Acme Rule Co., of Salisbury, Connecticut, as most surviving examples are marked with that company's name. Acme also was the maker of ivory rules on which thin ivory surface strips were riveted onto wood sticks, to save the cost of the expensive ivory (see IVORY RULES later in this chapter).

HAY'S "IMPROVED PRACTICAL MECHANIC'S RULE"

This is an English combination rule which was offered by Edward Preston & Sons at about the turn of the century. It is a broad four fold rule with a double arch joint and middle plates, marked, in addition to the usual scales of 8ths and 16ths, with special scales for the use of carpenters and mechanics.

The special scales marked on this rule are: (on the outside of the rule) a pair of 12 inch radius lines labeled "R^D," a "SIDE OF EQUAL SQUARE" line parallel to the upper radius line, a line of chords, approximately 8½ inches long, labeled "C," and (on the inside of the rule) a "CIRCUM" (circumference) scale. These scales can be clearly seen in the Preston catalog illustration on the next page.

This rule was invented by William Hay, of Dumbarton, Scotland, and was patented by him on June 11, 1867 (U.S. Patent No. 65,744, entitled "Improvement in Scale-Rules"). Presumably there was also an English patent as well, but that has not yet been found. The American patent was assigned to Robert Hay, of Mineral Point, Wisconsin, but no American-made examples have surfaced.

Figure 7-33: Hay's "Improved Practical Mechanic's Rule"

All known examples are of obviously English manufacture, and the only catalog listing offering them is that of Edward Preston & Sons, of Birmingham, England, where they are described as "Two Foot and Three Foot Four Fold Boxwood Rules Marked With Mechanics and Circumference Scales." Although both two and three foot rules were offered, no examples are known of the two foot version, and it may be that only the three foot version was ever actually manufactured and sold. It is worth noting that Preston was not, apparently, the actual maker of these rules, but instead contracted their manufacture to J. Smallwood, another Birmingham maker, and then sold them as their own product, marked with the Preston name.

According to the patent specification, the function of this rule was to "not only measure surfaces in the ordinary way," but to:

Ascertain the circumference of a circle of any given radius;

Find the side of a square equal to the area of a given circle;

Find the radius of a circle equal in area to a given square;

Divide a circle into any desired number of equal parts;

Find the side of the greatest square that can be inscribed within a given circle, and also the circle that will circum scribe a given square;

"... thus rendering the rule capable of much more universal employment in different kinds of work than those heretofore in use."

The radius lines are similar to the "lines of equal parts" (linearly divided lines originating at the joint pivot point) used for computation on the sector (see THE SECTOR, page 223, and the article "William Cox on the Sector," reprinted in Chapter III), differing only in that they are subdivided into 8ths instead of the usual 10ths. These radius lines, labeled R^D, were not for computation in the sense that the lines on a sector were, but were rather for geometrically solving (In conjunction with the chord scale) problems involving the subdivision of circles into equal parts. Considering that this rule carried ordinary graduations in 8ths and 16ths of an inch, dividing the radius lines to correspond seems reasonable.

The SIDE OF EQUAL SQUARE line parallel to the upper sector line was for converting between the dimensions of squares and circles of equal area. In this conversion the value on the R^D line represents the radius of the circle; the adjacent value on the SIDE OF EQUAL SQUARE line is the side of a square of the same area. Thus, referring to these scales, a circle of 5 inch radius will be found to have the same area as a square with a side of 8.9 inches.

The line of chords, C, is for setting the rule open to any desired specified angle. This operation was performed with the aid of a pair of dividers. Examination of the R^D lines reveals a pair of brass "gauge points" (small brass plugs with a tiny punched dimple in their center, set into the surface of the wood flush with the surface), similar to those found on Gunter's scales, at the 6 inch point in each scale. The dividers would be set to the distance representing the desired angle, 0 to 90 degrees, on the C scale; then the main joint of the rule would be opened until the divider points just rested in the dimples in the two gauge points, at which point the angle of the two radius lines would be the desired angle.

The other special scale on Hay's rule was the circumference scale. This scale had inches that were .3183 (equal to $1/\pi$) times the size of common inches; thus this scale, which extended the full length of the rule, was graduated into 75 (on the two foot rules) or 113 (on the three foot rules) circumference inches. This scale, and the common inch scale marked next to it, allowed easy conversion from diameter to circumference when laying out shapes on flat stock which would have to fit a given circle after being rolled and fastened.

It is worth observing that this design was first developed for use on two foot, two fold rules, as shown in the patent drawing. This explains why the two radius lines and the LINE OF EQUAL SQUARES

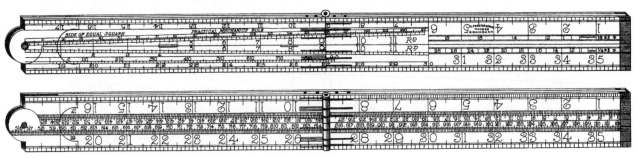

Figure 7-34: Illustration of Hay's rule in the Preston catalogue

are all only 12 inches long, and the line of chords only extends 9¼ inches down the rule on the three foot examples. Apparently when it was decided to put these scales on a three foot rule, they were not redesigned, but just put on as originally laid out, with some short drafting scales added to fill up the extra six inches.

The author wishes to acknowledge the contributions of Dr. Richard Knight, of Edgbaston, England in analyzing the scales on this rule and identifying their functions.

COOPERS' RULES

Coopers used a special hook rule to measure the width of the staves as they prepared them for making or repairing a cask. This rule, called a hook stave rule, was a nonfolding wood rule, six inches or so in length, with a brass hook on one end. This hook was large enough to allow the rule to measure the width of the stave on the outside when placed across the inside curve of the stave. Six inch examples have been noted by both Kerby and Lufkin, and longer, apparently special order, one foot ones by Lufkin.

CORDAGE RULES

Cordage rules are short nonfolding caliper rules used by riggers and seamen to measure the diameters of ropes and cables, and to provide them with essential information as to weight per foot, breaking strength, etc. These rules were extra wide, so as to have caliper jaws deep enough to span the large cordage with which they worked, and to provide space for the data tables containing the information. The most common patterns were large rules 6 inches long and 2½ inches wide, and small rules 4 inches long and 2 inches wide. The large rules could measure cordage diameter up to 4 inches; the small up to 3 inches.

Cordage rules were often custom ordered by rope or cable makers, such as the John H. Roebling & Sons or the Plymouth Cordage Company, for resale to their customers, and would be marked prominently with the ordering company's name as well as the maker's name.

The scales and tables marked on the body and caliper slide varied with the maker, and with the needs of the buyer ordering the rule. The slide was usually graduated to read diameter, but was sometimes also graduated to read equivalent circumference as well. The body usually bore a scale of inches and 16ths. The tables would be specific to one or two types of cordage (hemp, manila, sisal, steel wire, steel chain) and configuration (such as hawser or cable laid), and would give breaking strain and weight per unit length as a function of diameter.

These rules were fairly widely manufactured, with examples by Kerby, Stanley, and Belcher Brothers often turning up.

COUNTER MEASURES

A counter measure is a thin metal yardstick designed to be fastened to the surface of a store counter so as to facilitate the measurement of fabric and to prevent the yardstick from being mislaid. Graduated in fractions of a yard (see

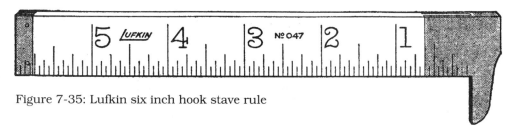

Figure 7-35: Lufkin six inch hook stave rule

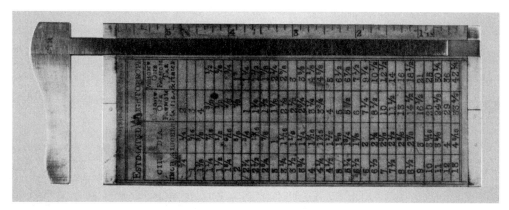

Figure 7-36: Stanley six inch cordage rule

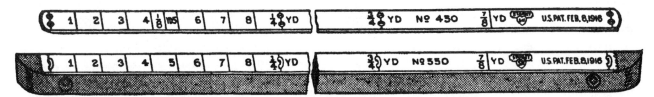

Figure 7-37: Stanley Nos. 450 & 550 counter measures

GRADUATIONS—FRACTIONS OF A YARD in the previous chapter), these measures were in common use in fabric and drapery shops prior to the invention of the rotary fabric gauge in the 1920s.

Three major American makers are known to have offered counter measures: Eagle Square, Stanley, and Lufkin. Major British makers offered these items as well; Rabone, Preston, and Mathieson all listed brass counter measures in their catalogs. The first American maker to offer a counter measure was the Eagle Square Company, of Shaftsbury, Vermont. This company, primarily a maker of framing squares, began offering counter measures some time in the 19th century.

Stanley acquired Eagle Square in 1916, and continued to offer its counter measure under the Eagle Square name for several years. Finally, in 1925 they redesigned it slightly, and introduced it in two varieties, the Nos. 450 and 550. The two measures were of almost identical design, both being nickel-plated steel, the major difference between them being that the No. 450 was a plain measure, designed to be screwed or stapled onto the surface of the counter, while the No. 550 was supplied stapled to a square stick for mounting on the edge of the counter.

Lufkin offered two different counter measures, the Nos. 68 and 68½, one graduated with fractions of a yard and inches/8ths, the other with fractions of a yard only, beginning some time before 1888. Both were made of brass, and were available either plain or nickel-plated.

Stanley discontinued production of its counter measures after 1932. Lufkin continued to offer its counter measures until about 1941.

DESK OR SCHOOL RULES

In the nineteenth century, as today, almost every desk, whether at home, in an office, or in a school, would have one or more short, relatively thin, nonfolding rulers somewhere in or on it, for the use of the occupant. These "desk" or "school" rules were typically about ¾ inch to 1 inch wide, 12 to 15 inches long (to allow carrying in a notebook or satchel, if necessary), and usually beveled on one edge. They were usually graduated in 8ths and/or 16ths of an inch, occasionally in 10ths of an inch, and in a few cases in centimeters and millimeters.

Over the years such rules were produced in a wide selection of woods, graduations, and construction details by many of the major American

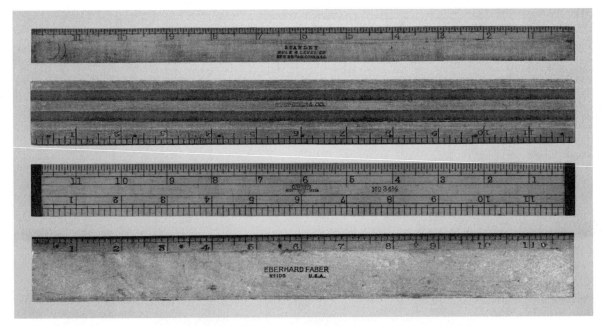

Figure 7-38: Four common 12 inch wood desk/school rules. Top to Bottom: Stanley No. 98, Stephens patent, Stanley No. 34½, Eberhard-Faber.

SPECIAL RULE TYPES AND USES 189

Figure 7-39: Starrett No. 367 nine inch steel desk rule with lifting knob

makers, including Stanley, Chapin, Stephens, and Belcher Brothers, to name a few. Metal ones were offered as well, particularly from the makers of machinist's tools, such as Brown & Sharpe, Starrett, and Lufkin. Most commonly they were beveled on only one edge (two edges if they had two different graduations). They were not equipped with the metal tear edge so common on desk rules today; this feature is a 20th century invention which only came into use after the major rulemakers had largely passed from the scene.

DRAFTSMEN'S RULES

In the interests of practicality architects, surveyors, and (to a large extent) engineers ordinarily prepare their working drawings to a different size (usually reduced, but sometimes enlarged) from that of the object depicted. The ratio of the dimensions of the object to the dimensions of the drawing is called the "scale" or "reduction" factor. Architects and surveyors usually expressed this ratio in terms of how many inches in the drawing represent one foot on the object (e.g.: 1/4 inch equals 1 foot, 1 1/2 inches equals 1 foot). Engineers more commonly expressed it as a simple fraction (e.g.: 1/4 scale, 1/2 scale).

To facilitate the preparation of these drawings and the taking of dimensions from them, the draftsman would employ a rule graduated to match the reduction factor used. These rules would be used by the draftsman at his drawing board to guide the pen while drawing straight lines and for measuring distances on the paper. They were made in a variety of shapes, sizes and materials, to suit the needs and preferences of the user, and graduated with scales appropriate to the type of drawings being prepared. Usually the rule itself was perhaps 1/2 inch longer than the scales, which thus began and ended short of the ends of the rule.

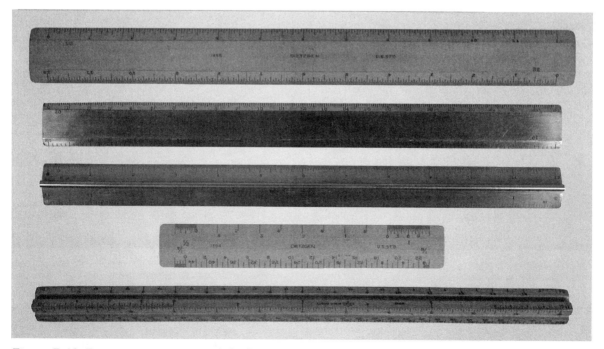

Figure 7-40: Representative assorted draftsmen's rules

The most common rule pattern is that most familiar to all modern students of engineering drawing. Typically triangular in cross-section, and 3/4 inch wide on each face, this rule would thus have six edges, each graduated with a different scale. Usually each face would have a semicircular groove down its face between the two scales to facilitate picking up and holding. The most common length was 12 inches, but shorter (3 inches and 6 inches) and longer (up to 18 inches) examples are not uncommon.

The second most common pattern was the flat rule with beveled edges. This was made in a number of variations: the two edges both beveled on the same side (two scales), the two edges beveled on opposite sides (also two scales), and both edges beveled on both sides (four scales). This pattern was also available in a wide range of lengths, from 3 inches to 18 inches, with 12 inches the most common.

Both patterns, flat and triangular, were made in a number of different materials. The most common was wood, sometimes with a celluloid or "ivorene" surface layer to show the graduations better. Other materials sometimes, but less frequently, used were metal (usually steel, German silver or nickel-plated brass), and (rarely) ivory.

Flat rules are sometimes encountered in the form of boxed sets, the box containing from 4 to as many as 12 rules, each with a different graduation. The most elaborate of these boxed sets will contain two rules for each scale factor, a full-length rule and a shorter, typically 2 inch, rule. These short rules, called "offsets," differed from their full-length counterparts in that the rule and the scales were exactly the same length, with no blank area beyond the scales at either end.

A third pattern of drafting rule frequently encountered is the flat 4 inch to 6 inch rules usually included in the oval "pocket case of instruments" which were popular almost until the end of the 19th century. These pocket cases would contain a small assortment of drafting instruments, a sector (for calculation), a linkage-type parallel rule, and a pair of 1½ inch to 2 inch wide scales for measurement and drawing straight lines. One of these scales would be graduated with a scale of inches, diagonally divided at one end into 100ths of an inch and at the other into 100ths of half an inch. A 180 degree protractor scale, centered in the middle of the bottom edge, was marked up the left end edge, across the top edge, and down the right end edge. The other rule was marked with as many as a dozen different open-divided architects' scales, covering its entire surface, and would have its ends shaped so as to form templates for drawing brackets of several different sizes. Since all of the scales on both rules were located away from the edges, distances were taken and transferred to or from the paper with dividers.

The scales put on draftsmen's rules would depend on their intended use. Rules for surveyors and civil engineers would be graduated with full-divided scales of chains and links for various scale factors (e.g. 1:1,000, 1:2,500). Rules for engineers would be graduated with full-divided inches/subdivisions. Rules for use by architects would be graduated with open-divided scales of feet and inches for various scale factors (¼ inch per foot, ⅜ inch per foot, etc.).

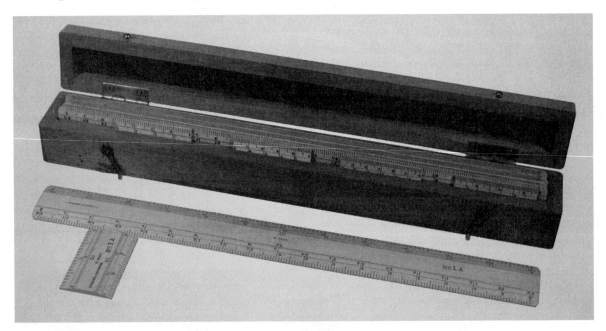

Figure 7-41: Boxed set of five drafting rules with offsets

SPECIAL RULE TYPES AND USES 191

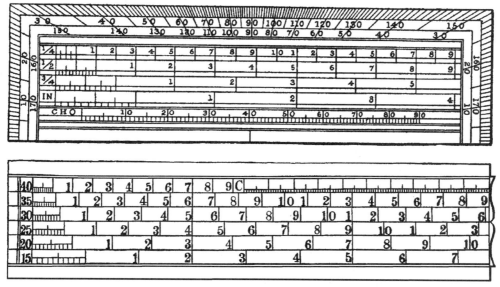

Figure 7-42: Six inch rules from a pocket case of instruments

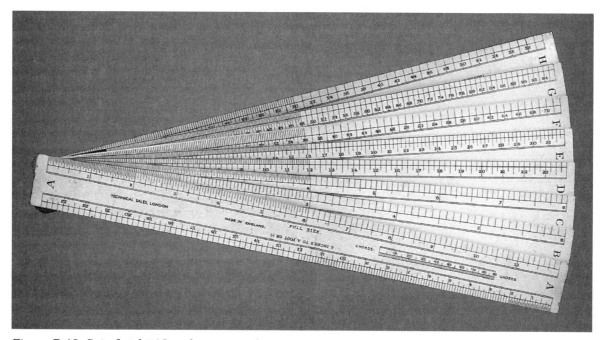

Figure 7-43: Set of eight 12 inch paper scales

Occasionally paper scales for use by draftsmen will be found. These were typically 1¼ inches to 1½ inches wide and from 12 inches to 20 inches long, and were printed on card stock or bristol board. The theory behind the use of paper scales is that they would expand and contract with the humidity to the same degree that the drawing paper would, thus automatically compensating for changes in the weather. Paper scales were usually sold in sets, in a paper or cardboard slip case.

The most common makers of drafting rules were the scientific instrument makers and the makers of instruments and supplies for draftsmen. The most frequently encountered names in the United States are those of Keuffel & Esser, Eugene Deitzgen, and Theodore Alteneder & Sons. In England the best-known makers were W. F. Stanley, Elliott, and Harling.

Folding Architects' Rules

A number of American rulemakers made folding rules which embodied, in a limited way, some of the features of an architects' drafting rule. These two foot, four fold rules had their inside edges (the edges brought together by closing the rule joint) beveled on the inside face to bring the scales on those edges closer to the surface being

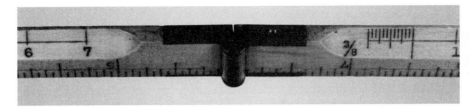

Figure 7-44: Beveled edges on an architects' four fold rule

measured when the rule was lying flat. This was only done for the central four inches of each leg; the last inch at each end being left unbeveled in order to allow the joints and tips to be fitted as on a conventional rule.

The scales marked on these four beveled edges were drafting scales (1/8, 1/4, 3/8, and 1/2 inch per foot) representing scale distances of 29' 12", 14' 12", 9' 12", and 7' 12" respectively. Bringing these scales closer to the surface being measured greatly improved the accuracy when either laying out work or taking off dimensions. Typical of this type of rule are the Stanley No. 53½, the Stephens No. 39, and the Lufkin No. 861A. Stanley, Chapin, and Chapin-Stephens also offered this pattern of rule in ivory.

ENGINEERS' (SLIDING) RULES

Unquestionably the most elaborately marked and graduated rules likely to be encountered by the collector today are the engineers' rules manufactured by a number of American and English makers during the 19th century.

These rules were made in a number of different patterns according to the views of the several inventors as to the needs of different trades and professions. Thus the collector will encounter Routledge's engineers' rule (the most common), Slater's cotton spinners' rule, Carrett's engineers' rule, and Wilkinson's spinners' version of Routledge's rule, to name the best known types.

All were mechanically identical, being two foot, two fold (occasionally four fold) rules with a logarithmic slide, differing only in the data tables, and (in the case of Slater's rule) comments marked on their surface.

These rules were an attempt to provide the user with not only the means for performing rapid calculations (the Gunter's slide), but also much of the physical data required to work out the common engineering problems of the day (the data tables), and some guidance as to how to perform particular classes of calculations (the comments). With this rule, a user could perform volume conversions, weight calculations, geometric analysis, steam engine and pump computations, and spinning machinery calculations.

Most patterns of engineers' rules were produced only in England by makers such as Rabone and Preston; only Routledge's rule was made by American makers as well. Examples and/or catalog listings exist for Routledge's rules by Stanley, Stearns, Chapin, Stephens, Chapin-Stephens, Belcher Brothers, Standard Rule, and Upson Nut.

Collectors wishing to find out more about engineers' rules are referred to the articles by Kenneth Roberts and Philip Stanley reproduced in Chapter III, and to the article by Knott (Ref. 2).

EXTENSION STICKS

A carpenters' extension stick is actually a pair of sticks which slide outward along one another and are used to measure or gauge surface-to-surface inside dimensions.

In the period before arithmetic skills were widespread, such sticks would not have been graduated. The user would have simply captured the dimension by scribing a line on one stick

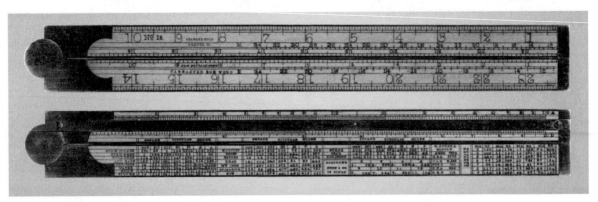

Figure 7-45: Routledge's engineers' rule

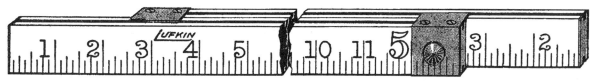

Figure 7-46: Lufkin 5 foot to 10 foot extension stick

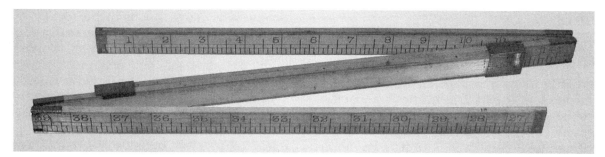

Figure 7-47: Kerby 2 foot to 4 foot folding extension stick

where the other one overlapped; later, back at the shop, by aligning the sticks to that mark the dimension could be recaptured and used.

With the spread of elementary education these sticks became more sophisticated. They were graduated (or made from a pair of yardsticks, etc.), so that they could be used to measure as well as gauge. Clips were added to keep them aligned when sliding, and a clamping mechanism held a setting once it had been found.

This is the form in which rulemakers began making extension sticks (or extension "rules," as they were sometimes called) in the late 19th century. Typically, these commercially made extension sticks would consist of a pair of maple sticks about 1 inch wide and 1/2 inch to 3/4 inch thick, with brass tips. The sticks were held together by a pair of brass brackets, one of which was equipped with a clamping thumbscrew, and were prevented from sliding completely apart by a round head screw driven into the tip of the back section.

A number of different configurations were available. The most common arrangement was to have two sliding sticks, with the length extended about twice the length collapsed. The usual lengths were 2 to 4 foot (that is 2 feet long when closed, and 4 feet long when fully extended), 3 to 6 foot, 4 to 8 foot, 5 to 10 foot, and 6 to 12 foot. One maker, Lufkin, also offered an extension stick with three sliding sections, which was 2 feet long when collapsed, but opened out to 6 feet when fully extended.

These sticks were graduated in 8ths of an inch, in such a way as to be direct reading. The front section was graduated one way, beginning at zero; the back section was graduated the other way, beginning at the collapsed length of the stick (on the 2 to 4 foot stick, for instance, the scale on the back section began at 2 feet). Thus as the stick was extended, the graduation on the back section directly opposite the end of the front section represented the overall length of the pair at that instant. These sticks were not graduated on the back.

Many of the major rule makers offered extension sticks as stock items. Catalog listings and/or examples have been found for Belcher Bros., Chapin-Stephens, Kerby, Keuffel & Esser, Lufkin, Stanley, and J. Watts, with the largest makers being Stanley and Lufkin. Two of these makers, Kerby and Chapin-Stephens, offered versions of their extension sticks made so they could fold up (when fully collapsed) to half their minimum length (i.e.: a 2 foot to 4 foot extension stick that would fold down to one foot). This was a very useful feature, allowing the stick to be easily stored in a carpenter's tool box.

One unusual extension stick that is sometimes encountered is the 18 inch to 34 inch extension stick patented by C. M. Mumford, of Springfield, Massachusetts, on April 9, 1889 (U.S. Patent No.

Figure 7-48: Mumford's patent extension stick

401,292, entitled "Extensible Measuring Stick"). In this rule, a metal rod slides out from a hole in the end of an 18 inch wooden stick. A longitudinal slot in the surface of the stick connects to the hole, allowing a moving pointer connected to the rod to move along a scale on the surface of the stick, thus indicating the total length of the stick/rod pair.

FOREIGN MEASURES OR COMPARISON RULES

A major problem in commerce during the 18th and for much of the 19th centuries was lack of uniform standards for linear measurement in Europe and elsewhere. Each political entity, of which there were hundreds, had its own definition of what constituted an inch, a foot, a cubit, a yard, an ell, etc. Within each political unit's own local area this presented no difficulty, but their use beyond the border in any kind of commerce created major problems. A merchant dealing in foreign goods in Europe had to be very aware of the existence and nature of the system of weights and measures used by his sources and customers. If he bought goods by the foot in England and sold them by the foot in France, for example, he would lose six percent on every transaction if he did not allow for the French foot (*pied du roi*) being almost ¾ inch longer than the English foot.

Nor did the introduction of the metric system beginning in 1815 instantly solve this problem. The adoption of the metric system in Europe was a gradual process, taking more than 100 years, and was slowed down every step of the way by the

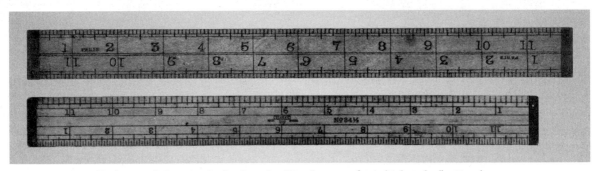

Figure 7-49: A Kerby *pied du roi* rule (top) and a Stanley one foot desk rule (bottom)

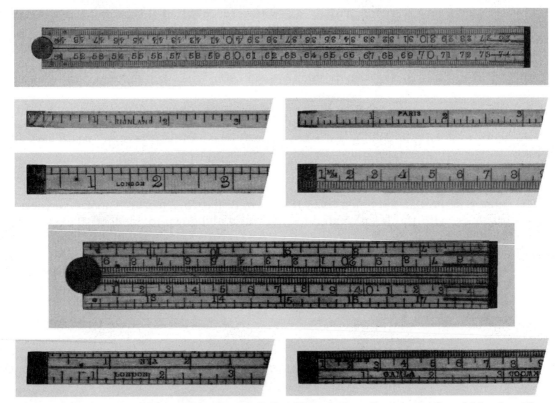

Figure 7-50: Two comparison rules by J & D Smallwood. Top: One meter, four fold rule with English, Paris (*pied du roi*), Rhineland, and metric graduations. Bottom: Two foot, four fold rule with English, metric, and Swedish (both "*nya*" and "*gamla*") graduations.

unwillingness of people to give up the familiar, no matter how awkward, and the innate resistance to change of many trades and crafts. ("What was good enough for my farther,. . . ," etc.)

This problem was dealt with in two ways. One was by the publication of encyclopedias of foreign weights and measures for the use of those involved in intercity or international trading. (see the references in the next chapter for examples) These were useful in an office environment, but could not be readily taken along into the field. Therefore, a more portable method of conversion had to be found.

The second solution to this problem was the development of "Comparison" or "Commerce" rules bearing two or more scales with differing graduations, which permitted direct on-the-spot comparison or conversion. These rules became increasingly common during the 18th century, and were offered by most major European rule-makers until nearly the end of the 19th century.

The most common non-English measures which are found on these multiple-graduated rules are the Spanish "Burgos" foot (*pie*) and inch (*pulgada*), the Rhineland foot (*fuss*) and inch (*zoll*), the French foot (*pied du roi*) and inch (*pouce*), the Swedish foot (*fod*) and inch (*tum*) (Sweden had two different systems of subdividing their *fod*: "*gamla*," where it was divided into 12 *tumme*, and "*nya*," where it was divided into 10 *tumme*), and the Russian *sajen* and *werschok*.

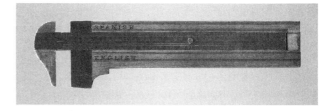

Figure 7-51: Buck & Hickman four inch caliper rule with English and Spanish ("Burgos") graduations

Some American makers also offered comparison rules, usually combining Spanish with English graduations. Examples exist of post-1921 Stanley zigzag rules graduated in English, Spanish, and Russian units.

A more extensive discussion of pre-metric linear measures in Europe, together with a complete table of these measures, giving their names,

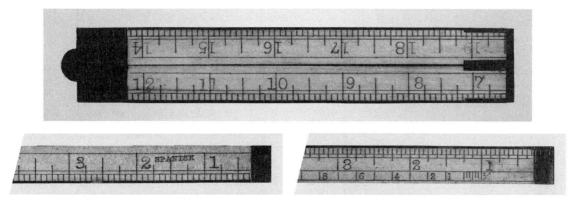

Figure 7-52: Chapin-Stephens rule with English and Spanish graduations

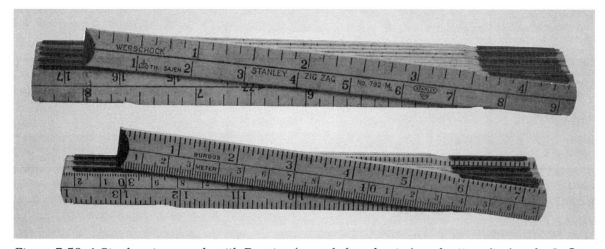

Figure 7-53: A Stanley zig-zag rule with Russian (*werschok* and *sajen*) graduations (top) and a Lufkin zig-zag rule with Spanish ("Burgos") and metric graduations (bottom). Both rules are graduated in inches/16ths on the back.

regions of use, and English and metric equivalents, is provided in Chapter VIII. A rule with unknown scales can usually be identified as to its intended locale of use by comparing it to the values in this table.

FREIGHT RULES OR FORWARDING STICKS

A freight rule is a long (four to six foot) nonfolding measuring stick used by freight agents and those responsible for forwarding (shipping) cargo to take the dimensions of crates, barrels, etc. for packing and billing purposes. Also referred to as forwarding sticks or cargo rules, they were usually graduated in feet and inches, and were equipped with a projecting hook at one end to make it easier to align the end of the rule with the far side of a crate or barrel when measuring. The example shown below by Kerby is also marked TIDEWATER S & W Co., presumably the name of the company for whose agents it had been acquired.

At least one maker (Belcher Brothers) is known to have offered a freight rule with a fixed and a sliding caliper jaw, to facilitate the measuring process. In addition to Belcher Brothers and Kerby, Stanley is also known to have offered Freight Rules.

GAUGING AND WANTAGE RODS

THE GAUGING ROD

The gauging rod and the wantage rod are the tools of the gauger, used to measure the capacity and contents of wooden casks. The gauging rod is used to estimate the capacity of the cask, the wantage rod to determine the number of gallons of liquid needed (wanted) to fill the cask.

Casks were made in various nominal sizes, from the half barrel of 16 gallons to the pipe of 120 gallons. The exact capacity of any cask, however, could vary slightly from its nominal value. Cooperage was hand made, (and still is, largely, even today) and any two containers of the same supposed capacity and shape, even when made by the same individual, could differ slightly in their proportions, and have different capacities. It was, therefore, necessary to measure each cask separately to determine its exact capacity.

The gauging rod was a simplified method of estimating this capacity without the elaborate measurements and calculations necessary for exact volume determination. This rod measured a single dimension of a cask, a dimension which was a function of head diameter, maximum diameter, and length, and then converted that dimension, by means of its graduations, into a capacity in gallons. The dimension used for this purpose was the distance from the bung hole, in the middle of the side of the barrel, to the opposite chine (the point at which the staves contact the head). This dimension was used for two reasons: it was capable of easy measurement with a rod inserted through the bung hole, and it represented a good working average of the length, head diameter, and bung diameter of the cask, thus compensating to a large extent for any manufacturing variation in the cask shape or dimensions.

The rod itself was three or four feet long, and 1/2 inch to 5/8 inch square, with one end tapered to a brass-plated wedge shape. It was calibrated from left to right, beginning at the tip, with two scales: the gauging scale and a scale of common inches. The three foot rod had a gauging scale for casks up to 120 gallons; the four foot rod had a

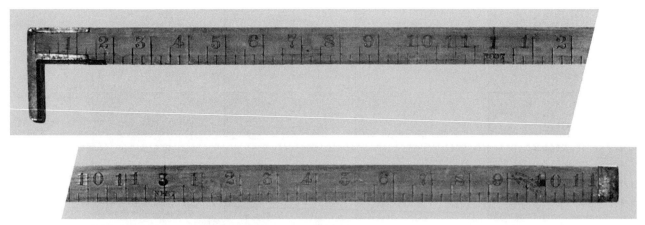

Figure 7-54: Kerby four foot forwarding stick with end hook

Figure 7-55: Stanley No. 45 gauging rod

gauging scale for casks up to 180 gallons. These gauging scales were nonlinear, and read directly in gallons, the graduations being scaled and marked in such a way as to include the conversion from cubic inches to gallons.

The method of using the gauging rod is clearly described in the following passage and illustration from a catalog of coopers' and gaugers' supplies published circa 1922:

"Placing the barrel, for instance, on its side, so as the bung hole will be at the top, insert the sharpened end of the gauge-rod in a slanting direction through the bung-hole until it reaches the place where the head fits into the stave directly opposite the one in which the bung-hole is, pushing it in as far as it will go, keeping the side on which the number of gallons is marked uppermost. Now notice exactly where the rod strikes the under side of the middle of the bung-hole, putting your finger on the place if you cannot see it before drawing out the rod. If the bung-hole was exactly in the middle of the barrel every time, that measure would be enough, but as it is not always so, it is necessary to reverse your rod and measure in the same way from the middle of the bung hole to the other end of the barrel and take the average. For instance, you measure one way and it shows 49 gallons; you measure the other way and it shows 47 gallons; then you take the average, which is 48 gallons, which is the capacity of the barrel. Be careful not to confound the inch scale on the rod with the gallon scale."

THE WANTAGE ROD

Once the capacity of a cask was known, the wantage rod was then used to measure its contents. As in the case of the gauging rod, the wantage rod depended for its accuracy upon the fact that casks were usually made with fixed proportions, and could thus have their contents found by a single linear measurement.

Commonly, wantage rods were 16 or more inches long, 5/8 of an inch square, and had a narrow brass lip or plate representing the zero point projecting slightly from one side. They were marked with either eight or twelve scales, each scale graduated for use with a particular nominal capacity cask.

The wantage rod was used as follows:

"With the barrel on its side, bung hole uppermost, insert the wantage rod vertically down through the hole until the brass plate is even with the underside of the stave. Then withdraw it, and note where the liquid mark intersects the scale on the rod appropriate to the cask being measured. The graduation corresponding to that mark indicates the number of gallons of liquid wanting to fill the cask."

Many rule makers sold gauging and wantage rods. Examples by Stanley, Belcher Bros., Kerby, and Lufkin are frequently encountered, with rods made by the smaller, lesser-known makers showing up occasionally.

COMBINATION INSTRUMENTS

Three interesting variations of these instruments, intended to minimize the number of tools needed by the gauger, are the gauging rod with wantage tables, the combined gauging and wantage rod, and the Prime & McKean combination gauging rod.

Gauging Rod With Wantage Tables

The Stanley No. 45 1/2 is a good example of the first variation. Four feet long, it was marked not only with the gauging and common inch scales, but with four wantage tables, for 16 (small barrel), 32 (full barrel), 64 (tierce), and 108 (hogshead)

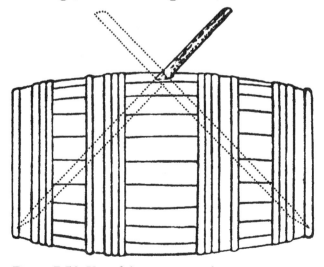

Figure 7-56: Use of the gauging rod

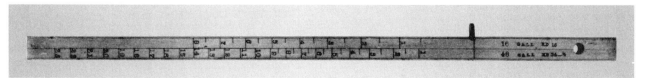

Figure 7-57: Stanley No. 44 wantage rod

gallon casks, equating the fluid level in each size of cask with the number of gallons it would take to fill it. The scale of common inches on the rod would be used to measure down to the liquid level, and then the table would be consulted to determine the wantage.

Combined Wantage and Gauging Rod

The second variation, the combined gauging and wantage rod was four feet long, and was configured as a gauging rod on one end, and as a wantage rod, complete with protruding brass lip, on the other. This rod also had the unusual feature of having two types of scales for determining how many gallons were in a cask. In addition to wantage scales, there were also content scales for indicating how many gallons remained in the cask, based on the distance from the bottom of the cask up to the liquid level. To distinguish the wantage scales from these content scales, the wantage scales were labeled "OUT," and the content scales were labeled "IN."

These combination rods were made by Kerby & Brothers, of New York City, and are usually marked with the name of the Hydraulic Press Manufacturing Company, of Mount Gilead, Ohio, a large maker of equipment and presses for cider mills.

Prime & McKean Combination Gauging Rod

The Prime & McKean combination gauging rod was the tool required by the IRS to be used by gaugers in determining the contents of barrels for the purposes of taxation from 1875 to 1895. It was an elaborate instrument, pointed at one end for use as a gauging rod, and equipped with caliper arms for measuring the length of a barrel and a short sliding fixture for measuring the head diameter. It was marked along its entire length with elaborate tables for determining the barrel capacity and contents.

Three versions are known to exist, a 44 inch model and a 72 inch model, both entirely made of metal, and at least one example is known with a metal rod and wood caliper jaws. The Federal Manual For Gaugers mentions these two types and also a third with both the rod and the caliper jaws made entirely of wood.

This instrument is the result of combining the features of two gauging rod patents. The first is that of Eli S. Prime of Baltimore, Maryland, granted on March 29, 1870 (U.S. Patent No. 101,309. entitled "Improvement in Gauge-Rod"); the second that of Edwin R. McKean of Nashville, Tennessee, granted on July 12, 1870 (U.S. Patent No. 105,352, entitled "Improvement in Cask-Gauging Instrument") By combining features from both patents the inventors were able to develop a single instrument which replaced five different gaugers' tools and met all of the requirement of a Federal gauger.

The Prime & McKean gauging instrument was made by the Stanley Rule & Level Co., and was

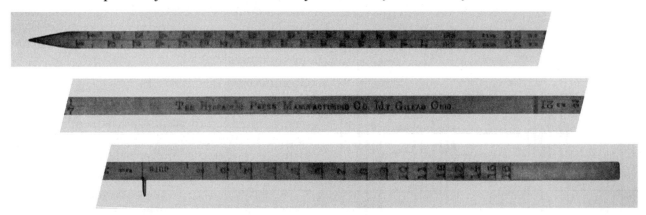

Figure 7-58: Kerby combined gauging and wantage rod

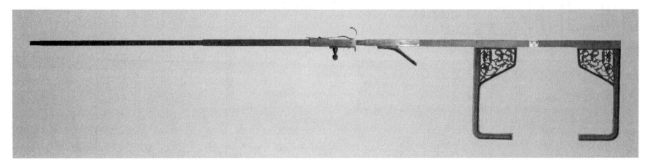

Figure 7-59: Prime & McKean combination gauging rod

marked with that company's name. A few unmarked examples have been observed, but the construction details correspond so closely to marked examples that there is no doubt that these are also of Stanley production. It was a special product, and was never listed in any Stanley catalog.

For more information on the history and use of the Prime & McKean gauging instrument, see the excellent article "The Stanley Cask Gauging Rod," by Gil & Mary Gandenberger (Ref. 3).

GEAR RULES

Designing the wooden patterns for gears, and then cutting them after they are cast, posed significant problems for designers, patternmakers, and machinists. Allowance had to be made for the difference between the tip (outside) diameter and the pitch (rolling) diameter of the gear, taking into account the number of teeth of the gear and the diametral pitch (the number of teeth per inch). The calculations involved were significant, and errors could cost time and money. A gear calculating rule is a measuring instrument for use by those involved in the gearmaking process to simplify these calculations and to eliminate the possibility of error.

Two versions of gear rules are known: Long's gear calculating rule, manufactured by the Stanley Rule & Level Co., and the steel gear rules offered by the Brown & Sharpe Company.

Long's Gear Calculating Rule

This rule was invented by Charles B. Long, of Worcester, Massachusetts, and patented by him on April 25, 1865 (U.S. Patent No. 47,436, entitled "Gear Cutting Rule"). It was intended to provide most of the data needed by designers, patternmakers, and machinists to plan gear systems, lay out the teeth on gear patterns, and cut the finished teeth in gear blanks. Long's rule was a two foot, two fold rule, similar in construction to the common patternmakers' folding shrinkage rule (that is, with a knuckle joint instead of the more common rule joint), and was brass bound. Its four edges were marked with scales of 10ths, 12ths, 14ths, and 16ths of an inch, and the intervening surface with the special scales for different gear tooth spacings.

There were eight of these scales, corresponding to diametral pitches of 5, 6, 7, 8, 10, 12, 14, and 16 teeth per inch of pitch diameter. Each scale was identified at its origin with a fraction indicating its addendum (a technical term for the distance from the pitch circle of a gear to the tips of the teeth, equal to one divided by the diametral pitch), and was marked along its length with figures showing the number of teeth of that pitch which could be cut in a gear of the corresponding outside (tip to tip) diameter.

Since these scales included the allowance for pitch diameter/addendum, they could be used directly, without the calculations usually accompanying gear design. A patternmaker, for example, preparing a pattern for casting a 26 tooth gear with a diametral pitch of 6, could lay his try square across the rule at the figure 26 on the 1/6 scale, and find the required diameter of the blank pattern, $4 \frac{2}{3}$ inches, on the nearest inch scale (in this case, the 12ths scale).

"Full directions for the practical use of the Tables ..." were provided with each rule. These constituted five pages of text and diagrams, including a page of testimonials by satisfied users.

These rules were at first made (from about 1865) in the Stanley factory (formerly the factory

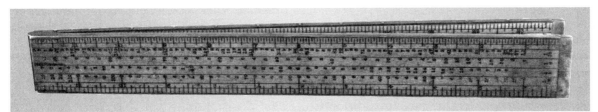

Figure 7-60: Long's gear calculating rule

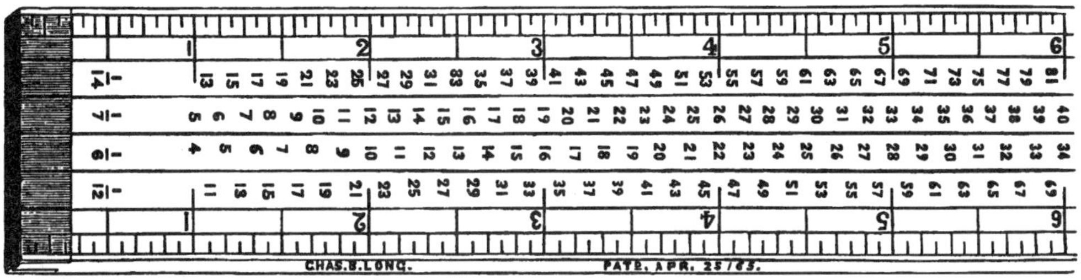

Figure 7-61: Tables on the gear calculating rule

of E. A. Stearns & Co.) in Brattleboro, Vermont, and were marked E. A. STEARNS & CO. This manufacturing site was selected because this rule was similar in construction to the two fold shrinkage rules which were part of the Stearns product line. The Brattleboro factory was closed and the production of the two fold shrinkage rule, the gear rule, and a few others transferred to the Stanley factory in New Britain in 1879, and the marking changed at that time to THE STANLEY RULE & LEVEL CO. It is not known exactly when this rule was discontinued.

BROWN & SHARPE STEEL GEAR RULES

These were one foot steel rules, 1¼ inches wide, marked with scales similar to those of the Long rule. Two versions were offered, the No. 61, and the No. 78, differing slightly in the way their scales were organized and marked. The No. 61 was marked with eight scales, four on each side, which ran the length of the rule. The No. 78 was graduated in inches for its entire length on both sides, and was marked in the first and last inch on each side with 20 scales (five at each end on each side). These rules were slightly more difficult to use than the Long rule, requiring the user to manually add in the compensation for the difference between the root diameter and the tip diameter, a step that the Long pattern did automatically.

These steel gear rules were manufactured for many years by Darling, Brown & Sharpe, beginning in 1868, and later the Brown & Sharpe Manufacturing Co. The early examples will be marked: J. R. BROWN & SHARPE; those made later BROWN & SHARPE MFG CO. or just BROWN & SHARPE. In 1913, as part of an extensive renumbering of the entire Brown & Sharpe rule line, these two rules were given new product numbers; the No. 61 became the No. 377, Style 1, and the No. 78 became the No. 377, Style 2.

No. 5 Graduations

At various times Brown & Sharpe offered their steel rules in 14 different patterns of graduations, to suit the needs and preferences of their customers. Their catalogue noted that, in addition to their Nos. 61 and 78 gear rules, rules with No. 5 graduations could also be used to size gears.

The No. 5 graduations were only offered on certain 12" or 24" rules, and consisted in dividing each inch on each edge differently, as follows:

No. 5 graduations:
1st corner   16, 32, 64
2nd corner   11, 14, 15, 17, 18, 19, 20, 21, 22, 23, 24, 25
3rd corner   26, 27, 28, 29, 30, 31, 33, 34, 35, 36, 37, 38
4th corner   39, 40, 41, 42, 43, 44, 45, 46, 47, 48, 49, 50, 100

By selection of the inch containing the appropriate subdivisions, and the use of dividers and some simple rules, the machinist could use these scales to calculate tip and root diameter for any desired number of teeth and teeth per inch.

GLAZIERS' RULES

A glass-cutter needed a special configuration of rule to measure and cut glass for windows from the large sheets supplied by the makers. The rule had to be long enough to measure from one edge to the other for the largest sheet expected to be cut (up to 10 to 12 feet for store display windows), and wide and thick enough so it could guide the glazier's diamond without flexing as he scored the glass prior to breaking it.

These rules could be plain, with simple tips like a carpenters' folding rule, or be equipped with a shallow hook on one end (on the flat side) to aid in measuring across wide sheets. Usually made of maple, they could be anywhere from 3 to 12 feet long, 2 to 3 inches wide, and up to ⅜ of an inch thick.

These rules were also made in the form of squares, to aid in cutting the glass at exact right angles to the edge. One such configuration resembled a drafting T square with a 1 to 1½ foot head; another resembled a diagonally-braced carpenters' try square with a 2 to 2½ foot stock. In both configurations, the face of the head/stock would be plated with brass to reduce wear from the hard, sharp edge of the glass.

The largest maker of glaziers' rules and squares was the Lufkin Rule Company, which offered these rules in all of the configurations described above from the 1920s to the 1960s. Other makers were Kerby & Brothers and Belcher Brothers.

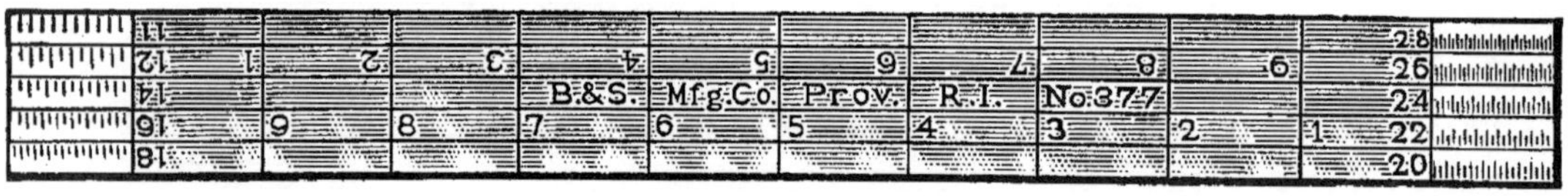

Figure 7-62: Brown & Sharpe No. 377 (style 2) gear rule

SPECIAL RULE TYPES AND USES 201

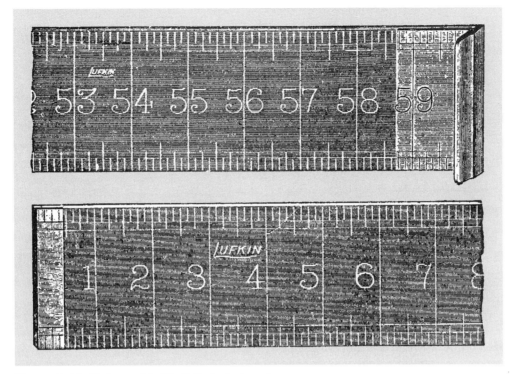

Figure 7-63: Plain and "with lip" glazier's rules

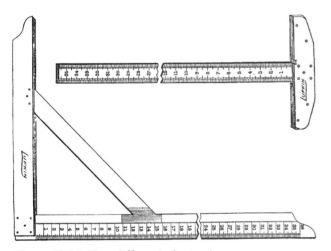

Figure 7-64: Two different glaziers' squares

GUNTER'S SCALES

The Gunter's scale is a large rule marked with elaborate scales which could be used in conjunction with a pair of dividers and a chart or map to solve complex problems in trigonometry and spherical geometry. Invented circa 1625 by Edmond Gunter of England, this tool was used up until the end of the 19th century by mathematicians and surveyors, and, most extensively, by navigators (seamen referred to it colloquially as a "Gunter").

In its most common form the Gunter's scale was a flat rule, typically 1½ to 2 inches wide, ¼ to ⅜ inch thick, and either 1 or (more commonly) 2 feet long. One edge was usually beveled, and the more elaborate specimens would have the tips and one or both edges brass bound. These rules were graduated to a standard pattern.

On the top edge of the front surface was a scale of inches/10ths, graduated right to left. The bottom edge was beveled and not marked with any scale.

The front surface was marked with the same scales which appeared on the front and back surfaces of the simpler plain scale (see PLAIN SCALES later in this chapter) as follows: on the left half of the front surface was marked an 11 inch scale divided into half inches, with the left half inch diagonally divided into 100ths of a half inch and the two right hand half inches diagonally divided into 100ths of an inch. The half inches in between were not subdivided. On the right half of the front surface were marked the so-called "plain" or "natural" lines of numbers: scales representing various functions from trigonometry and spherical geometry. Four of these started near the right hand end of the scale and extended left. These were: a Line of Numbers/Line of Equal Parts, labeled LEA; a Line of Rhumbs, labeled RUM; the Meridian Line (from Mercator's chart); labeled M*L; and a Line of Chords, labeled CHO. Five others started near the center of the scale and extended right. These were: a second Line of Rhumbs, labeled RUM; a second Line of Chords, labeled CHO; a Line of Sines, labeled SIN; a Line

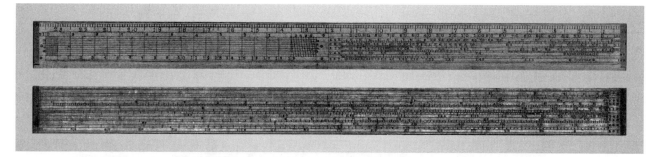

Figure 7-65: Front (top) and back (bottom) of a Merrifield Gunter's scale

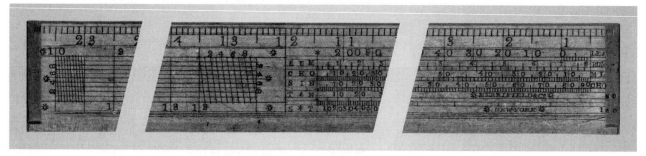

Figure 7-66: Scales on the front of a Gunter's scale (details)

of Tangents, labeled TAN; and a line of Sines of Tangents, labeled S*T.

On the back surface were marked the "logarithmic" or "artificial" lines of numbers, 24 inch scales representing the logarithms of various mathematical functions. These scales, all of which started near the right hand end of the scale and extended left, were: logarithms of the Sines of Rhumbs, labeled S*R; logarithms of the Tangents of Rhumbs, labeled T*R, logarithms of Numbers (two cycles in length), labeled NUM; Logarithms of Sines, labeled SIN; logarithms of "Versed" (reversed) Sines (cosines), labeled V*S; logarithms of Tangents, labeled TAN; logarithms of the Meridian Line (from Mercator's chart), labeled MERID; and a Line of Equal Parts (a linear, non-logarithmic scale), labeled E*P.

Set flush into the surfaces of the scale were a dozen tiny brass plugs, each with a small dimple punched in its face. These plugs were located at the origins of the various scales, or at key values in the graduations, and were placed there to pro-

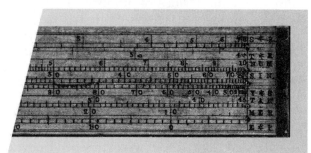

Figure 7-67: Scales on the back of a Gunter's scale (details)

vide a good location for the tip of a pair of dividers when setting them or reading their setting. They also served to protect the surface of the wood from damage due to repeated piercing by those tips.

All calculations were performed on the Gunter's scale geometrically, physical distances being transferred between the chart and the scales, and from one scale to another, with the aforementioned dividers. Addition, subtraction, and the taking of trigonometric functions were performed on the plain scales; multiplication, division, and powers and roots on the logarithmic ones.

More information on the use of the Gunter's scale can be found in the two articles "Some Notes on the History and Use of Gunter's Scale" and "Further Notes on the Operation of Gunter's Rule," which are reprinted in Chapter III.

Considering its nearly 300 year history, the Gunter's scale remained remarkably unchanged. One might suppose that it would have evolved or been modified significantly over time, but that is not the case. A few variations were suggested over the years, but none were generally adopted, with the result that the instruments sold in the late 19th century were essentially the same as those made 250 years earlier. A remarkable testament to Gunter's planning and vision!

The Gunter's scale eventually fell into disuse toward the end of the 19th century. Its method of operation admitted of significant errors, and it was eventually superseded by pencil and paper calculations using tables of logarithms, and by the logarithmic slide rule as that device became more sophisticated and accurate.

HATTERS' RULES

A hatters' rule, also known as a hat size gauge, is a short nonfolding rule with an extension slide, used by hatters to measure the inside diameter of customers' old hats preparatory to fitting them with new ones. Usually five inches long, and extending to almost nine inches, these rules were graduated in 8ths of an inch, in both directions, on the front surface and from left to right on the slide, and had a table on the back relating hat size, head circumference, and hat circumference.

This table had three columns; the first, Size, for the nominal hat size, the second, Head, for the circumference of the head, and the third, Hat, for the circumference of the hat (outside the band), all in inches. For more information on this table, see the section TABLES in Chapter VI.

Typically these rules would have a captive slide (one which could not be fully removed from the body), with the scale figured to be self indicating. The graduations on the slide would begin at five at the tip, and progress to 9 near the right end; as the slide was extended to the left, the graduations and figures on the slide at the point where it left the body would give the total overall length.

Hatters' rules were almost invariably not marked with the maker's name, making it difficult to determine which makers offered them. It is known that Stanley made them; a few examples have the demonstrable provenance of having originally come from the Stanley factory. As to the other major American makers, we can only speculate.

HORSE MEASURES

It was, and still is, customary to measure the height of a horse at the highest point of its back between the two shoulders. In order to perform this measurement a special type of rule was developed, capable of reaching past the swell of the horse's body to the measurement point. These special rules are called "Horse Measures."

Typically, a horse measure consists of a stick five or six feet long, with a square or rectangular cross section perhaps 3/4 of an inch on the side. On this long stick slides a right-angle arm about two feet long. This arm is usually provided with a level glass to aid in keeping the arm horizontal and the long stick vertical while making a measurement. The long stick is graduated in inches, with every fourth inch numbered sequentially, thus reading directly in "hands," the usual method of describing the height of a horse.

In use, the long stick is placed vertically on the ground next to the horse, and the sliding arm, kept horizontal, is slid up or down until it just touches the horse's back between the shoulders. The measured height in hands/4ths is indicated by the reading where the underside of the arm touches the long stick.

There are some interesting variations of this instrument. Sometimes they were made such that the long stick folds in half and the arm folds down, allowing for ease of carrying/storage. Another variation was to build the functionality into a walking cane, with a sliding extension from the tip of the cane to make it long enough, and a folding arm housed in the cane body near the handle.

Occasionally a horse measure will be encountered that appears too small, with a 2½ foot stick and a 1 foot arm. This is actually a dog measure, intended for measuring the height of dogs.

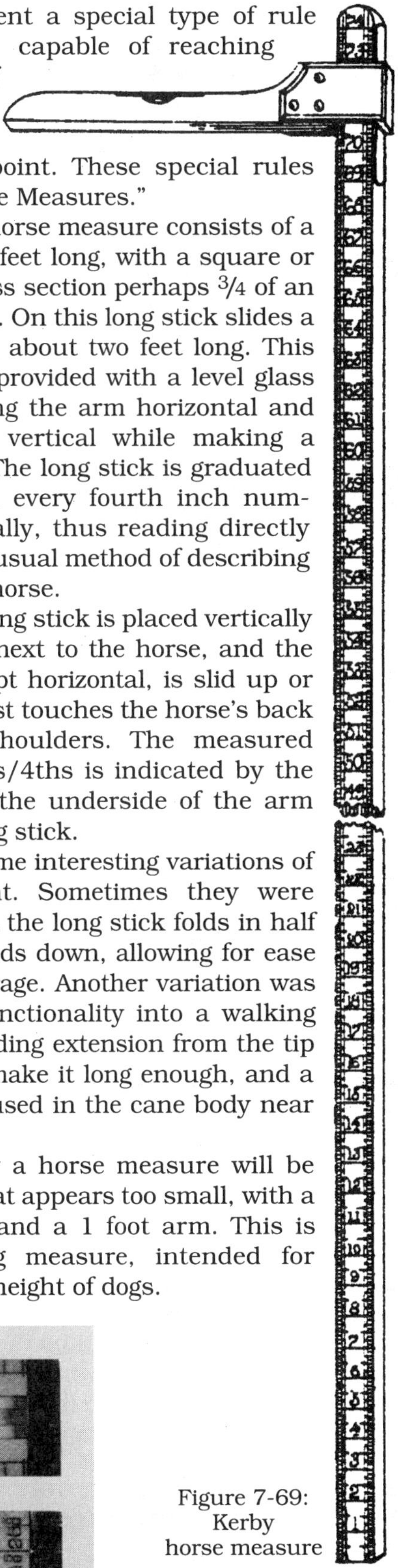

Figure 7-69: Kerby horse measure

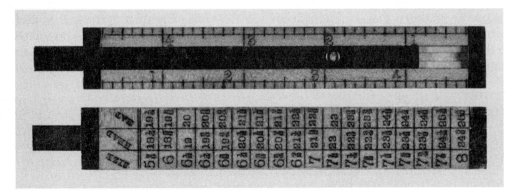

Figure 7-68: Front (top) and back (bottom) of a Stanley hatters' rule

Most horse measures encountered in this country are of English origin. The only known American maker is Kerby & Bros., of New York.

IRONMONGERS' RULES

"Ironmongers' Rule" is the English name for a folding rule with a sliding caliper, marked with tables giving the weights of various shapes of steel and wrought iron plates and bars. The most common patterns were either one foot, three fold, or two foot, four fold, in either boxwood/brass or (less frequently) ivory/German silver. The tables, which ran down the center of the rule between the scales on the edge, would be for such quantities as SQUARE IRON PER FOOT, FLAT IRON PER FOOT BY ¼ IN THICK, or ROUND IRON PER FOOT for widths/diameters from ¼ inch up to four inches. Frequently a CWT (hundredweight) table, relating cost per pound to cost per hundredweight (112 pounds), would also be provided.

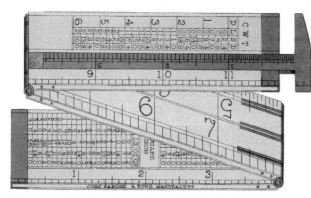

Figure 7-70: One foot, three fold English ironmongers' rule

The ironmongers' rule was almost exclusively an English product. With the exception of one rule produced by Stanley in the 1860's no American maker offered an equivalent product. That rule, the six fold Stanley No. 58½ was literally covered with no fewer than 18 tables of weights for various shapes of steel, iron, and other materials, covered to the point that there was no space remaining for marking the rule number or the maker's name. For more information on this extreme example of the ironmonger's rule, see the article "The Stanley No. 58½ Rule," reprinted in Chapter III.

IVORY RULES

Rules made of ivory, with their clear black graduations on the smooth white ivory surface, and their German silver trim, present an exceptionally fine appearance, and were as highly prized by users in the 19th century as they are by collectors today. An ivory rule in "like new" condition, crisp and white, is a pleasure to see and hold, and is usually the centerpiece of an owner's collection.

Most ivory folding rules were of one of three types: six inch, two fold; one foot, four fold; or two foot, four fold. One American maker, Belcher Brothers, did offer two foot, two fold ivory rules, but this was the exception. Joints and trim were usually German silver (brass with a 20% admixture of nickel); prolonged contact with yellow brass tends to gradually stain ivory green and the German silver looked better with the ivory anyway. Both plain and caliper rules were offered.

It is interesting to note that ivory rules were more subject to expansion and contraction with changes in humidity than was boxwood, and as a result were less accurate. This was noted by Charles A. Strelinger & Co. in its 1895 catalog (Ref. 4):

> "Ivory rules are desirable for gifts, and where the eyesight is poor, as the black markings on white are more easily read. We do not consider them as reliable as the Boxwood, as they are more apt to shrink. (We) have seen a first-class ivory rule that had shrunken nearly one-eighth of an inch in two feet."

This sensitivity to humidity also accounts for the large number of bound ivory rules where the ivory is cracked. Under dry conditions the ivory in these rules would attempt to shrink, but would be constrained from doing so by the binding; the resulting stress would cause the ivory, which is actually fairly brittle, to crack across.

Ivory rules were fairly expensive to manufacture, compared to their boxwood equivalents. Ivory rose markedly in price during the last half of the 19th century, and actually tripled in price between 1870 and 1905. Rule makers managed to keep their prices from rising by improved manu-

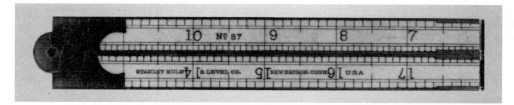

Figure 7-71: Stanley No. 87 two foot, four fold ivory rule

facturing methods and by cutting profit margins, but this situation could not continue. As labor and material costs continued to rise during and after World War I, ivory rule manufacture became unprofitable, and by 1925 had ceased entirely in America.

Due to the cost of the material, a great deal of ingenuity was devoted to saving ivory in rule manufacture, and at least two patents were issued for inventions in this area.

HASTINGS PATENT RULE

Glover S. Hastings of Granby, Connecticut invented a composite wood and ivory rule which was patented by him on March 16, 1875 (U.S. patent No. 160,904, entitled "Improvement in Measuring Rules"). This was a rule where the sticks were wood in the center, with ivory glued on at the edges, where the graduations were marked. The patent claimed a "savings in ivory" as one of its advantages. This patent was assigned one half to a Frank S. Johnson, of Plainfield, Connecticut. These rules were made (in ebony and ivory) by the Standard Rule Co., of Unionville, south of Granby. Known examples are all one foot, four fold caliper rules, and are marked with the patent date and the number "113."

HOGARTY'S PATENT RULE

Michael Hogarty of Salisbury, Connecticut invented a different type of composite wood & ivory rule which was patented by him on December 22, 1891 (U.S. patent No. 465,664, entitled "Rule"). Hogarty's invention was a rule where the sticks were wood, plated with a thin riveted-on layer of ivory. The patent claimed a "savings in ivory" as its primary advantage. These rules were made by the Acme Rule Co., of Salisbury, Connecticut. Known examples are all one foot, four fold rules, either with or without calipers, and are marked "PATENT APPLIED FOR."

The Acme Rule Co. was also the maker of a two foot, "three" fold wood rule that could also be used as a drafting T square (see COMBINATION RULES earlier in this chapter).

LINING OR ACCOUNTANTS' RULES

In the early 19th century persons keeping account books had frequent need for a rule. It served as a placeholder when setting an account book aside, it guided the pen when striking out an item, and it was a necessity when drawing the vertical and horizontal lines used to separate columns and rows on the page. This last was due to the fact that, prior to the invention of ruling machinery, account books were supplied unlined, and the bookkeeper had the responsibility for drawing those lines prior to making the entries on each page.

The first two uses could be satisfied by any common rule, but the third, the lining of the blank pages, required a rule which could draw a series of parallel equally-spaced lines, and a number of ingenious rules and devices were invented to perform this task. Some provided a means for keeping the lines parallel, but left the spacing up to the user. Others had a vertical scale to space the lines, leaving it to the user to keep them parallel by eye alone. The most elaborate handled both tasks, not only keeping the lines parallel, but also ensuring that they were equally spaced at the interval selected by the user.

Properly speaking, lining rules are only peripherally measuring instruments, and it could be argued that they have no place in a book dealing with such. Although most were graduated, their primary function was that of a straight edge, and they would have functioned equally well if the

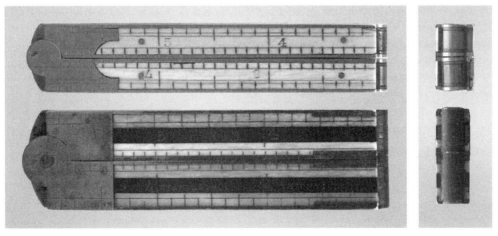

Figure 7-72: Patented wood and ivory rules. Top: Hogarty's patent (Acme Rule Co.).
Bottom: Hastings patent (Standard Rule Co.).

graduations had been omitted. Despite this, they have been included here for two reasons: because they were frequently the product of the makers of conventional measuring rules, and to illustrate the wide variety of these rules which were available.

Simple Accountants' Rules

Any desk rule could be used by an accountant to line pages, providing it were wide enough to hold down with one hand while drawing the line with the other. The spacing and the parallelism of the lines would depend on the skill and eye of the user. However, it was useful if the rule had a metal edge which remained straight under use, and which could be used to hold down a piece of paper while tearing off a strip. It was also useful if the shape allowed the user to lift the drawing edge cleanly from the paper after marking the line, thus avoiding blots or smears. Many rules which fit these requirements exist, and two examples are shown here.

Tingley's Rule

This rule, made by Tingley & Co., of New York City, possessed both of the special features mentioned above. It had a metal tear edge, and, additionally, was slightly hollow on the bottom surface, to allow the user to press the edge more firmly on the paper while tearing. Also, the back edge of the bottom surface had a narrow rebate running its entire length; pressing that edge down with the fingers after drawing a line would cause the front edge to pivot straight up away from the paper.

Gibbs/Rowe/Mann Pattern Rules

Another interesting accountants' rule is that which had the form of a wooden cylinder with a flat metal blade protruding from it. This configuration allowed the user to "roll" the metal edge of the blade up away from the paper after drawing a line. The example shown below was made by F. G. Rowe, of Boston, Massachusetts. It is 14 inches long, and made of rosewood with a brass blade. Other examples, from 9 inches to 18 inches in length, made by Rowe and others, have bodies of walnut, ebony, cocobolo, and other fancy woods, and blades of German silver and nickle-plated steel.

The three known makers of this type of rule are Rowe, G. E. Gibbs (also of Boston), and John E. Mann, of Cleveland Ohio.

The basic configuration of this rule was apparently never patented. The only known patent relating to this type of rule is that of Mann, granted on June 10, 1884 (U.S Patent No. 299,993, entitled "Ruler"), and that patent was concerned not with the basic configuration, but with minor construction details. Gibbs examples are sometimes marked "PATENT APPLIED FOR" or "PATENTED," but no patent by Gibbs or assigned to him has ever been found. Mann's examples are marked "PAT'D June 10 '84."

Rules to Draw Parallel Lines

Simple Rolling Rules

The simplest form of rule for keeping successive lines parallel was a simple cylindrical rule long enough to span the paper being lined. In use it would be rolled down the page, being periodically stopped by one hand while drawing the line with the other. Sometimes instead of being round it would be octagonal, or have octagonal ends, to space the lines evenly.

More Elaborate Rolling Rules

A number of inventors developed fairly elaborate devices to make drawing the parallel lines faster and easier. Here are a few typical examples:

Carrington's Rule

One of the earliest such inventions was Carrington's parallel ruler, patented by James Carrington, of Wallingford, Connecticut on April 14, 1832 (U.S. Name&Date Patent No. X6998, entitled "Ruler For Counting Houses, &c."). This rule had a solid wood body from 12 inches to 18

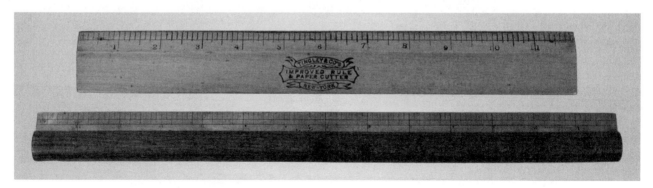

Figure 7-73: 12 inch desk rule by Tingley & Co., and 14 inch Rowe's patent accountants' rule

SPECIAL RULE TYPES AND USES 207

Figure 7-74: Three simple rolling lining rules. Top to bottom: round, octagonal, round with octagonal ends.

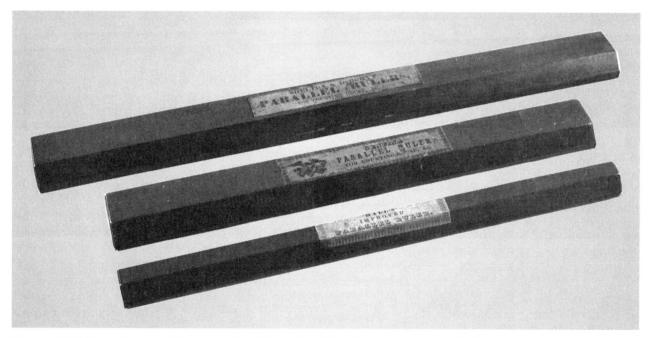

Figure 7-75: Three forms of Carrington's rolling rule: 18 inch with two pairs of rollers, 14 inch with a single pair of rollers, and 14 inch with a single roller.

inches long housing two parallel wood rollers held in place by pewter end caps, an arrangement which made the ruler easier to handle and hold steady while marking. In the 18 inch examples, the single pair of rollers was replaced by two pairs of 9 inch rollers side by side. Another variation on the basic design has been observed where the ruler was markedly narrower, and employed only a single roller.

This ruler was apparently a very successful invention. Judging from the number of surviving examples it must have been widely sold for a considerable period of time. It is interesting to note that these rulers were made for sale under other names besides Carrington's. Examples have been found marked "Shelton & Osborne," "Eagle," "Hall's Improved," "Parallel Ruler for Counting Houses" (no maker's name), etc., etc.

Appleton's Patent

This lining rule is a different advance on the basic rolling rule. The roller was linked to a second, non-rolling round stick by plates at the end. When the roller was rolling, the second stick would slide; when drawing the line, pressure on the second stick would keep the ruler from moving and spoiling the line.

RULES TO DRAW AND SPACE PARALLEL LINES

An interesting rule for spacing lines as they were drawn and keeping them parallel is the example shown on the next page made by the Eberhard Faber Co. The two rectangular slots in the rule have scales on beveled brass plates at their outer edges (the brass plates on the inner edges are for picking up the rule). The scales indicate the distance in 16ths of an inch from the

Figure 7-76: Appleton's patent lining rule

Figure 7-77: Slots with scales on an Eberhard Faber accountant's/lining rule

lower edge of the rule. To draw the lines, the rule would be placed on the paper with the beveled edge toward the bottom of the page, and the upper edge used to guide the pen. The spacing between the lines would be controlled by aligning the previously drawn line with the desired graduation line on the two vertical scales. With care, the successive lines could be maintained reasonably parallel, although the two slots were too close together to give really good results.

A more elaborate example is the Bostock & Pancoast Lining Rule, shown below. This device was designed to both space the lines uniformly and to keep them parallel while doing so. The bevel-edged rule had a brass plate extending from its bottom edge, with a captive flat block sliding on it. The distance the block was allowed to slide, and thus the spacing of the lines, was set by the position of a thumbscrew in a slot below the block. After drawing one line, the block was slid down against the thumbscrew, and then pressed down against the paper while the rule was moved down until it touched the block, at which point the next line could be drawn.

This rule was invented by Edward Bostock, originally of Philadelphia, Pennsylvania, but later of Albany, New York, and patented by him on August 6, 1867 (U.S. Patent No. 67,487, entitled "Adjustable Ruler"), and on May 23, 1871 (U.S. Patent No. 115,019, entitled "Adjustable Parallel Ruler"). It was also available in a more elaborate version where the rule was connected to the brass plate by a protractor head, allowing it to be set to any desired angle to the direction of motion. This second version was probably intended for use as a section liner, and not as an accountant's lining rule (see SECTION LINERS, opposite).

The foregoing is only a sample of the types of lining rules which were devised and made over the years. The number of different rules, and the diversity of their features, is such that it would be possible to build an entire collection just of these instruments alone.

Figure 7-78: Bostock and Pancoast lining rule

(Section Liners)

It is possible that some devices which seem to be intended for lining account books were actually intended for use by draftsmen to draw the section lines in technical drawings showing cross sections, or by engravers for drawing the finely spaced lines which give shading to woodcut and copperplate pictures. The applications are similar, the major differences being that section or shading lines are usually closer together, and are almost always drawn at an angle to the horizontal. As a rough rule of thumb, any lining rule which has a protractor head, or in which the line of the rule can be set at an angle to the direction of motion is probably actually a section liner.

LOG STICKS, LOG CANES, AND LOG CALIPERS

One of the most persistent problems which faces the buyer or seller of saw logs is that of estimating the amount of lumber which a log will yield when taken to the mill. Over the years a number of methods have been developed to systematize the estimating process, methods usually referred to as "log rules."

A log rule consists of a measuring method, formula, or data table, or some combination of the three, which allows one to estimate the amount of sawn lumber which can be realized from a log. One takes the dimensions of the log, applies the rule to it, and the result is the number of board feet which it will yield when cut up. Over the years a number of log rules have been developed, from rules as simple as the 400 year old "quarter-girth" formula to modern data tables derived from measurements performed on a large number of logs.

Developing such rules was a complex matter, involving a large number of variables and near-intangible factors. Allowance had to be made for the shape of the log, with taper and sweep taken into consideration. Bark thickness had to be taken into account. Wastage during the sawing process also had to be allowed for, with saw kerfs, slabs, and shrinkage included. Finally, the varieties of trees and the sizes of the timbers and boards being cut at a particular mill had to be taken into account.

Log rules were developed in three different ways. Some, like the "quarter-girth" rule or the Doyle rule (1825), were based on mathematical formulas. Others, like the Scribner rule (1846), were based on carefully drawn diagrams. Still others were derived from tabulated yield data from a large number of actual sawn logs. (see Nonlinear Graduations in the previous chapter)

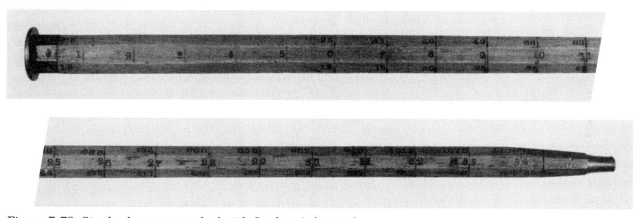

Figure 7-79: Stanley log cane marked with Scribner's log scales

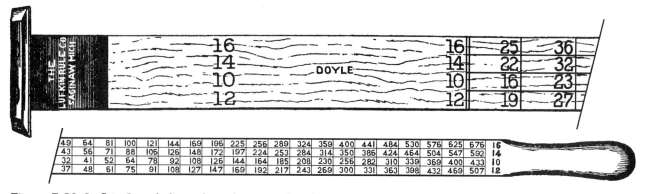

Figure 7-80: Lufkin 8-scale log rule with square head

In actual use these rules were converted into graduated scales which were marked on measuring sticks for use by the seller/sawyer. Such a stick would be marked with a number of scales, one for each different length of log (8 feet, 10 feet, 12 feet, etc.); by simply laying the stick across the butt of the log the board foot estimate could be read directly on the appropriate scale.

The usual method of marking these log scales on a stick was to space the graduation lines one inch apart, and to place the board foot value for that diameter next to each line. This same line spacing was used for all the log scales on a stick, only the board foot figures differing from one scale to the next. This method of graduating log scales is in contrast to the method of graduating board sticks, where each adjacent line represents an increment of one board foot, and the line spacing is consequently different for each scale. This difference is useful in distinguishing log sticks from board sticks; if the lines on all scales are equally spaced, it's almost certainly a log stick; if each scale has a different line spacing, it's almost certainly a board stick.

The most common form of log sticks were flat, flexible log sticks. Rectangular log sticks, with and without caliper jaws, and log canes, were also made.

Log Canes

Log canes were three feet long, and made of hickory for strength. They had a round brass head, and were tipped with a brass ferrule to reduce wear. The round brass head served a double duty, providing a good grip when used as a cane, and acting as a shallow hook to aid in placing the head of the cane at the far edge of the log when measuring. The cane itself was octagonal, with a single scale on each of its eight faces, and was tapered from 3/4 inch thick at the head to 1/2 inch at the ferrule. Log canes were usually marked with scales for either Scribner's or Doyle's log rules, the most common rules in use in the 19th century.

Flat, Flexible Log Sticks

Flat, flexible log sticks were usually hickory, three or four feet in length, and had a swelling teardrop-shaped handle for a good grip. The body of the rule itself was 1, 1 1/4 or 1 1/2 inches wide, depending on whether it was marked with six, eight or ten log scales, and was thin enough to be somewhat flexible. A wide variety of heads were available, from a plain cap to a single or double hook, in either steel or brass.

The largest producer of flat, flexible log sticks was the Lufkin Rule Co., but examples by the Cleveland Rule Co. are also frequently encountered. Lufkin would mark their log sticks with scales based on any of eleven log rules as requested by the customer, and examples with other scales beyond these eleven have been noted.

Log Calipers

Log sticks were sometimes made in the form of large calipers, with a fixed and a moving jaw, allowing the log to be measured while still standing, before felling the tree. This was extremely useful for estimating the amount of lumber to be

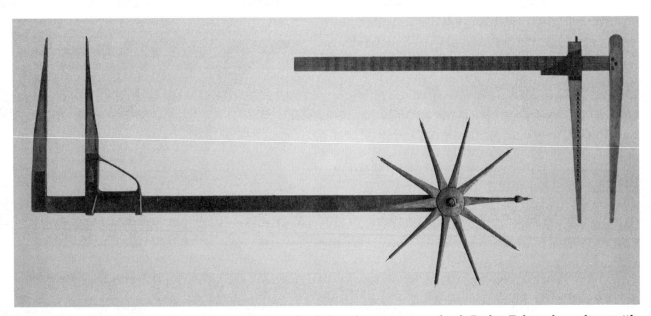

Figure 7-81: Left: Greenleaf log caliper with "pinwheel" length measuring wheel. Right: Fabian log caliper with tally holes in jaw.

realized from a woodlot, prior to any cutting. The bar of the caliper would be typically three or four feet long, and the fixed and moving jaws perhaps 18 inches, allowing the caliper to measure tree trunks up to three feet in diameter. Some were simple, with iron fastenings and a plain bar; others are large and spectacular measuring instruments, with the bar edged with brass to prevent wear, and brass castings fastened to/sliding on the bar to carry the jaws. Usually, the bars on these calipers would be 3/4 inch thick and 2 1/2 inches to 3 inches wide, and would be marked with up to a dozen scales on the two sides and (sometimes) on the edges as well.

More elaborate log calipers would be equipped with a ten-pointed, pinwheel-shaped measuring wheel, mounted on the opposite end of the fixed jaw. The sharp points on the spokes of these wheels were six inches apart; thus the wheel could be "stepped" down a log to measure its length, with each two steps representing one foot. One spoke (the "starting spoke") was weighted with a lead collar, to cause that particular spoke to always be on the bottom when the wheel was placed on the log ; and every other spoke around the pinwheel was marked with stripes indicating how far it measured from the starting spoke, to facilitate keeping track. These wheels could be ordered from the caliper maker as either standard or as an option, or could be purchased later from a forestry supply dealer and attached by the owner. Thus it is not uncommon to find a caliper by one maker equipped with a wheel by another.

MACHINISTS' RULES

The development of machine work in the 19th century to the point which required accuracy in the range of thousandths of an inch fostered the development of special measuring tools to achieve this accuracy. The first of these was the machinists' steel rule, produced commercially just prior to the Civil War. These rules, graduated frequently up to 100ths of an inch, allowed machinists to work to closer precision than was possible with the wood or ivory rules hitherto available.

Nor did the subsequent development of more accurate measuring tools, the vernier caliper and the micrometer, render the machinists' rule obsolete; for much rough work accuracy to a 50th or 100th of an inch is adequate, and only in special cases must the rule be set aside and the more precise instrument used.

The machinists' rule is made of steel, and can be anywhere from 1 inch to 4 feet long, although 6 inches, 12 inches, and 24 inches are the usual lengths. Steel is used for durability, rules being often subject to rough usage on the shop floor, and because it can be graduated with the fine scales needed by the users. The most common types are flat steel rules, anywhere from 1/4 inch to 1 1/2 inches wide, and from 1/8 inch to 18 gauge thick. Brown & Sharpe also offered machinists' steel rules with a square cross section in lengths up to 6 inches, and with a triangular cross section in lengths up to 12 inches.

Figure 7-82: Flat, square and triangular steel machinist's rules

The earliest makers of steel machinists' rules were Joseph R. Brown, of Providence, Rhode Island, beginning in 1850, and Samuel Darling, of Bangor, Maine, beginning in 1852. In 1853 Brown went into partnership with Lucien Sharpe, making rules marked J. R. Brown & Sharpe. Darling, on his part, was in partnership first with Edward H. Bailey, and subsequently with Michael Schwartz, and his rules are marked either D&B (very rare) or D&S. In 1868 the two makers merged, and rules made after that time are marked Darling, Brown & Sharpe.

Other early makers of steel machinists' rules were L. S. Starrett, of Athol, Massachusetts beginning in 1882; J. Wyke & Co., of Boston, Massachusetts, beginning in 1885; the Sawyer Tool Co., of Athol, Massachusetts, beginning in 1896; and the Lufkin Rule Co., of Saginaw, Michigan, beginning about 1920.

Machinists' rules were graduated with scales divided as finely as inches/100ths, the practical limit of taking off dimensions with a pair of dividers with the aid of a magnifying glass. Usually each edge of the rule would have a different graduation divided upon it, and sometimes, in extreme cases, a single edge would have different inches divided differently. As an example, one edge could have its first inch divided into 100ths, its second inch into 50ths, its third inch into 20ths, and the remaining inches into 10ths. By 1900 it had become common to offer rules graduated across their ends as well as along their long edges, making it possible to measure the width and depth of narrow grooves.

Both L. S. Starrett and Brown & Sharpe, the two largest makers of steel machinists' rules, had standard patterns of graduations which they offered on their rules. Starrett offered sixteen different patterns of scales, Brown & Sharpe nineteen. These combinations were usually the same for both makers. That is, the Brown and Sharpe No. 4 graduation pattern (its most common) was the same as the Starrett No. 4 pattern.

No. 4 Graduation:

	Brown & Sharpe	Starrett
1st corner	8	64
2nd corner	16	32
3rd corner	32	16
4th corner	64	8

This was true for at most patterns, but there were exceptions. Starrett offered no equivalent of the Brown & Sharpe No. 9 pattern; Brown & Sharpe offered no equivalent of the Starrett No. 16 pattern.

Brown & Sharpe No. 9 Graduation.

1st corner	10,	20
2nd corner	16	
3rd corner	32,	64
4th corner	50,	100

Starrett No. 16 Graduation.

1st corner	32
2nd corner	64
3rd corner	50
4th corner	100

The typical width of machinists' rules depended on the length, from 1/2 inch (for a one inch rule), up to 1 1/4 inches wide (for two to four foot rules). Thicknesses varied from about .045 inches, for semi-flexible rules, to about 1/10 of an inch, for heavy rules to be used at the bench.

MASONIC RULES

The Freemasons use a number of "operative" tools in their lodge rituals to symbolize certain aspects of their beliefs and principles. One of these is a two foot, three fold rule called a "gauge." This rule, about one inch wide, is assembled with face-to-face joints like the middle joints in an ordinary four fold rule. It is graduated on one side from 0 to 24 inches, and on the other from 0 to 8 inches on each of the three sticks. In Masonic symbolism the 24 inches represent the 24 hours of the day, and the three 8 inch segments represent the proper allocation of the day's time: 8 hours for sleep, 8 hours for work, and 8 hours for family and civic activities.

The joints on these rules would be either plain or plated with arch plates, like the middle joints of a four fold rule with a double arch joint. The usual materials for the sticks were either boxwood or ivory, although at least one example exists made of solid German silver, silver plated. If made of boxwood or ivory, the joints & trim would have been either brass or German silver, as appropriate.

The two makers known to have made boxwood or ivory Masonic rules are Stanley and Chapin. It is not known who was the maker of the solid German silver example mentioned above, but judging by the fine decorative engraving on the

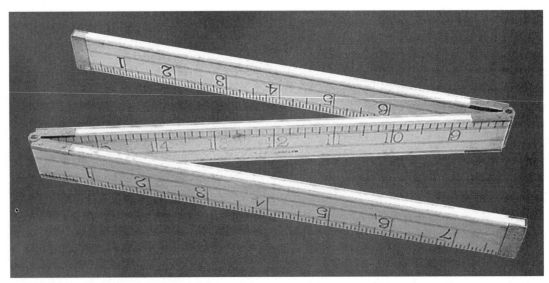

Figure 7-83: Ivory Masonic rule

SPECIAL RULE TYPES AND USES 213

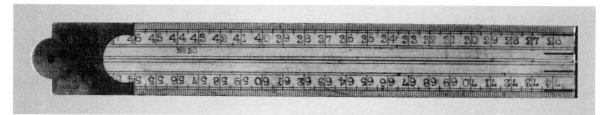

Figure 7-84: Stanley No. 30 1 meter, 4 fold rule

edge of the rule it was probably one of the larger silver firms such as Gorham or International.

METRIC RULES

The marketing of metric rules by American makers apparently began some time after the Civil War. The reasons for the appearance of metric graduations and rules on the American scene at this time were twofold. First, during the previous thirty years the metric system had become almost universally adopted, country by country, in Europe; many immigrants from those countries had become used to that system and preferred it. Second, there was an organized effort at this time by industry and the scientific community in the United States to secure the adoption of the metric system here (Ref. 5). Metric weights and measures had become legal, but not required, in 1866; the American Meteorological Society had founded an "American Metric Bureau" in 1876 to produce and market metric system teaching aids, including metric rules, for schools. Ultimately, these efforts came to nothing, and the United States chose not to abandon English measures. For twenty years, however, a number of American makers offered metric rules either as standard products or options. Hubbard & Curtis, of Middletown, Connecticut offered metric graduations as an option in 1872, Stanley listed metric rules as standard products in their 1877 catalog, Chapin offered metric graduations as an option in their 1882 catalog, and examples have been observed made by the Standard Rule Co. dating from 1889 or earlier.

Stanley was the most active in making metric rules, offering both standard length and metric length rules with metric graduations. Apparently Stanley was the maker of choice to supply rules for sale by the American Metric Bureau. The list of metric length rules available from that Bureau (Ref. 6) closely matches the different patterns known to have been offered and made by Stanley. None of the other makers mentioned above went so far as to make rules in metric lengths, but contented themselves with simply offering metric graduations on their standard pattern rules.

From that time to the present, all major makers have offered metric rules and graduations, usually as options on standard rules, available on special order, but occasionally as standard products. Meter sticks are frequently encountered, particularly those made by Stanley and Lufkin. Standard pattern two-fold and four-fold rules in 1, 2, and 3 foot lengths with metric graduations, while not common, are also found from time to time.

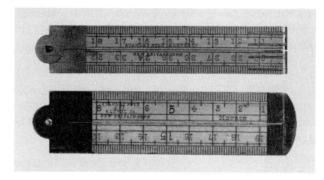

Figure 7-85: Two small Stanley metric rules

MILK CAN GAUGES

Dairy farmers used to deliver fresh milk to wholesale dairies in large wood-staved or galvanized steel cans. To measure the amount delivered, the buyer and seller would use a dip stick

Figure 7-86: Kerby milk can gauge

called a milk can gauge to determine the amount of milk in each can. This gauge was similar in physical construction to the more familiar coopers' gauging rod, being a 30 inch square stick with a brass-plated wedge-shaped tip.

Because it was used as a dip stick to measure the content of the can, and not to gauge the capacity, the milk can gauge was graduated and used differently. The scales began at zero at the tip, and proceeded uniformly up the length of the stick, ending at the point which represented the full liquid level of the can. Each of the sides of the stick was graduated with a scale for a different type of milk can, so that a single stick could deal with four different sizes of cans. These gauges were largely a custom product, ordered by the local dairy wholesaler, marked with scales suitable to the milk cans in use in his area. The example shown below is marked with four scales labeled: 20 Q^t MILK CAN, 30 Q^t MILK CAN, 40 Q^t MILK CAN IRON CLAD, and 40 Q^t MILK CAN L&G.

In use, the stick was dipped vertically into the can of milk until the tip touched the center of the bottom; the graduation line (of the scale appropriate to that size of can) where the milk surface intersected the scale represented the number of quarts contained in the can.

The most common maker of milk can gauges was Kerby & Brothers, in New York City, although Stanley is also known to have made them. The Stanley gauges are identical to their No. 45 gauging rods, differing only in their length and graduations; the Kerby gauges had a thicker 5/8 of an inch cross section and the end of their wedge-shaped tip was flattened.

MULTIPLE SLIDES

An interesting category of folding rule is that of rules with more than one slide. Most known examples would appear to be experimental or trial pieces, made either at the whim of a workman or to try out some arrangement which was abandoned before the commencement of serious production. Below are some examples.

STEARNS TWO FOOT, TWO FOLD RULE WITH THREE SLIDES

This is a very unusual design, and only one example of this rule is known to exist. It is a two foot, two fold rule with an arch joint, significantly wider (two inches) than usual. It is equipped, however, with no less than three slides. An ordinary Gunter's slide is located in the usual location for a carpenters' sliding rule. The two other slides are located in the other leg, on the opposite side of the rule. These two slides run side by side (the rule is thicker than normal (1/4 inch) to accommodate this structure), and both are extension slides. The difference between them is that the upper one is graduated in inches and 8ths, and the lower in inches and 10ths. All three slides are made of brass.

Why it was felt desirable to have two extension slides, one reading in 8ths, and the other in 10ths is not clear. It does not appear to be a very useful feature. It must be one of those "solutions in search of a problem"!

It is probable, but not certain, that this rule was an experimental product, and was never produced commercially. The size of the joint in proportion to the width of the rule does not feel right, nor does the size of the curve in the arch joint plates. This would indicate that the rule was made by adapting the joint plates of an ordinary 1 1/2 inch two fold rule. Surely Stearns would have put a better-proportioned joint on this rule if it had gone into production. It is definitely a company-made piece, however, and not some employee's whimsey; the use of the E. A. STEARNS & CO., BRATTLEBORO, VERMONT stamp is evidence of that.

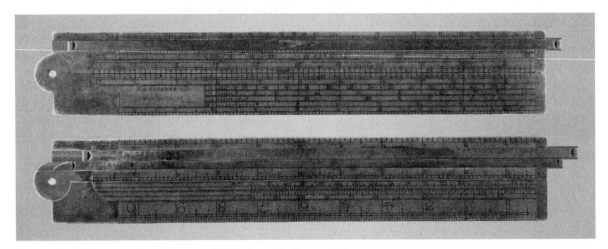

Figure 7-87: Stearns two foot, two fold rule with three slides

Stanley Two Foot, Two Fold Rule With Two Slides

Several examples of Stanley two foot, two fold rules with two slides are known to exist. The example shown below is typical. In all of these rules, the slide in the customary location is a Gunter's slide. The second slide, which is a simple extension slide, graduated in inches and 8ths, is located on the same side of the rule as the Gunter's slide, but in the other leg.

Other than the presence of the second slide (and the absence of the drafting scales whose place it took) these rules in all respect conform to the normal configuration of a Stanley carpenters' (sliding) rule.

It is not clear whether these Stanley two-slide rules were an experimental product, or were just whimsies turned out by workers for their own amusement. Both bound and unbound versions (both with arch joints) have been found, and both types are invariably marked with the Stanley number appropriate to their trim (i.e., the unbound ones are marked "No. 12," and the bound ones marked "No. 15."). This would suggest that the extra slide was added to a production rule by some worker. However, the leg containing the extension slide is nicely marked with double embellishment lines on each side of the slide. This is more likely to have been done if the rule was an experimental model being considered by the company.

Six Inch Cordage Rule With Two Slides

Another interesting rule with multiple slides is the cordage rule shown in Figure 7-89. This rule was made with two slides, the ordinary caliper slide graduated with the usual circumference scale, and on the other half of the face a six inch Gunter's slide for doing computations. The presence of this slide made it possible to greatly simplify the table of cordage parameters on the back of the rule. Instead of all combinations of all parameters, it was only necessary to include one basic gauge value for each parameter. Given that basic gauge value, the desired variation for the cordage in question could be calculated easily using the Gunter's slide.

This rule, which is probably English or European in origin, is not marked with a maker's name. It was probably made in at least limited quantities, however; the design and level of finish is too high for it to be a one-of-a-kind item.

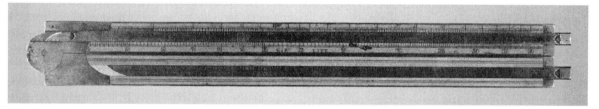

Figure 7-88: Stanley two foot, two fold rule with two slides

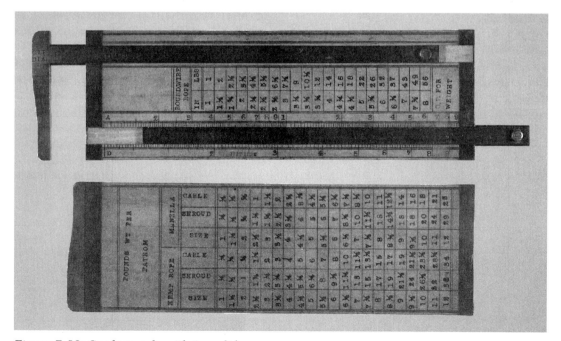

Figure 7-89: Cordage rule with two slides

ENGLISH TWO FOOT, TWO FOLD DIALING RULE WITH TWO SLIDES

This two foot, two fold rule is a carpenters' (sliding) rule with a second slide to extend the functionality of the Gunter's slide. This second slide, which is made of wood, is in the same leg of the rule, just below the brass Gunter's slide, and adjacent to it (the rule is thicker than normal (¼ inch) to accommodate this structure). The D scale is marked on the upper edge of this second slide. Below this second slide, on the body of the rule adjacent to it, are a series of calibration points, 12 in number, marked at various locations along this edge. Each of these points is marked with a number representing its value (816, 359, 232, etc.). In operation, the second slide is moved to the right until either the left hand end of the D scale or the value 10 in the center of the D scale is aligned with the calibration point appropriate to the calculation to be performed. The main slide is then operated in the usual way to perform the calculation; the offset of the second slide automatically introduces the selected calibration value into the calculation. A complete set of printed instructions, written by a John Browne, presumably the inventor of this rule, was available, explaining the use of the calibration points and the second slide.

The rule is also marked on the other leg, below the slide with four special scales, IN MERID (Inclinations of Meridians), CHORDS, LATITUDES, and HOURS. These scales were included so that the rule could be used to lay out vertical and horizontal sundials. The printed instructions also explained the use of these scales in the process of designing sundials.

On the other side of the rule was marked a pair of sector lines divided from 0 to 30, for simple calculations involving multiplication, division and proportion. Above the upper line was a parallel line divided from 0 to 40. This combination of lines (a 30 division sector with an extra line with 40 divisions) is called "Scamozzi's Lines" after the renaissance architect Vicenzo Scamozzi. Scamozzi had proposed a rule with these sector lines for proportioning the orders and elevations of buildings in his book, *The Mirror of Architecture*, and English carpenters had developed methods of using the same lines for proportioning the rafters in framing buildings. Below the lower sector line was a pair of lines labeled CHOR (chords) and TANG (Tangents) which were used with the sector lines for setting out angles and computing offsets. These scales were also explained in Browne's instructions.

This rule is an English pattern which saw significant production. Several examples have been found marked SAMPSON ASTON, and one marked Z. BELCHER (before, or shortly after he emigrated to the United States). Unmarked examples also exist. All are unmistakably of English manufacture.

PLAIN SCALES

The plain scale is a one foot rule marked with measuring and trigonometric scales which could be used in conjunction with a pair of dividers to solve problems in plane and spherical trigonometry. It is not known when this scale was invented. The scales, their development, and their method of use are described extensively in Bion (Ref. 7), Book I, Chapter 6, so its invention must predate 1750. It probably dates from about the the beginning or middle of the 17th century, as does the Gunter's scale, to which it is closely related.

In form the plain scale is highly standardized. It is a flat rule, typically 1½ to 2 inches wide, ¼ to ⅜ inch thick, and 1 foot long. The bottom edge is beveled, sometimes on the front, sometimes on the back. All known examples of plain scales are unbound, and do not have metal tips.

The graduations were also highly standardized. Near the top edge of the front surface was a scale of inches/10ths and a second scale of 10ths/100ths of a foot, both graduated left to

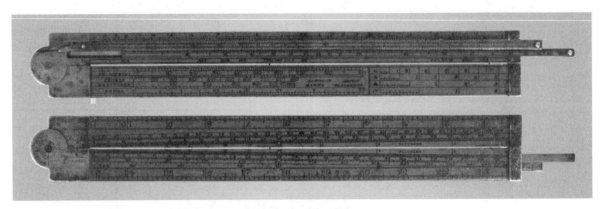

Figure 7-90: Two foot, two fold dialing rule with double slide

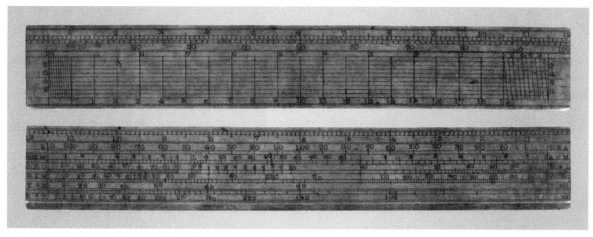

Figure 7-91: Front (top) and back (bottom) of a plain scale

right. Below that on the remainder of the front surface was marked an 11 inch scale divided into half inches, with the left half inch diagonally divided into 100ths of a half inch and the two right hand half inches diagonally divided into 100ths of an inch. The half inches in between were not subdivided.

On the back of the rule were marked the so-called "plain" or "natural" lines of numbers; scales representing various functions from trigonometry and spherical geometry. Four of these started near the right hand end of the scale and extended left. These were: a Line of Numbers/Line of Equal Parts, labeled LEA; a Line of Rhumbs, labeled RUM; the Meridian Line (from Mercator's chart); labeled M*L; and a Line of Chords, labeled CHO. Five others started at the left end of the scale and extended right. These were: a second Line of Rhumbs, labeled RUM; a second Line of Chords, labeled CHO; a Line of Sines, labeled SIN; a Line of Tangents, labeled TAN; and a line of Sines of Tangents, labeled S*T.

These scales, and the method of developing them geometrically, are discussed in the section NONLINEAR GRADUATIONS in Chapter VI.

Set flush into the surfaces of the scale were a dozen tiny brass plugs, each with a small dimple punched in its face. These plugs were located at the origins of the various scales, or at key values in the graduations, and were placed there to provide a good location for the tip of a pair of dividers when setting them or reading their setting. They also served to protect the surface of the wood from damage due to repeated piercing by those tips.

All calculations were performed on the plain scale geometrically, physical distances being transferred between the working drawing and the scales, and from one scale to another, with dividers.

Although most of the plain scales encountered are probably of English origin, it is known that American makers made them as well. Examples marked BELCHER BROS & CO. have been found; as has also one marked JACOB LEVY & BROTHER, NEW YORK.

PRINTERS' RULES

Printers' rules were specially graduated rules used by printers in composing and setting type. They were not so much concerned with inches as with line width and line spacing, expressed in type-sized units, or in column length, and they used rules graduated in those units when preparing to set type, or lay out a page.

The traditional form of printers' rule was a 3/4 inch square stick, 1 foot long, marked with eight scales, one on each edge of each surface. Other forms were also made; one foot, four fold rules marked with four or six scales, and one foot flat metal rules marked with four scales (these last were usually provided with a small hook at the end for measuring linotype slugs, ornamental borders, etc.)

The scales placed on these rules would be those for measuring the most commonly used type sizes: 6 point (called NONPARIEL), 7 point (called MINION), 8 point (called BREVIER), 9 point (called BOURGEOIS), 10 point (called LONG PRIMER), 11

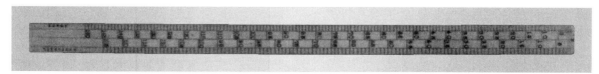

Figure 7-92: Stanley No. 186 one foot printers' rule

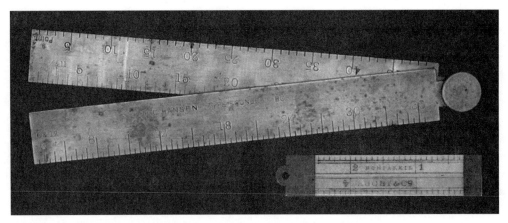

Figure 7-93: Folding printers' rules; top: German silver; bottom: ivory

point (called SMALL PICA), and 12 point (called PICA) (1 point equaled approximately .0138 of an inch; see LINEAR, 1:1 GRADUATIONS, Chapter VI). Scales for the extremely small type sizes, such as the PEARL (5 point) and RUBY (5½ point), and the very large type sizes, ENGLISH (14 point) and GREAT PRIMER (18 point) were omitted; it was apparently assumed that the printer/typesetter could use the LONG PRIMER SMALL PICA, MINION, and BOURGEOIS scales by doubling or halving the count.

One other scale was frequently included, a scale calibrated in a unit called AGATES (1 agate equaled 1/14 of an inch). This was a unit used by printers for measuring column length for gross estimating and billing purposes.

Occasionally printers' rules will be found which are concerned with paper size, and not type size. These rules will be marked with the tables of the dimensions of the various sizes of paper used by printers in the 19th century. These tables can be recognized by the names (similar to the names given to type sizes) assigned to the more than 25 different paper sizes (Ref. 8). Typical names/sizes are: FOOLSCAP (13¼" x 16½"), DEMY (15½" x 20"), ROYAL (19" x 24"), ELEPHANT (23" x 28"). Such rules with tables of paper size are usually English.

ROLLING MEASURES

An interesting variation on the ordinary straight rule is the rolling measure. This is an instrument with a circular wheel of a known diameter which can be rolled across a surface and which will indicate the distance traversed. Sometimes, as in the case of the Young's patent "Revolving Measuring Wheel," the wheel is only part of the instrument, with the indicating mechanism occupying the rest. Other examples, such as Brackett's "Protractor" combine the two functions, turning the entire instrument into the rolling wheel.

MAP MEASURES

The most common example, still made today, is the rolling map measure. This is a small instrument for tracing a route on a map and indicating the distance from start to finish in miles or kilometers. In its usual form it is an instrument resembling a pocket watch, with a small protrusion on the bottom edge ending in a small wheel. This wheel is geared to a hand under the crystal like the minute hand of a watch. As the wheel

Figure 7-94: Swiss-made K&E map measure

rolls, the hand advances. One or more circular scales are graduated on the face of the instrument under the hand, each scale, with its markings, corresponding to a different scale of map. Sometimes the instrument will be equipped with two hands, one counting the revolutions of the other, thus permitting longer measurements. It is customary in these instruments to provide some means of resetting the hand(s) to zero before starting a measurement.

Here are two other examples of rolling measures:

YOUNG'S PATENT ROLLING MEASURE

This is another example of a rolling measure in which the wheel is separate from the indicator. Invented by Louis Young of New York, New York, and patented by him on November 20, 1855 (U.S. Patent No. 13,833, entitled "Revolving Measuring Wheel"). It is about 5½ inches long, and has a measuring wheel at one end and an indicating wheel at the other. The measuring wheel has a circumference of six inches, and is graduated in inches/8ths. A pointed indicator mounted next to it can be set to indicate the position of the zero mark at the beginning of the measurement.

At the other end is a housing holding an indicator wheel which counts the revolutions of the measuring wheel. The two wheels are connected by a sliding rod and pawl arrangement which advances the indicating wheel one count for each revolution of the measuring wheel, the count being indicated by a scale on the wheel that advances past a line on the edge of a viewing window. There is a small reset lever in the indicating wheel housing which resets the indicator wheel to zero when beginning a measurement.

The distance traversed (in inches) is found by multiplying the revolution count by six, and then adding the residual distance indicated by the pointer adjacent to the measuring wheel.

The construction of this instrument is in the best tradition of scientific instrument making. It is made entirely of German silver and brass, assembled with small screws, and with elegant curved spokes on the measuring wheel and the indicating wheel housing.

BRACKETT'S PATENT ROLLING MEASURE

This is a rolling measure invented by C. W. Brackett, of Jordan, New York, and patented by him on October 13, 1885 (U.S. Patent No. 328,281, entitled "Protractor"). This instrument is an example of a rolling measure in which the entire instrument constitutes the measuring wheel. It is circular in shape, with a diameter of nearly 4 inches, and has a circumference of exactly 12 inches. The circular edge of the wheel is graduated in inches/16ths.

The construction of Brackett's measure is nowhere near as fine as that of Young's. It is made from nickle-plated, stamped out sections of sheet steel, and assembled with rivets.

Brackett's measure is used by grasping the central hub, which is free to rotate, between the thumb and index finger, and rolling it across the

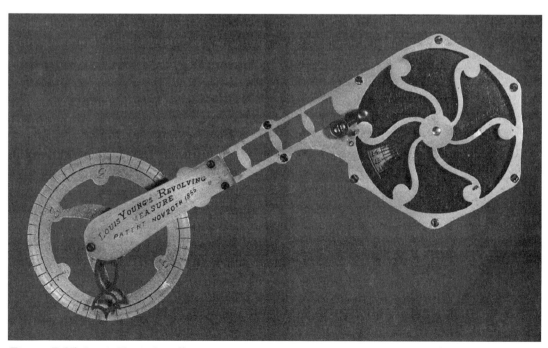

Figure 7-95: Louis Young's rolling measure

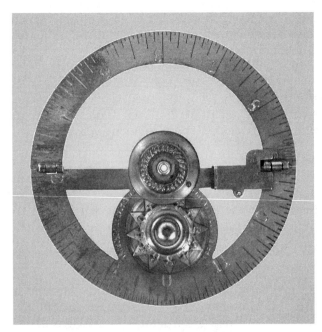

Figure 7-96: Brackett's rolling measure

surface. The hub is geared to a counter wheel on the body of the rule, causing it to advance one count for each full revolution.

There is no provision for marking the starting point on the circumference, nor for resetting the counter wheel. The measurement must be started with the zero point in contact with the surface, and the revolution count is determined by subtracting the starting count from the ending count. The distance traversed (in feet and inches) is the revolution count (equal to feet) plus the indicated inch value at the point of contact with the surface at the end of the measurement.

One interesting feature of Brackett's measure is that it can be folded in half for storage when not in use. The periphery is fitted with two hinges, allowing the top to be folded over for placing in the tool kit or pocket. Adjacent to one of the two hinges is a sliding latch which, when actuated, will prevent folding from occurring while the measure is being used.

As originally described in the patent, Brackett's rolling measure was graduated in degrees on the back edge of the upper half, thus making it possible to use it as a semicircular protractor (hence the title of the patent) when it was folded up. This feature has not been found in any known examples of this instrument.

SADDLERS' RULES

The saddler's rule, also known as a harness-maker's rule, was used by leather workers to measure and cut leather. It was three or more feet long, and extra wide, as much as 2 inches. The length was dictated by the size of the leather pieces which had to be measured and cut. The width was necessary to resist the side force of the knife when guiding it with the edge of the rule.

SEALERS' RULES

During the 18th and 19th centuries it became increasingly common for national governments to require that municipal authorities oversee the use of weights and measures in their locality, to ensure accuracy and to prevent fraud. A town or county would be required to employ a sealer of weights and measures to test the scales and weights, linear measures, and volume measures used by the local merchants for accuracy. This individual would be provided with reference standards and equipment to be used in carrying out his task, and merchants would be required to regularly submit their measures used in trade for him to compare with these standards and mark as complying. This activity is still performed today.

Two types of linear measures were routinely used by sealers: reference standards for checking linear measures used in trade, and dry measure gauges for testing the dry volume measures used in trade for items such as vegetables or grain.

REFERENCE STANDARDS

The most common reference standard of checking linear measures was the standard yard. The simplest (and earliest) examples of this instrument were usually thick lengths of brass or bronze (or wood, if lightness was paramount) exactly 36 inches in length, marked on one side in feet, with the first foot subdivided into inches/8ths, and on another side in 8ths of a yard, with the first 8th subdivided into four half-nails (1 nail equals 2¼ inches).

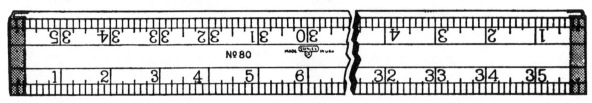

Figure 7-97: Stanley No. 80 saddler's rule

Later examples were more elaborate, being about 40 inches long, with brass or steel stops exactly 36 inches apart fixed to the surface at each end. This version had the advantage of not requiring elaborate comparison procedures; if the yard measure being tested fitted easily, but not loosely, between the stops its overall length was acceptable. These go/no-go standards were also graduated in inches/8ths and in 8ths of a yard/nails between the stop blocks, so that the graduations could also be checked.

A certain amount of variation in design will be found in these standard yards. Examples have been observed that are hinged in the middle and fold up to only 20 inches in length, thus facilitating their transport to the merchant's place of business. Some will have a stop at one end only, to locate one end of the yardstick being checked while comparing it's far end to the graduations marked on the standard surface.

Two examples of reference standards are shown below, Figure 7-98. The upper standard is a gauge for checking yardsticks, the distance between the two protruding sections being exactly 36 inches. It is brass, 40 inches long, and 1/4 inch thick. This gauge is not marked in any way, and may have been made by some local craftsman for use by the sealer in his home town.

The lower standard, a reference standard yard, was used by the sealer of weights and measures in Waywanda, New York, a new town created when Chester, New York was subdivided in about 1840. It is brass, 36 inches long, and 1 inch by 3/4 inch in cross section, and is clearly professionally made. It is stamped with the eagle mark which appears on many reference standards from about this time, but is not marked with the maker's name. It is possible (based on a comparison of the shapes and sizes of the number stamps) that this instrument was made by Joel Andrews, a scientific instrument maker known to have been active in the region of Albany, New York at this time.

In addition to individual instrument makers, two business concerns are known to have manufactured standards for use by sealers, W. & L. E. Gurley, makers of surveying instruments, and the The Fairbanks Company, makers of scales. Both offered a complete line of tools and standards for performing length, volume, and weight testing by local sealers, including standard yards and gauges as described above.

Dry Measure Gauges

A sealer's dry measure gauge is a measuring instrument used by the sealer to determine if the volume measures (e.g., peck, 1/2 bushel) used by a merchant were of legal size, that is, not over- or undersized. Three types of these sealers' rules, known as "dry measure gauges" are known, that made by F. E. Marsh and the two made by W. & L. E. Gurley. All three gauges operated on entirely different principles.

The Marsh Dry Measure Gauge

The Marsh dry measure gauge consisted of a Stanley No. 62½ two foot, four fold rule with 13 elaborate scales stamped onto the brass binding on the edge of the rule, a DIAMETER scale with

Figure 7-98: Sealers' standards. Top: yardstick checking gauge; bottom: standard yard from Waywanda, New York.

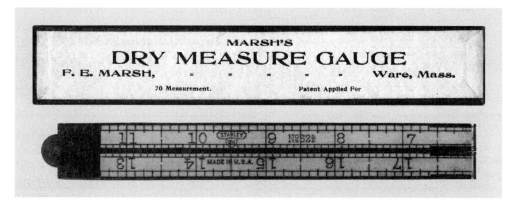

Figure 7-99: Marsh's dry measure gauge (rule is a Stanley No. 62½)

Figure 7-100: Scales on the edge of the dry measure gauge

graduations labeled A through L, and a corresponding set of 12 depth scales with graduations indicating measure capacity.

The DIAMETER scale was used to measure the diameter of the measure being tested. The marking on that scale matching the diameter would then determine which of the other scales would be used to measure the depth.

These rules were manufactured by F. E. Marsh, the local sealer of weights and measures in Ware, Massachusetts from about 1905 until his death in 1938. He would purchase stock No. 62½ rules from Stanley, graduate and stamp the markings on the brass binding, and then market the rules under his own name, packaged in a labeled cardboard box, with accompanying instructions.

Gurley Dry Measure Gauge (Airy's Pattern)

The Airy's pattern dry measure gauge made by W. & L. E. Gurley was also a two foot, four fold rule. However, it had only two scales: a DEPTH scale on one side of the rule, and a DIAMETER scale on the other. Both scales were logarithmic, but with different starting points and different scale factors. In application the scales were used, as indicated, to measure the depth and diameter of the measure being tested, with the two resulting numbers added together. That sum indicated the capacity of the measure, in accordance with a table also marked on the rule, as follows:

30	½ PINT	70	4 QUART
40	PINT	80	PECK
50	QUART	90	½ BUSHEL
60	2 QUART	100	BUSHEL

This use of logarithmic scales to calculate measure capacity was invented by George Biddell Airy, at one time Astronomer Royal of England (Ref. 9). Gauges based on its principles were in use in that country as early as 1894, in the form of solid brass rules nearly two feet long.

The scales on the Gurley dry measure gauge differed from those devised by Airy in one respect: the scale factors and constants used to develop them were specifically selected so that the depth scale by itself could also be used to calculate the volume of rectangular measures. This improvement over Airy's pattern was invented by Edward D. N. Schulte, of Troy, New York, and patented by him on July 6, 1915 (U.S. Patent No. 1,145,706, entitled "Improvement in Computing-Rules").

Gurley was offering its folding wood version of Airy's gauge as early as 1900, in its *Handbook for Sealers* (Ref. 10). The company did not manufacture the gauge itself, however, but purchased it ready-made and graduated from a commercial manufacturer of wood folding rules. Early examples were made by the Stanley Rule & Level Co., and can be identified as such by the five-sided head on the main joint pin. Other examples have been found with the characteristic arch joint shape which marks them as having been made by the Lufkin Rule Co. (These must have been made sometime after 1924, since Lufkin did not commence rule manufacture until that date.)

Gurley Dry Measure Gauge (Go/No Go)

This was a gauge in the form of a brass two foot, three fold rule, marked with the diameters and corresponding depths of the commercially-made dry measures in common use in the first half of the 20th century. It was adequate for a quick check of a standard measure, but could not be used to test a measure whose depth/diameter ratio was nonstandard.

Given the fact that Lufkin was one of the makers of the four fold wood dry measure gauge for Gurley, it is probable that it was the maker of this

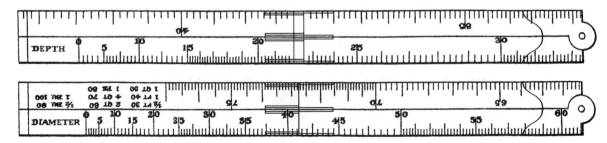

Figure 7-101: Scales on the Gurley dry measure gauge (from the patent illustration)

SPECIAL RULE TYPES AND USES 223

Figure 7-102: Gurley folding brass Go/No Go dry measure gauge

go/no go gauge as well. This is supported by the fact that the gauge was similar in construction to the three fold brass blacksmiths' rules with a locking pivot joint made by the Lufkin Rule Co.

(It is interesting to note that Gurley in its *Handbook* sold not only the two dry measure gauges which were its own products, but also the Marsh pattern dry measure gauge, which was not.)

SEAMSTRESS OR WORKBASKET RULES

These were small rules, usually 12 or 24 inch, folding to 3 or 4 inches, made of boxwood or ivory intended for use when making small measurements while sewing. They were usually graduated in 16ths of an inch, and were sometimes equipped with a sliding caliper. These elegant little rules were apparently a popular form of lover's gift; a number of examples have been found marked with the name of a recipient and (sometimes) the giver.

When a workbasket rule incorporated a sliding caliper, to facilitate the measurement of button diameter when preparing button holes, the caliper usually would be graduated in 40ths of an inch, the increments by which buttons were sized (see BUTTON GAUGES, earlier in this chapter).

A few workbasket rules are known made of mother-of-pearl, with sterling silver trim. These elegant examples are one foot, four fold caliper rules, similar in configuration and size to the Stanley No. 40/Chapin No. 77 ivory caliper rules, and are marked with the name of the Gorham Silver Co.

THE SECTOR

Properly speaking, the sector is not a measuring instrument *per se*, but a calculating instrument; however, its configuration is so rule-like that it has been included here so that rule collectors can recognize and understand one when they see it.

The sector takes the form of a two-fold rule, marked with a special set of scales which can be used in conjunction with a pair of dividers to perform multiplication, division, interest calculations, and solve problems in navigation.

The sector rules usually encountered by rule collectors are the ones which were included in the seven inch high oval pocket cases of drawing instruments so common in the 18th and 19th centuries. This pattern of sector, called "Gunter's sector" after the mathematician who perfected its

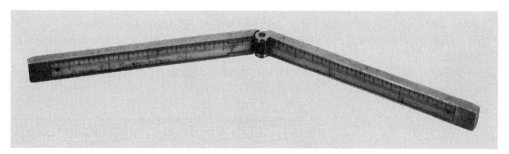

Figure 7-103: Six inch, two fold ivory workbasket rule marked "*Anna L. Spaulding*"

scale arrangement, is a one foot, two fold rule of either boxwood or ivory, with a round brass joint. It is marked with two sets of scales (which Gunter called "lines"): the sector scales, which all radiate from the center of the pivot point to the end of the rule, and the ordinary scales, marked along the inside and outside edges of the rule parallel to the edge (see Figure 7-104, below). The sector scales are: the "lines of lines or lines of equal parts" (the primary sector lines), the "lines of superficies," the "lines of solids," the "lines of quadrature," the "lines of segments," the "lines of inscribed bodies," the "lines of equated bodies and metals," and the "lines of sines and chords." All of these lines are put on the rule in pairs (one on each leg). The ordinary scales are: the "lines of sines and chords" (duplicated on each leg), the "line of tangents," the "line of secants," and the "line of meridional parts." Two of these lines, those of sines and chords and of secants, are six inches (one leg) in length; the other two are marked on the outside edges of the rule, spanning both legs, and are thus twelve inches in length.

The basic principal upon which the sector operates is that "the corresponding sides of similar triangles are proportional." That is, for any matching pair of sector scales, and referring to Figure 7-105 opposite, it will be seen that AE:DE as AC:BC. By making use of this proportionatility, and setting one of the sides equal to 1, multiplication can be rapidly performed graphically, with nothing more than the sector and a pair of dividers.

For example, to multiply 1.3 by 2.5 the procedure is:

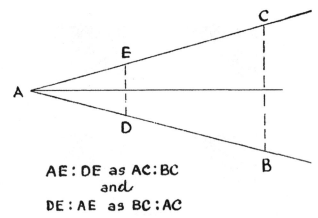

AE : DE as AC : BC
and
DE : AE as BC : AC

Figure 7-105: Proportional triangles

1. Set one point of the dividers at the pivot point, and the other to the value 1 on either line.
2. Without changing the setting of the dividers, set one point to the value 1.3 on one line, and open the sector until the other point rests on the value 1.3 on the other line. (this sets AE = 1.3 and DE = 1)
3. Without changing the setting of the sector, set one point of the dividers at the pivot point, and the other to the value 2.5 on either line. (This sets BC = 2.5)
4. Without changing either setting, slide the points along the two divergent lines until each point falls upon the identical value on each line. This value, seen to be 3.25, is the product of 1.3 and 2.5.

This can be validated as follows:

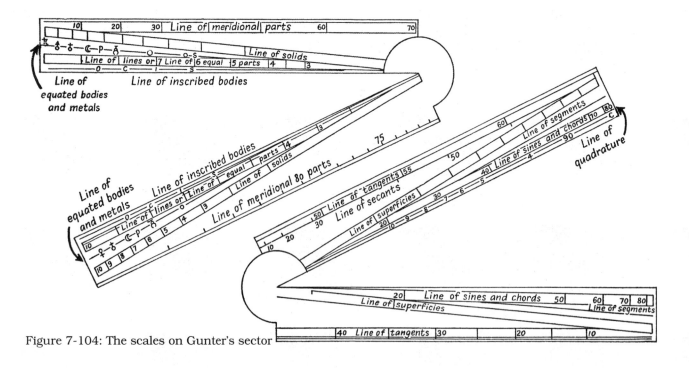

Figure 7-104: The scales on Gunter's sector

If	AE:DE as AC:BC
Then	AC x DE = AE x BC
But if	DE = 1
Then	AC x DE = AC x 1 = <u>AC = AE x BC</u> ...

which is the multiplication just performed.

The theory behind the sector, and its basic configuration, was developed independently by two mathematicians, Galileo Galilei, in Italy, and Thomas Hood, in England, in about 1597 (Refs. 11 and 12). Galileo initially saw it only as a way to solve military problems relating to cannons and fortifications. Only after several years did he even partially appreciate its use as a general calculating instrument. Hood developed his version primarily as an instrument for taking sights and calculating the results in land surveying. Neither realized that its most valuable feature was its use as a general purpose calculating instrument.

The individual who developed the final configuration for the instrument was Edmund Gunter, an Englishman. In about 1606 he proposed an instrument in which all of the features of Hood's sector which were not related to calculation were eliminated, leaving only the lines of numbers, and that other lines of general use be added. These were the lines of superficies and the lines of solids (to facilitate dealing with areas and solids), the lines of quadrature, the lines of segments, the lines of inscribed bodies, and the lines of equated bodies and metals (to facilitate dealing with other classes of problems), and the lines of sines and chords, the line of tangents, the line of secants, and the line of meridional parts (placed on the rule so it could solve problems in navigation).

Equally as important as the development of his sector was Gunter's publication in 1623 of his instructions for its use in general calculation and navigation (Ref. 13). This book, one of the most important English books ever published on calculation, was reprinted many times, widely referenced, and publicized the sector as no prior effort ever had. Largely due to Gunter's work, the sector became the most commonly used calculating instrument, and remained so until supplanted by the slide rule more than 250 years later.

SHRINKAGE RULES

The shrinkage (or contraction) rule is an important item in the set of specialized tools used by the patternmaker in the production of patterns for foundry use.

Such patterns, originally made primarily of wood, but more recently also of metal and plastic, are models of objects which are intended to be cast from molten metal. The pattern is used to form a hollow mold by packing damp sand around it, and then withdrawing it. The hot liquid metal is poured into the resultant cavity, where it cools and solidifies into a copy of the pattern.

As the metal in the mold cools and solidifies, it also shrinks, with the result that the copy is measurably smaller than the pattern. To compensate for this shrinkage, the pattern is deliberately made oversize by a percentage appropriate to the melting point and expansion coefficient of the particular alloy being cast and the size and shape of the casting. This shrinkage allowance can be as little as $1/2$ of a percent ($1/16$ of an inch per foot of length) or as great as almost 6 percent ($11/16$ of an inch per foot of length).

Selecting the appropriate shrinkage allowance is complicated by the fact that there is frequently an intermediate stage in the pattern-to-casting sequence. Often the original pattern will be used to cast a more durable metal pattern of, say, iron or aluminum, which will in turn be used to make the final casting. In such a case, two shrinkage allowances, added together, would have to be used in making the original pattern (e.g., if the intermediate pattern were iron ($1/8$ of an inch per foot), and the final casting brass ($3/16$ of an inch per foot), then the original pattern would have to be made with a shrinkage allowance of 5/16 inch per foot. Typical shrinkage allowances for different alloys and alloy combinations are shown below.

To compute and apply this shrinkage allowance to all dimensions during the construction of the pattern would be a tedious and error-prone task. Instead, the patternmaker uses a "shrinkage" rule, a rule whose graduations are slightly oversize to include the shrinkage allowance, to

Alloy or Alloy Combination	Shrinkage Allowance
Aluminum	3/16" per foot
Aluminum/Brass*	3/8" per foot
Bismuth	5/32" per foot
Brass	3/16" per foot
Britannia Metal	1/32" per foot
Cast Iron	1/8" per foot
Cast Iron/Brass*	5/16" per foot
Cast Iron/Cast Iron*	1/4" per foot
Copper	3/16" per foot
Lead	5/16" per foot
Malleable Iron	1/8" per foot
Steel	1/4" per foot
Tin	1/12" per foot
Zinc	5/16" per foot

Figure 7-106: Typical shrinkage allowances

dimension the pattern. Such rules are graduated as if they were their nominal length, but are actually physically longer by the shrinkage allowance. For example, a two foot shrinkage rule for cast iron would be graduated into exactly 24 "inches," but would actually be 24 1/4 standard inches long.

Shrinkage rules were identified in different ways. In the United States the most common method was to mark them with the shrinkage allowance expressed in fractions of an inch per foot (the method used in the table above). An alternative method less commonly used was to mark them with a figure indicating the shrinkage allowance relative to that for cast iron (i.e., 1 "shrink" equals 1/8 inch per foot, 1 1/2 "shrinks" equals 3/16 inch per foot). The continental custom was to mark the rule with a proper fraction representing the allowance in "feet per foot" (or "meters per meter") (i.e., 1/96 equals 1/8 inch per foot, 1/64 equals 3/16 inch per foot).

A patternmaker would have a whole set of such rules, each with a different shrinkage allowance, to address all possible requirements imposed by different jobs.

Prior to the middle of the 19th century, shrinkage rules were not commercially available, and each patternmaker was forced to make his own. A method for doing this by scaling from a standard rule is given in Chase (Ref. 14). Even after rulemakers began to offer shrinkage rules as standard items this was still done if the required allowance was not available.

Typically, a shrinkage rule would be made of boxwood, (nominally) two feet long, and about 1 1/2 inches wide. They were made both nonfolding, and in a two fold form, and were invariably graduated from left to right along the top of each of the four edges.

The folding shrinkage rule was somewhat unusual in its construction. It did not have the edgewise round, square, or arch joint usually used for two foot, two fold rules, but instead had a knuckle joint, and folded so that the sticks were face to face, instead of edge to edge. The purpose of this was to allow the patternmaker to place it on edge on the work while marking, without any interference from a protruding joint. This meant that the rule could not be laid flat on one side, of course, but apparently this was felt to be less of a problem.

This knuckle joint was the type known as triple plated; i.e.: it had two extra pairs of plates set into the wood in addition to the plates on the edge. The working environment in a pattern shop was such that a rule could receive rough usage, and it was necessary to make the joint as strong as possible.

The most common graduations were 8ths and 16ths of an inch, but 10ths and 12ths were also used, particularly after about 1890. Between about 1850 and 1910, shrinkage rules were made in an increasing number of different contraction allowances by Stanley and other makers such as D. H. Stephens & Co., the Lufkin Rule Co., and the Keuffel & Esser Co. As a result, the practice of making one's own rules became less common, and most patternmakers' tool kits from post-1900 will contain few, if any, craftsman-made shrinkage rules.

After World War I, pattern making became increasingly a precision trade, assuming aspects similar to those of a machinist or diemaker, and the precision achievable using wooden measuring instruments graduated in 16ths of an inch was no longer adequate (a Lufkin catalog from ca. 1945-1950 describes wood shrinkage rules as being suitable for "rough work," but recommends steel rules for "close work," as being more durable and more finely divided) As a result, the wood shrinkage rule was gradually supplanted by its steel equivalant, graduated in 32nds or 100ths of an inch, produced by machinist tool makers such

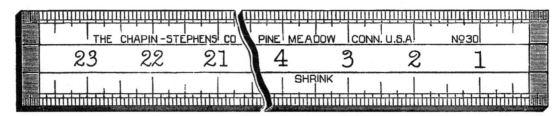

Figure 7-107: Chapin-Stephens non-folding shrinkage rule

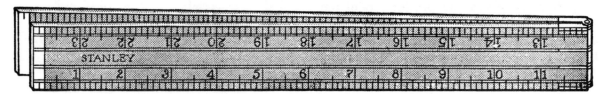

Figure 7-108: Stanley folding shrinkage rule

as the Brown & Sharpe Co., The Lufkin Rule Co., or the L. S. Starrett Co.

An unusual wood shrinkage rule is sometimes encountered which has an extremely large shrinkage allowance, in the neighborhood of around 15%. No combination of metals would require a shrinkage allowance this large. These extreme shrinkage rules were actually intended for sizing the patterns for casting ceramic objects such as sinks, drinking fountains, etc., which would be fired in a furnace after being cast and dried. It is a characteristic of such ceramics to exhibit major shrinkage in the process of drying and being fired.

The flat-folding construction of American shrinkage rules differed significantly from the standard pattern in England and Europe. In these markets it was traditional to construct folding shrinkage rules in the same manner as ordinary two fold and four fold rules, with an edge joint, and, in the latter case, two knuckle joints.

SIX FOLD RULES

The two foot, six fold rule, measuring only four inches when folded, was very easy to carry in the pocket, and had much less tendency to drop out when the workman leaned over. It was also an attractive rule, with pleasing proportions and plenty of nice-looking brass or German silver joints.

The number and arrangement of these joints posed some problems, however. The rolls for the four middle joints protruded on both sides of the rule, as compared to one side only on four fold rules; thus the six fold rule could only be laid flat on one edge while measuring or marking. Also, because these middle joints folded both ways, this rule was somewhat more difficult to handle; it had a tendency to sag when held unsupported in a horizontal position.

A third problem, pointed out by Crussel (Ref. 15), was that some carpenters, instead of using the figures on the rule, calculated the distance by counting inches from the nearest joint. It must have been confusing to such a user to use a six fold rule, where the joints were not at 6, 12, and 18 inches, but at 4, 8, 12, 16, and 20 inches instead.

It is not known when six fold rules were first offered commercially in the United States. Their five separate joints made them almost twice as expensive as four fold rules, which had only three. It is probable that they only became economically practical with the introduction of machinery to stamp out and file joint plates in the middle of the 19th century.

The earliest known price lists offering these rules was that of the Stanley Rule & Level Co. in February 1859, and of Hermon Chapin in July of the same year. E. A. Stearns & Co. began making them some time prior to 1861. It is not certain when Stephens or Belcher Brothers first offered them; the scarcity of catalogs or price lists for these makers makes it very difficult to date the introduction of any product exactly. In any event, from about 1870 on these rules were commonly available from all of the major makers.

During the period when they were offered, these rules were available in at least four different patterns, and in both boxwood with brass and ivory with German silver. The most widely offered, and the most frequently encountered today, are the boxwood rules with an arch main joint and edge plates. They were also available full bound, however, and one maker (Belcher Brothers & Co.)

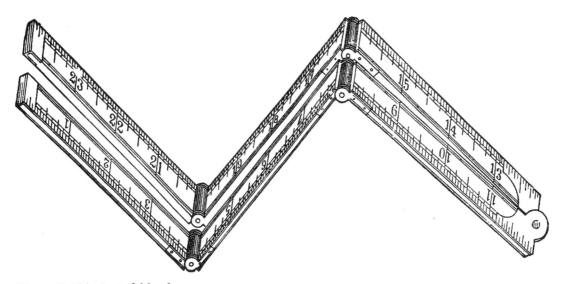

Figure 7-109: A six fold rule

also offered them with middle plates. As was customary, the wood rules were trimmed with yellow brass, and the ivory with German silver (white brass).

Four different examples of six fold rules are shown in Figure 7-110 below. These are, from top to bottom, a Standard No. 58 (unusual in that it is 7/8 of an inch instead of 3/4 of an inch wide); an E. A. Stearns No. 60 (ivory & German silver), an early Stanley No. 58½ (completely covered with various weights of metals tables), and a late Stanley No. 58½ (no tables, but full bound).

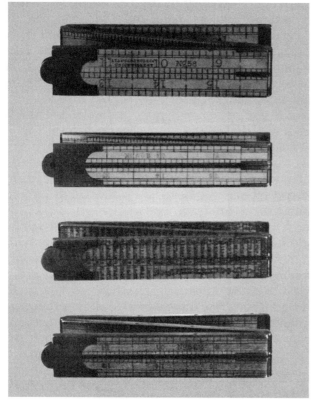

Figure 7-110: Different patterns of six fold rule

Because of the extra work involved in their manufacture, these rules always sold at a premium. In 1879, for example, The Chapin six fold No. 96 was listed in its catalog at $13.00 per dozen, while the four fold No. 17, otherwise similar, was listed at only $8.00 per dozen.

These rules continued to be made right up until the eve of World War I, when, like many other low-volume products, they were dropped due to labor and material shortages. Stanley offered them for the last time in 1915, and presumably Chapin-Stephens dropped them about the same time.

STATIONERS' GOODS

During the middle years of the 19th century some rulemakers offered products outside of their normal product line of measuring instruments. These items were either specialty measuring instruments, such as hatters' rules, or wood items without graduations that might concievably be sold through stationary stores. Three such companies were the Stanley Rule & Level Co., Hermon Chapin, of Pine Meadow, Connecticut, and Belcher Brothers & Co. of New York City.

Stanley, in particular, offered a number of such disparate items in its catalogs. From 1862 to 1865 its line of such goods was so extensive as to warrant a special section in the catalog labeled Stationers' Goods (see illustration opposite). Among other things, one could buy specialized tools for use by navigators, watchmakers, and accountants, checkermen, and geometrical models for use in schools. (This was a period when Stanley had not yet focused its efforts on tool manufacture exclusively, and was still offering buttons, toy pistols, and even roller skates.) Chapin and Belcher Brothers did not offer quite as extensive a selection as Stanley, including them under Miscellaneous in catalogs instead of listing them separately.

By about 1870 most of these non-tool items had been eliminated from these makers' product lines. As the companies grew and expanded their operations, these unusual items were dropped as being outside their main focus and marketing effort. The button gauges were the last to go, remaining in the catalogs until about 1880, after which they also were eliminated.

STEAM ENGINEERS' RULES

The variations in pressure in the cylinder of a steam engine at different times during the stroke of the piston provides a great deal of information to the engineer operating the engine as to engine horsepower and efficiency, the timing and proper functioning of the intake and exhaust valves, and correct steam pressure. To measure and record these pressure variations, the engineer would use an instrument known as a Steam Engine Indicator to plot pressure versus position for an entire cycle on a small 2 inch by 4½ inch card.

This instrument was connected by a string to the engine crankshaft or connecting rod, the back and forth motion of the piston thus causing a drum on the indicator to rotate back and forth. A card was wrapped around this drum, held in place by two spring clips. A piston in the indicator, connected to a vertical stylus, was connected by a steam tube to the engine cylinder. The changing pressure in the cylinder during its stroke, balanced by a spring behind the piston, would move a stylus up and down on the surface

STATIONERS' GOODS.

11

No.		Per dozen.
160,	Rulers, Mahogany, Flat, Plain Finish, Bevel Edge, Assorted, 12 to 15 inches, - - -	$0.40
162,	Rulers, Mahogany, Flat, French Polish, Bevel Edge, Assorted, 12 to 15 inches, - - -	0.80
164,	Rulers, Mahogany, Octagon, French Polish, Assorted, 12 to 15 inches, - - - -	1.50
166,	Rulers, Rosewood, Flat, French Polish, Bevel Edge, Assorted, 12 to 15 inches, - - -	1.50
168,	Rulers, Rosewood, Round, French Polish, Assorted, 12 to 15 inches, - - - -	2.50
170,	Rulers, for Scholars' use, Boxwood, Bevel Edge, Extra Finish, Graduated to Eighths and Sixteenths, 1 foot, - - - - -	1.00
172,	Drafting Scale, Boxwood, with Scale for Drafting Twelfths and Twenty-fourths of an inch, 1 foot,	2.00
174,	Architects' Drafting Scale, Boxwood, with Scale for Drafting for $\frac{1}{8}$, $\frac{1}{4}$, $\frac{3}{8}$, $\frac{1}{2}$, $\frac{5}{8}$, $\frac{3}{4}$, $\frac{7}{8}$, 1, $1\frac{1}{4}$, $1\frac{1}{2}$, $1\frac{3}{4}$, 2, $2\frac{1}{4}$, $2\frac{1}{2}$, $2\frac{3}{4}$, 3 inches to the Foot, - -	5.00
176,	Gunter's Scale, Boxwood, - - -	9.00
186,	Printers' Rules, Satin Wood, Square, 1 foot, -	2.00
188,	Printers' Rules, Boxwood, 4 Fold, 1 Foot, -	3.00
204,	Checkermen, Boxwood, Plain, Assorted Sizes, put up in neat boxes, containing 1 Set, per gross Set,	18.00
205,	Checkermen, Boxwood, Polished, Assorted Sizes, put up in neat boxes, containing one Set, per gross Set,	30.00
206,	Dissected Cube, for illustration of Cube Root, Boxwood, per dozen Set, - - - -	6.50
208,	Cubical Blocks, 1 Inch Square, containing 64 cubical Blocks, - - - - -	10.00
210,	Button Gauge, Boxwood, - - -	4.00
212,	Hatters' Rules, Boxwood, - - -	4.00
214,	Watch Glass Gauges. - - -	1.50

Figure 7-111: List of stationers' goods from an 1863 Stanley catalog.

of the card. The result, a roughly kidney-shaped closed curve drawn on the card, would be measured for height, width, and area, and that information, together with the engine speed and clues provided by the shape of the curve would tell the engineer what he needed to know about the operation of the engine.

The area of the curve was measured using a planimeter, a clever instrument which could determine the area of any closed shape with reasonable accuracy by simply tracing around its periphery. The height and width were determined using short three or four inch flat bevel-edged boxwood scales graduated decimally into 20, 30, 40, etc. divisions to the inch, the particular scale used being a function of the ratio between piston stroke and card width and the ratio between inlet steam pressure and card height.

This second ratio was determined by the strength of the spring balancing the stylus piston, and a specific spring would require a specific scale. Since these indicators were supplied with a set of springs of various strengths for use with different steam pressures, they would also be supplied with a corresponding set of scales.

Sometimes a set of these scales was supplied with the indicator or could be purchased separately from the maker. Some types of planimeters had detachable scales which could be used either on the instrument when measuring area, or detached when measuring height and width.

When found, these scales will usually be marked with the name of the maker of the steam engine indicator with which they were originally

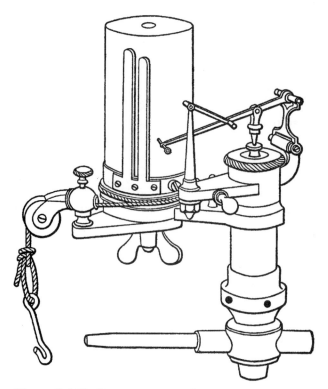

Figure 7-112: Steam engine indicator

supplied, the two largest makers being the Ashcroft Valve Company and the Crosby Steam Gauge Company.

TAILORS' RULES

Tailors face four tasks: taking the measurements of the subject, drafting a pattern that conforms to those measurements (ending with the various pieces drawn in chalk on the fabric), cutting out the pieces, and sewing them together. Of the four the second, drafting the pattern, was by far the most difficult. Measuring required care, as the final pattern depended on the measurements for proper dimensions; cutting required a moderate amount of skill; sewing required more time than either one, but required much less knowledge and training. Drafting the pattern, on the other hand, was a complex task requiring a high degree of skill. Forming a flat, basically two-dimensional material like cloth into a three-dimensional covering for the person, a covering adhering to the styling dictates of the day, was more an art than a process, one that was very difficult to master.

Over the years many individuals have attempted to systematize the process of measuring and drafting, to make the process accessible to more than a gifted few (Ref. 16). The records of the U.S. Patent Office show many patents on tailoring systems and implements, with no fewer than 44 such patents being granted prior to the U.S. Civil War, and many more thereafter.

The instruments associated with these patents are frequently encountered by collectors, and in their variety could form a field of study all by themselves. Roughly speaking they fall into three categories: measuring sticks and sleeve sticks, squares, and bias squares.

MEASURING STICKS AND SLEEVE STICKS

Measuring Sticks

Measuring sticks are ordinary straight rulers, from one to three feet in length, and about 1½ inches wide. They were sometimes graduated into

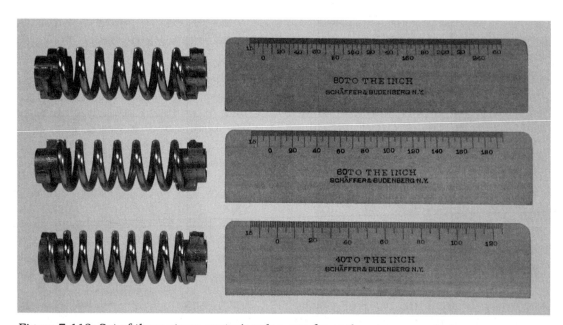

Figure 7-113: Set of three steam engine's rules, one for each pressure spring

inches/8ths, but more frequently with special graduations for use in the tailoring system for which they were designed. Probably the most common special graduations were to have the stick marked with scales of reduced-size inches: halves, quarters, thirds, and sixths. By using measurements made with these reduced size graduations the tailor could proportion the different parts of the pattern appropriately. Other special scales would sometimes be included for applications such as spacing buttons or marking cloth on the bias.

Another class of straight stick was the "ell" measure, used for measuring cloth as yard goods and for large layout tasks such as skirts and coats. The English and American ell is 45 inches in length, although prior to the 19th century this varied somewhat with time and locale. This 45 inch dimension can be seen reflected in the 1 1/4 yard tailors' sticks sold until recently by the Lufkin Rule Co. Ell measures made prior to the middle of the 19th century were often square, slightly tapered rules with a turned handle on the large end; more modern ones are simply 1 1/2 inches by 5/16 of an inch with bevels on both edges. Typically ell measures would be graduated in inches/8ths, 8ths of a yard and (on early examples) nails for the first 9 inches, one nail being equal to 4 1/2 inches.

Sleeve Sticks

Sleeve sticks were used by tailors for drawing the curved lines when drafting the patterns on the fabric. The name "sleeve stick" comes from the curvature required in drafting the arms of a pattern, but almost every other line in the pattern had to be more or less curved as well. Few of the lines in the human body are perfectly straight, and the edges of the pattern pieces have to curve to correspond. A tailor drafting a pattern would have to have a number of these sticks, to allow him to draw a variety of curves. As a result they were frequently made in sets of 10 or 12, with each stick in the set having a different curve.

TAILORS' SQUARES

The tailors' square is a light wooden square perhaps 3/16 of an inch thick, with a brass reinforcing plate (and sometimes a brace) at the corner. The most commonly encountered examples are 14 inches by 24 inches, but the overall size and the proportions vary widely. Examples as small as 4 inches by 6 inches (presumably for making children's or doll's clothing) have been found.

Tailors' squares are usually graduated in inches/8ths on one side and with the same special graduations which appear on tailors' sticks on the other.

These squares were used in both the measurement process and in drafting the pattern. Placing one blade in the armpit allowed accurate simultaneous measurement of the arm heights and thickness at the shoulder, for example. During drafting the right-angle blade helped establish a reference line from which other dimensions could be marked off.

Take-down or folding tailors' squares are sometimes encountered. In the 19th century it was not uncommon for a tailor to go to the client's home or place of business to measure him or her for a new garment. A square in which one leg could be disconnected from or folded down against the other was much easier to carry

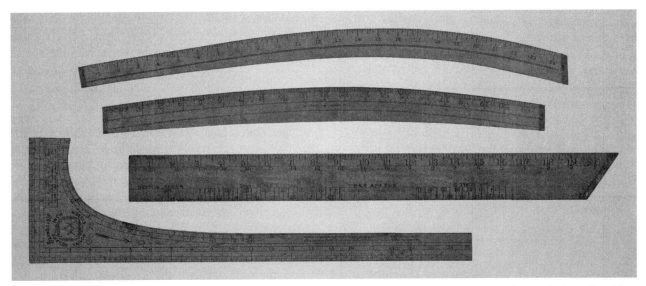

Figure 7-114: An assortment of tailors' rules. Top to bottom: Lufkin sleeve stick, Watts sleeve stick, rule with 45 degree tip to aid in cutting on bias, combined rule, curve, and square.

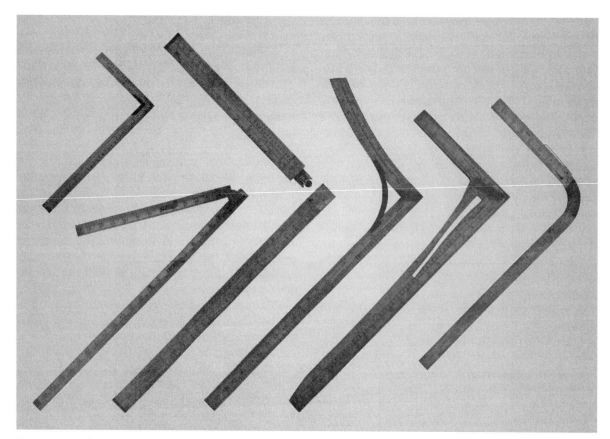

Figure 7-115: A sampling of tailors' squares

through the streets. Examples of each of these two types are show in Figure 7-115. Tailors' rules which folded double or came apart into two pieces were also made, for the same reason.

BIAS SQUARES

Cloth cut at a 45 degree angle to the threads of the fabric, (the "warp" and "weft") has different stretch and movement properties than cloth cut in line with these threads. Tailors would sometimes cut one or more pieces of the garment in this way ("on the bias" as they termed it) to take advantage of these properties in making these pieces work with the rest of the pattern and ensure that the garment would move and drape properly.

To meet the needs of tailors special tools were developed to simplify marking a 45 degree line on the fabric either for cutting, or as a base line to allow the piece being laid out to be skewed by this amount. These instruments were made in a variety of forms. Squares were offered with a pivoting blade that could be set and locked at 45 degrees. Several different makers offered 45 degree right triangles with legs about 16 inches long. One manufacturer offered an instrument resembling a queen post truss with two parallel legs of unequal length joined by short sections at 45 degrees.

These distinctions between the different types of tailors' tools are not ironclad, however. Inventors would frequently find ways to combine two or more functions in a single instrument. In figure 7-114, for instance, the bottom stick is designed to also serve as a curve stick and as a square. In the same figure, the stick directly above it has its tip cut at 45 degrees to allow it to mark bias lines. In Figure 7-115 several of the squares include curves in their design to be used in marking curved lines while drafting. In Figure 7-116 the upper instrument is a folding square with a brace allowing it to be locked at 45 degrees and mark bias lines.

THE MITCHELL SHOULDER SQUARE

One tailoring instrument that does not fit into any of the above categories is the so-called Mitchell Tailor Square. This is a hybrid instrument developed specifically for the purpose of measuring the upper torso. It would be placed under the armpit of the subject with the bar pointing forward and the sliding jaw then moved in to contact the front of his arm at the shoulder, after which the attached tape measure would be used to measure his chest circumference. This square was apparently named after its designer, but who Mitchell was, and whether he ever

SPECIAL RULE TYPES AND USES 233

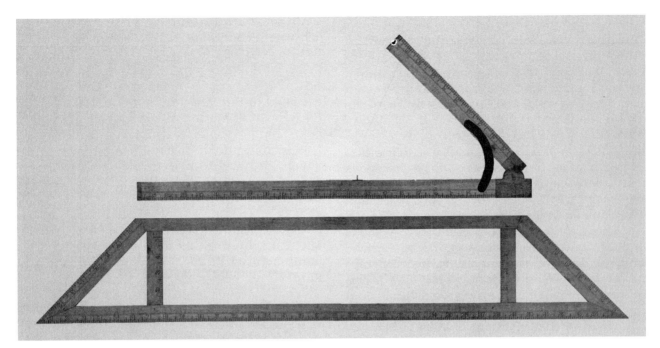

Figure 7-116: Two tailors' tools for cutting on the bias

patented it is not known. It is shown here only because examples are not uncommon, and its design is unusual enough to puzzle a collector who encounters one.

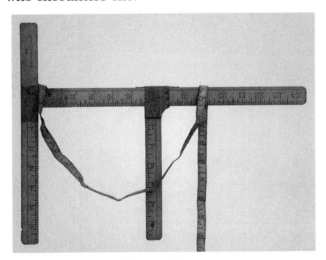

Figure 7-117: Mitchell shoulder square

TEXTILE INDUSTRY RULES

Two types of specialized rules were used in the textile industry: rules for measuring and grading the manufactured goods, and rules which aided in the design of the mill and the integration of the machinery.

MEASURING AND GRADING INSTRUMENTS

Cotton Staple Gauge

One criterion for grading the output of a cotton or woolen mill was the number of threads per inch in the woven fabric; the more threads per inch, the heavier the cloth and the better the quality.

The Stanley Rule & Level Co. manufactured a special rule for measuring this parameter, called a "Cotton Staple Gauge." This was a one inch wide by three inch long rule with two beveled edges, graduated in both English and metric divisions. Its small size allowed it to be carried in the vest pocket of the foreman or textile purchaser. Officially, Stanley only offered this rule in its 1929 catalog, but it must have been part of its product line for many years before that; both boxwood and ivory examples have been found with trademarks indicating pre-1900 manufacture.

Figure 7-118: Stanley No. 299 cotton staple gauge

English makers offered more elaborate rules and devices for performing this function, ranging from rules with blades that would cut strips exactly 1/2 inch wide from the fabric for subsequent thread counting, to precision instruments that would traverse a magnifying glass a controlled distance across a fabric sample as the counting was performed.

Putnam's Bolt Measures

By the late 19th century the customary method of packaging yard goods was to form them into a "bolt," that is, to wrap them around a flat board made of pasteboard about ½ inch thick, 8 inches wide, and 20 inches to 24 inches long. If the fabric was wider than the length of the board, it was folded lengthwise prior to wrapping. This form of packaging made the fabric easier to ship, store, and dispense in the dry goods shop. As lengths were required for sale they could easily be cut off the end of the wrapped fabric and the remainder, still on the board, returned to the shelf. This did create a problem for the dry goods merchant, however, of estimating the amount of fabric remaining on the bolt. Completely unwrapping the fabric from the board to measure it and then rewrapping it was a slow process and could stretch, wrinkle or soil the cloth.

In 1898 Alfred E. Putnam of Milan, Michigan invented a special measuring instrument for performing this estimating in a quick and simple manner. This instrument was a 20 inch sliding caliper with 5 inch jaws made of maple with nickel-plated sheet steel joints. The bar of this caliper was one inch square and was marked with 47 scales, one for each number of "rounds" (turns) of cloth which could be on the board. In use, the number of rounds of cloth on the board were counted, using a slightly-blunted wood stylus, and the result used to select which of the 47 scales to use. Then the caliper was used to measure the width of the bolt including the cloth; the position of the moving jaw would indicate on the selected scale the number of yards in the bolt.

This caliper (the upper instrument in Figure 7-119) was marked: PUTNAM'S CLOTH CHART, COPYRIGHTED 1896, PATENT PENDING. The patent referred to has not been found, however, nor has any information relating to the claimed copyright.

Eleven years later Putnam invented a second version of his "Cloth Chart" and was granted a patent for it on May 14, 1907 (U.S. Patent No. 853,262, entitled "Computing-Calipers"). This caliper, which operated on the same principal as his original caliper, was made entirely of nickel-plated steel, and instead of having its scales marked on its surface had them printed on a long cloth chart which was rolled up inside the bar of the instrument. This bar had a long window slot, and by turning a large knob adjacent to the fixed jaw the desired scale could be moved to the window for use. The middle instrument in Figure 7-119 illustrates this second version.

This all-metal caliper is marked: PUTNAM'S IMPROVED CLOTH CHART, A. E. PUTNAM MFR., WASHINGTON IA. and PAT'D MAY 14 MAY 21, '07. The second date is for a patent for a similar device granted on May 21, 1907 to William C. Fogle, of Williamsburg, Kansas (U.S. Patent No. 854,476, entitled "Measuring Device"), and assigned to Putnam. It is not clear what relation Fogle's patent has to Putnam's all metal caliper; it bears only a vague resemblance to it. Perhaps this second patent covered some minor feature which Putnam desired to incorporate into his caliper, or perhaps Putnam simply purchased the patent to avoid competition.

It is interesting to note that although in both patents Putnam's residence is given as Milan, Michigan, by the time this second version was manufactured he had relocated to Washington, Iowa.

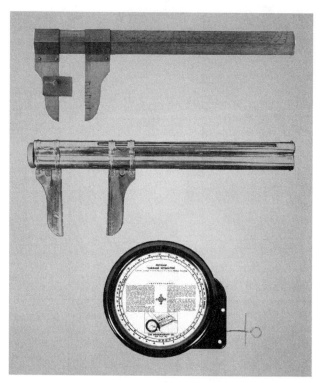

Figure 7-119: Three generations of Putnam's yardage estimator

By the 1920's three of Putnam's sons, Howard, Nathan (Clyde), and John had joined the firm, which then became known as A.E. Putnam & Sons. Clyde was an active partner, devoting his full time to the business, while Howard and John were less involved.

In 1937, at 73 years of age, A.E. Putnam invented and patented a third and radically different version of his "Cloth Chart." In this new instrument, which he referred to as his "Wonder Cloth Measure," he abandoned the use of calipers to measure the bolt. Instead he used a steel tape linked to what resembles an eight inch diameter tape measure in a bakelite case (see the bottom instrument in Figure 7-119). Pulling out the tape, which was only 17 inches long at its maximum extent, caused a circular dial on the instrument's face to rotate. This dial was marked on its circumference with a logarithmic scale representing yards. Surrounding this dial was a fixed logarithmic scale representing rounds of fabric on the board. To use this instrument, the rounds would be counted as before, and then the tape pulled out far enough to span the bolt including the rounds. Opposite the number of rounds on the outer scale would be the number of yards in the bolt on the inner scale. this greatly simplified the measuring process, and eliminated the problem of selecting the appropriate scale.

All known examples of this instrument are marked "PATENT PENDING," but a search of all likely patent classifications turned up no patent. Further research will have to be conducted to identify this patent if, in fact, it was ever issued.

Following his father's death in 1939, Clyde continued to operate the company and manufacture the "Wonder Cloth Measure" until his own death in 1947. At that time the company was purchased by the Measuregraph Company of St. Louis, Missouri, a firm specializing in instruments and machinery for folding and measuring fabric. The author has been informed by its management that Measuregraph continued to make this instrument under its own name until well into the 1950s. Examples from this period are marked "PUTNAM YARDAGE ESTIMATOR," with Measuregraph named as the manufacturer.

TEXTILE SLIDE RULES

The second type of rule used in the textile industry were rules incorporating Gunter's (logarithmic) computing slides and data tables for performing the calculations necessary to connect up the various machines. For example, the machines doing the carding had to produce exactly enough carded cotton to supply the machine converting it into roving; if they ran too slowly, the roving machine would be starved for stock; if too fast, the carded cotton would accumulate faster than it could be processed. Since all machines in the early mills were run from a common drive shaft, it was necessary to size the driving/driven pulleys so that each machine was working at the appropriate speed.

The engineers' rule, with its Gunter's slide and data tables was often used for this purpose, and many 19th century books of instructions on the use of the slide rule dwelt at some length on textile industry problems. In addition to the ordinary engineers' rule, other specialized rules were developed, both in America and in England, adapted to this class of problem. Hogg's rule, developed in America from Routledge's engineers' rule is one such; Slater's rule, invented in England by William Slater, is another.

Hogg's Textile Slide Rule

Hogg's rule is a slide rule developed for use in the textile industry, invented some time prior to 1886 by James Hogg, of Lawrence, Massachusetts. Hogg was a designer or inventor connected with the textile industry (during his career he patented at least four other inventions relating to spinning and carding machinery). He developed his slide rule by rearranging the four scales on the Gunter's slide of the engineers' sliding rule, duplicating two of them, and augmenting those six with two new ones: a linear scale (which we would call the L scale today) and a reciprocal logarithmic scale (which we would call a CI scale today). At three locations on the C scale (at the values 5.5, 10, and 11.4) he inserted tiny brass plugs, similar to those found on Gunter's scales

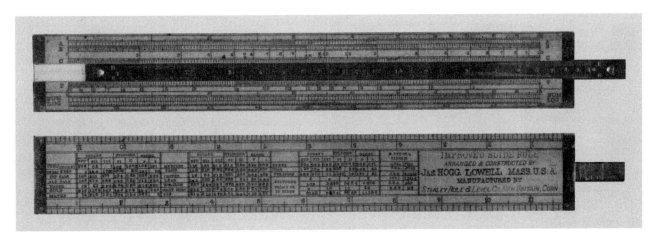

Figure 7-120: Hogg's textile slide rule

(see the section GUNTER'S SCALES, above), to mark three gauge points which were frequently used in textile-related calculations.

On the back of his rule Hogg placed most, but not all, of the same tables which were marked on the back of the engineer's sliding rule, omitting only the table of gauge points for pumping engines. He used the empty space thus created to place his name and the name of the actual maker of the rule. The full message reads as follows: IMPROVED SLIDE RULE / *ARRANGED AND CONSTRUCTED BY* / Jas. HOGG, LOWELL MASS, U.S.A. / MANUFACTURED BY / *Stanley Rule and Level Co., New Britain, Conn.*

Hogg also published a book of instructions for use with his rule in 1887 (Ref. 17). This book, 163 pages long, was divided into two parts. The first part, consisting of 72 pages, dealt with the instrument's use as an ordinary engineers' sliding rule, and was similar in content to many other 19th century instruction books for carpenters' and engineers' sliding rules. The second part was devoted entirely to the problems and calculations peculiar to textile manufacture (e.g.: "To Find the Draft-Gear Constant for Each Operation," "To Find the Twist Change-Gear Constants.")

This rule is frequently referred to as "Hogg's Patent" rule, but this is apparently a misnomer. There is no record of Hogg ever receiving a patent on this rule, and he never made such a claim in his instruction book or any advertisement for it.

Slater's Engineers' Sliding Rule

This is a version of the engineers' sliding rule developed by William Slater, of Bolton, England, for use by engineers and mechanics in textile mills. On this rule the tables of gauge points found on Routledge's rule (see ENGINEERS' (SLIDING) RULES, above) are replaced by written-out formulas for solving the common problems encountered in setting up textile machinery. By marking the rule in this way, Slater eliminated the need for a separate instruction book to accompany the rule.

A typical formula:

REV OF SPINDLES PER INCH.
MULTIPLY TURNS BY THE REV.
OF SPINDLES FOR 1" OF RIM.
DIVIDE BY THE N° OF INCHES PUT UP.

As far as can be determined, Slater's rule was never made by any American manufacturer. All known examples are marked either Edward Preston & Sons or John Rabone & Sons, both English makers.

TINSMITHS' RULES

These were steel rules, 1 1/4 inches wide and 3 or 4 feet long, widely used by sheet tin and copper workers. They were used both to mark and measure, and also provided the basic data needed in this type of work.

Tinners rules are thin (1/16 inch) and made of hardened steel, making them flexible enough to bend around gentle curves and conform to the surface being marked/measured. They were supplied with a hanging hole at one end.

On the front these rules were graduated with two scales, a common inches/16ths scale and a "circumference" scale. This second scale had "inches" which were $1/\pi$ (equal to 0.3183) times the size of common inches, allowing easy conversion between diameter and circumference when laying out shapes on flat sheet stock prior to cutting and forming. On the back these rules were marked with tables giving the dimensions of common standard containers (flat top cans, flaring dry measures, etc.) and basic equivalents for converting gallons to cubic inches, etc.

Examples of this rule which retain the circumference scale but omit the data tables are sometimes encountered. This second version was for use by general sheet metal workers, who had no need for the information in those tables.

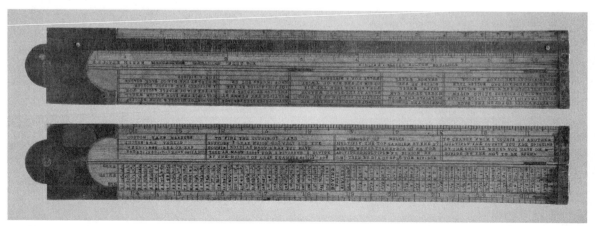

Figure 7-121: Slater's engineer's slide rule

SPECIAL RULE TYPES AND USES

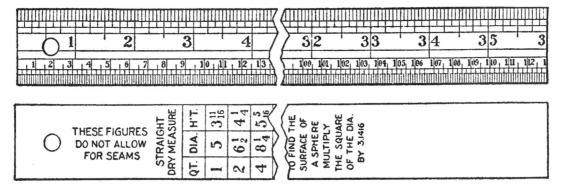

Figure 7-122: Lufkin three foot tinner's rule

At least two companies, the Lufkin Rule Co., and Peck, Stow, and Wilcox, (a major manufacturer specializing in tinsmiths' machinery and supplies) offered these rules commercially until quite recently.

WATCH GLASS GAUGES

Replacing watch glasses (or "crystals") was one of the most common tasks for a watchmaker, and the most critical part of the job was selecting exactly the correct size of crystal from his stock of repair parts. Crystals were not mounted by upsetting the metal of the case over their edge, but were cemented in place. The effectiveness of this cementing depended largely on the proper sizing of the crystal to the groove. After determining the diameter of the groove which receives the crystal, the repairman would use a watch glass gauge to measure the diameters of various crystals from his stock until he found one exactly the right size.

The sizes of watch glasses is expressed in either Geneva or Lunette measurement, two different standard developed in Switzerland. Geneva measurement is expressed in units and 16ths, where a unit is approximately equal to .0825 of an inch. Lunette is expressed in units and 4ths, where a unit is approximately .0864 of an inch. Thus a watchmaker would say that a Waterbury watch takes a 18 1/16 Geneva or 17 1/4 Lunette crystal.

To measure a crystal he would use one of the gauges shown in Figure 7-123. Both gauges are flat pieces of wood or metal, with a transverse step fastened across them near the handle. The crystal being measured would be placed on the surface of the gauge, its lower edge resting against the step, and its diameter determined by what graduation line it touched at its upper edge. On the Lunette gauge all lines, both the whole numbers and the 4ths were marked on the surface and the diameter read off directly. On the Geneva gauge sixteen vertical lines labeled 0 through 15 intersected the horizontal graduations, and the horizontal graduations sloped slightly, each line being one unit higher at its right end than at its left. The size of the crystal was shown by which horizontal graduation it touched and which vertical lines intersected that horizontal line at the point of intersection. As an

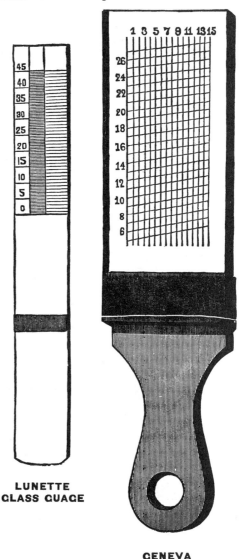

Figure 7-123: Two European watch glass gauges

example, a crystal whose edge touched the 19 line at the point where the vertical 11 line crossed it had a diameter of $19^{11}/_{16}$ Geneva measure.

WOODCUTTERS' RULES

These were long (four to six foot) measuring sticks used to measure stacked firewood. Wood for fuel is traditionally bought and sold in quantities called "cords," where a cord of wood is defined as 128 cubic feet. An 8 foot stack of wood 4 feet deep and 4 feet high, for example, would contain exactly one cord. A woodcutters' rule would have two scales: one in feet and inches for measuring the length and width of the pile, the other for measuring the height of the pile. This second scale would be graduated to read the contents in cords of a pile 16 feet by 4 feet (the usual way wood was stacked); thus each foot of this second scale would represent $1/2$ a cord. These stick were usually $3/4$ or 1 inch square in cross section, and had flat brass end plates.

YARD STICKS

The yard stick is the rule most commonly encountered in household use, and has been for many years. Its three foot length, and its graduation into inches/fractions and fractions of a yard adapt it to almost all measuring requirements.

Yard sticks vary widely in their construction details. Most were flat, and either $3/4$ of an inch or 1 inch wide, but some makers also offered them with a $1/2$ inch square cross section. In England they were available with a round cross section. Most, but not all, had brass tips. The majority were made of maple, but hickory was also used, particularly for the thinner sticks, and cheap ones of softwood abound.

It is the scale of fractions of a yard which distinguishes the yard stick from any common three foot rule or stick. This scale divides the stick into 8ths of a yard, with additional markings at $1/16$ yard and $1/32$ yard. On early yardsticks the $1/8$, $1/16$, and $1/32$ yard markings were sometimes labeled as "2 NAILS", "1 NAIL", and "$1/2$ NAIL"; (the nail, equal to $4^{1}/_{2}$ inches, is an early name for $1/8$ of a yard which fell out of use around the middle of the 19th century).

The fraction of a yard scale was for measuring fabric for tailoring or upholstery purposes. Sewing patterns usually specify the amount of fabric required in yards, and that is the way it is customarily priced and sold. Many stores dealing in "yard goods" would not only have several yard sticks loose on hand, but would also have one or more fastened to the counter, to facilitate use and prevent loss (see COUNTER MEASURES).

Both scales on a yard stick were usually graduated from left to right. Stanley was the one exception to this rule; that company graduated their yard sticks from right to left well into the 20th century, only changing to the more common pattern in the mid-1920s.

A number of makers also made yard sticks in the form of three foot, four fold boxwood rules, marked on one side with yard graduations and on the other with inches/16ths. Some put the yard graduations on the inside, others on the outside, purely as a matter of choice.

The yard graduations on these folding yardsticks differed in two ways from those on the common sticks: the scale was graduated in both directions, with markings and figures running from both left to right and from right to left, and two equally spaced embellishment lines were added. The right to left graduations were above the upper line and the left to right graduations were below the lower line; the space between them was used for decorative asterisks emphasizing the graduation lines. Both sets of figures used the graduation lines as the separators for their fractions.

Figure 7-124: Common yard stick

Figure 7-125: Yard graduations on a four fold boxwood yard stick

REFERENCES:

1. Henry Coggeshall, *Timber Measure by a Line of More Ease, Dispatch, and Exactness Than Any Other Way Now in Use, By a Double Scale ...* (London: publisher unknown, 1677).
2. John V. Knott, "Joshua Routledge, 1775-1829," *Journal of the Oughtred Society*, 4:2 (October 1995): 25.
3. Gil Gandenberger, and Mary Gandenberger, "The Stanley Cask Gauging Rod," *Stanley Tool Collector News*, 16 (Winter 1995), 14-17.
4. —, *A Book of Tools, Machinery and Supplies* (catalogue) (Detroit, Michigan: Charles A Strelinger & Co., 1895).
5. Peggy Aldrich Kidwell, "Publicizing the Metric System in America from F. R. Hassler to the American Metric Bureau." *Rittenhouse*, 5:4 (August 1991), 111-117.
6. —, "Metric Measures of Length," *Metric Manual for Schools: the Decimal System of Measures and Weights, with Exercises and Problems*, 4th ed. (Boston, Massachusetts: American Metric Bureau, 1877). (available from the American Metric Bureau)
7. Nicolas Bion, *The Construction and Principal Uses of Mathematical Instruments*, trans. Edmund Stone (London: J. Richardson, 1758).
8. William Edgar Garrett Fisher, "PAPER," *The Encyclopedia Brittannica*, 11th ed., vol. XX (New York: Encyclopedia Brittannica, Inc., 1911), 725-736.
9. T. N. Sear, "Airy's Dry Measure Gauge for Checking Trade Volume Measures," *Bulletin of the Scientific Instrument Society*, no. 45 (June 1995): 28-29.
10. —, *A Handbook For the Use of Sealers of Weights and Measures*, 4th ed. (Troy, New York: W. & L. E. Gurley, 1912).
11. Galileo Galilei, *Operations of the Geometric and Military Compass*, trans. Stillman Drake (Padua: Pietro Marinelli, 1606, Washington: Smithsonian Institute Press, 1978).
12. Thomas Hood, *The Making and Ufe of the Geometricall Inftrument, Called a Sector* (London: John Windet, 1598). (Reprinted as part of the *English Experience* series, New York: Da Capo Press, 1973).
13. Edmund Gunter, *The Description and Use of the Sector, the Cross-Staffe, and Other Instruments* (London: William Jones, 1624). (Reprinted as part of the *English Experience* series, New YorK: Da Capo Press, 1971).
14. I. McKim Chase, M.E., *The Art of Pattern-Making* (New York: John Wiley & Sons, 1903).
15. E. H.Crussell, *Jobbing Work for the Carpenter* (New York: The David Williams Co., 1914).
16. Claudia B. Kidwell, *Cutting a Fashionable Fit* (Washington: Smithsonian Institution Press, 1979).
17. James Hogg, *A Practical Course of Instruction with Hogg's Improved Slide-Rule in Arithmetic and Mensuration* (Boston: Rand Avery Company, 1887).

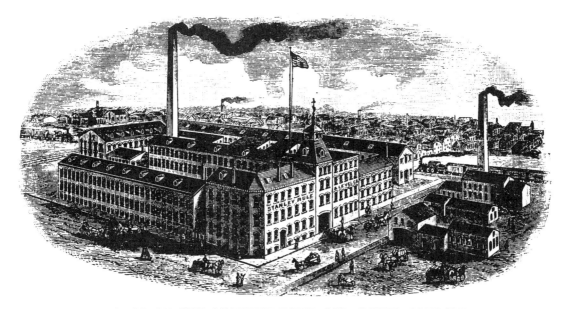

WORKS OF THE STANLEY RULE AND LEVEL COMPANY.

VIII
Foreign Units of Length

One of the more interesting facets of the study of linear measuring instruments is the number of different standard units of length which are encountered when examining the instruments themselves. There are scales divided in Rhineland *fuss* and *zoll*, in French *pied du roi* and *pouces*, in Spanish *burgos* measure, or in Russian *sajen* and *vershok*. These, and hundreds of others not mentioned, are all units of length which were in use in Europe at the end of the 18th century, and their origins, and their gradual decline in use, are worth studying.*

ORIGINS

Units of linear measurement had their origins in human development some time after counting and writing came into use. Three factors led to their development: astronomy, with the need to take observations to develop a calendar for the identification of religious holidays; construction, and building design, with the need for the designer to communicate with those doing the actual building; and commerce, with the change from the first stage of the barter economy ("this pile for that pile") to the second stage ("so much of this for so much of that").

The earliest measures were almost certainly based on the dimensions of body parts. They were convenient, and their size could be readily visualized by the user. The measures known to have been used (together with their usually accepted values) are:

- The DIGIT (the width of the first joint of the finger), equal to 3/4 of an inch;
- The THUMB (the width of the first joint of the thumb), equal to 1 inch;
- The PALM (the width of the palm), equal to 3 inches;
- The HAND (the width of palm plus the thumb), equal to 4 inches;
- The FINGER (the length of the middle finger), equal to 4 1/2 inches;
- The LONG PALM (the distance from the heel of the palm to the tip of the middle finger), equal to 9 inches;
- The SMALL SPAN (the width of the palm plus the extended thumb), equal to 6 inches;
- The GREAT SPAN (the distance from the tip of the thumb to the tip of the little finger with the fingers fully spread), equal to 9 inches;
- The FOOT (the length of the foot), equal to 12 inches;
- The CUBIT (the length of the forearm from the elbow to the tip of the middle finger), equal to 18 inches;
- The YARD (the distance from the tip of the nose to the tip of middle finger with the arm outstretched), equal to 36 inches;
- The ELL (the length of the arm/the distance from one shoulder to the end of the other palm with the arm outstretched), equal to 45 inches;
- The FATHOM (the distance from middle finger tip to middle finger tip with the arms outstretched), equal to 72 inches.

As the use of such units of measurement became more common, the issue arose of *whose* dimensions would be used, and what to do if that designated person was not present.

Early political units, with their ruler-oriented governments, often used the king's body as the source of their units (well into the 19th century the French foot was known as the *pied du roi*, literally the "foot of the king"). This would work, after a fashion; it certainly eliminated ambiguity, but it had the disadvantage that it would change each time a new ruler succeeded to the throne.

Sometimes, in an effort to compensate for individual variation, recourse was had to averaging, as when the rod was defined as the sum of the lengths of the left feet of 12 randomly selected men. In other cases, some other apparently invariant basis was selected, as when the inch was defined as the sum of the lengths of three barleycorns "neither too large or too small, from the middle of the ear."

---

*This chapter is based on a talk delivered by the author at the October 7, 2000 meeting of the Tool and Trades History Society in Dartford Bridge, Swanley, Kent.

Some units of length implied a conversion of a sort from other activities or fields. In ancient Persia the parasang (approximately 3 miles) was defined as an hour's walk for a horse. The ancient Roman mile was defined as 1000 paces (literally "*milia passuum*"). The problem of making a measurement in the absence of the king, or other primary standard, was solved by cutting pieces of wood or forging lengths of metal to exactly replicate that standard. These, the first linear measuring instruments, could be widely distributed and referenced.

Human nature being what it is, there was frequently a temptation to "adjust" a measure in favor of one party or another. If cloth is being sold by the yard, the seller would obviously favor a shorter yard than the buyer. In a more recent example, American sawmills preferred to use Scribner's log rule when buying wood; it read lower for a log than the other popular rule, Doyle's, and they could thus buy their logs more cheaply.

Because of their importance to commerce, and because of their being subject to tampering, primary linear standards were sometimes cut in stone, and deposited in the local temple, or placed on open display in the market place where they could be referenced by all.

HISTORICAL MEASURES

We have some knowledge of the standards in use in the ancient world. This comes from three sources: literary references, still-existing standards (either carved in monuments, or as separate standards), and analysis of existing objects to determine what measure was used to calculate their various dimensions.

The best example of this third method is the Great Pyramid of Egypt. It has been determined, based on careful measurement, that all of its dimensions, both overall and of its various components (doorways, etc.), are multiples of a single unit, a cubit of 20.63 inches, subdivided into 28 digits.

Other known examples of ancient measures are the cubit used in Babylonia and Asia Minor, which was the same as the Egyptian cubit but subdivided decimally, and the cubit of 18.23 inches, subdivided into 25 digits, which was used in later Egypt.

There was also a Judean cubit of 25.1 inches (later reduced to 21.6 inches), a Carthaginian cubit of 22.2 inches, a Mesopotamian cubit of 20.0 inches, and a Persian cubit of 19.2 inches.

The ancient Greeks used a whole series of linear measures, exactly related to one another. The basic unit was the *dactylos* (digit) of 0.76 Inches. Some others in the series were the *dichas* (span) (equal to 8 *dactylos*) of 6.07 inches, the *pugon* (foot) (equal to 18 dactylos) of 13.65 inches, the *pachys metrios* (cubit) (equal to 24 *dactylos*) of 18.21 inches, and the *bema* (yard) (equal to 40 *dactylos*) of 30.34 inches. They also used an Olympic *pous* (foot) (equal to $16 1/3$ *dactylos*).

The Romans used two inches, the *digitus* of 0.75 inches, and the *uncia* of 0.97 inches. Their larger measures were all based on multiples of the *uncia*. These were the *palmus* (palm) (equal to 3 *uncia*) of 2.91 inches, the *bes* (equal to 8 *uncia*) of 7.77 inches, the *spithama* (span) (equal to 9 *uncia*) of 8.74 inches, the *pes* (foot) (equal to 12 *uncia*) of 11.65 inches, the *cubitus* (cubit) (equal to $1 1/2$ *pes* which was equal to 18 *uncia*) of 17.47 inches, and the *gressus* (yard) (equal to $2 1/2$ *pes* which was equal to 30 *uncia*) of 29.12 inches.

The Roman system was related to the Greek system through the Roman *digitus*, which was equal to the Greek *dactylos*. The Roman *pes* of 11.65 inches was equal to 16 *digitus*; the Greek *pous* of 12.14 inches was equal to $16 1/3$ (equal to $1/6$ of 100) *dactylos*.

Little information is available about linear measures during the dark and middle ages. After the collapse of the Roman empire, the uniform Roman measures gradually fell into disuse, replaced by local standards which gradually drifted apart, due to a lapsing of commerce, and the reintroduction of lapsed earlier Celtic and Germanic measures. Most existing standards today date from the Renaissance or later.

MEASURES AFTER THE RENAISSANCE

With the spread of commerce after about 1000 -1400 AD political units began to enforce the use of standard measures throughout their territory. The number of different measures in use was largely determined by the degree of political unity prevailing at this time. For nations such as England, which was essentially unified by the end of this period, a relatively small number of units were in use. France and Spain, two recently unified nations, had more. Germany and Italy, both fragmented into dozens of city-states, had literally hundreds.

After 1400, in spite of the progress in political unification, varieties of linear measures continued to proliferate. This was the result of two main causes.

First, new standards did not displace the old ones. The establishment of national standards by newly centralized governments did not result in the elimination of the old local standards, but only in the addition of a new set of standards to those already existing.

Second, new standards were constantly being created. A locally-used container, such as a particular pail or tub, would become a standard. Sometimes the standard applied outside the walls of a city (and thus outside the control of the city guilds) would become different than that inside the walls. Fractions and multiples of existing standards would become standards in their own right. Different standards would develop for different varieties of a given material (examples of different measures for woolens, silks, and linens are common), or for the type of transaction (wholesale versus retail). Units of account, for which no actual measure existed, evolved. The last of wine was defined as 12 barrels, but no actual quantity in gallons was specified. Capacity measures based on a container could be defined three ways: heaped, struck (leveled off), or shallow (short of the edge).

The table below is a list of the units of linear measurement which had come into use in Europe by the end of the 19th century. It covers the range of units from about 1 inch in length to about 6 feet, and is sorted first by Country or Region and second by length, and for each unit gives the length expressed three ways: in inches and 64ths (expressed in fractional form), in inches and 100ths, and in centimeters and 20ths (both expressed in decimal form).

This table is derived primarily from 19th century sources. A number of tables and guides were published in the 1800s for use by merchants in dealing with this proliferation of weights and measures, and I have used them extensively to generate this table. A list of the documents consulted will be found at the end of the table.

It is hoped that this table will be a valuable resource in identifying the country of use for unlabeled measuring instruments.

LARGELY ELIMINATED BY THE METRIC SYSTEM

This proliferation of standard measures of length was one of the primary reasons for the development of the metric system of weights and measures. This system, first proposed in 1791 during the wave of rationalism which followed the French revolution, was formally adopted by France four years later, and spread across Europe forcibly by Napoleon's conquests over the next 20 years.

Although frequently resisted (most nations reverted to their old standards as soon as France was defeated), it continued to gain acceptance, and was gradually adopted across Europe during the ensuing 50 years.

The Netherlands and Belgium adopted the Metric system in 1817, Greece in 1836, the German *Zollverein* (customs union) in 1840, and the German states themselves in 1868, Italy (partially) in 1845, and wholly in 1863, Switzerland in 1851, Denmark in 1852, Spain and its colonies in 1859, Portugal in 1864, Sweden in 1865, Norway 1866, and Austria in 1876, when it joined the *Zollverein*.

This adoption was resisted by conservatism and tradition in most nations, and many governments had to repeatedly pass laws outlawing the old units. Sometimes, to soften the blow, a country would attach the names of the old units to the closest equivalent metric unit.

Such is the force of tradition, however, that some old measures persisted well into the 20th century. The French *pied du roi* was observed in use up until World War I; the Spanish *burgos* foot and the Russian *sajen/archine* were both still in occasional use as late as 1921. An interesting discussion of this persistence of the old measures can be found in Kennelly, listed in the bibliography following the table.

EUROPEAN UNITS OF LENGTH BEFORE THE METRIC SYSTEM

Country/Region	Unit Name	Comments	Inches/64ths	Inches/100ths	Cm/20ths	Subdivisions
Austria						
	Zoll		1-2/64	1.04	2.65	
	Fuss		12-28/64	12.45	31.60	= 12 Zoll
	Klafter		74-42/64	74.66	189.65	= 6 fuss
Innsbruck	Elle		31	31.00	78.75	
Salzburg	Elle	silk	31-36/64	31.56	80.15	
Salzburg	Elle	linen	39-38/64	39.59	100.55	

Country/Region	Unit Name	Comments	Inches/64ths	Inches/100ths	Cm/20ths	Subdivisions
Austria (continued)						
Tyrol	Elle		31-42/64	31.66	80.40	
Vienna	Elle	silk (legal)	30-42/64	30.66	77.90	
Belgium						
	Talie	old measure	1-46/64	1.71	4.35	
	Palme	legal	3-60/64	3.94	10.00	
	Pied	old measure	10-55/64	10.86	27.55	
	Pied		11-52/64	11.81	30.00	=10 pounces
	Aune		47-15/64	47.23	119.95	
	Vaem		72	72.00	182.90	
Antwerp	Pied		11-15/64	11.24	28.55	
Antwerp	Elle		22-18/64	22.28	56.60	
Antwerp	Elle	for wool	26-61/64	26.96	68.50	
Antwerp	Elle	for silk	27-21/64	27.34	69.45	
Brabant	Elle		27-25/64	27.39	69.55	
Brussels	Pied		11-29/64	11.45	29.10	
Liege	Pied		11-20/64	11.32	28.75	
Liege	Pied St. Lambert	land	11-34/64	11.52	29.25	
Liege	Pied St. Hubert	builder's	11-41/64	11.64	29.55	
Liege	Elle		21-45/64	21.71	55.15	
Liege	Toise		66-1/64	66.02	167.70	
Mechelen	Pied		11-2/64	11.03	28.00	
Mechelen	Aune		27-15/64	27.23	69.17	
Namur	Elle		26-7/64	26.11	66.30	
Ostend, Ypres	Elle		27-34/64	27.53	69.95	
Oudenarde	Elle		26-18/64	26.28	66.75	
Tournay	Elle		24-26/64	24.40	61.95	
Ypres	Pas		26-61/64	26.96	68.50	
Bohemia						
	Zoll		62/64	0.97	2.45	
	Fuss	old measure	11-43/64	11.67	29.65	
	Fuss	imperial	12-29/64	12.45	31.60	
	Elle		23-25/64	23.39	59.40	
Prague	"Foot"		12-20/64	12.31	31.25	
Prague	Elle	of Bohemia	23-25/64	23.39	59.40	
Prague	Elle	of Moravia	31-8/64	31.13	79.05	
Presburg	Elle		21-62/64	21.97	55.80	
Bulgaria						
	Lak't		25-38/64	25.59	65.00	
	Archin	tailor's	26-49/64	26.77	68.00	
	Archin	mason's	29-54/64	29.84	75.80	
Cyprus						
	Pic		26-29/64	26.45	67.15	
Denmark						
	Linie		6/64	0.09	0.20	

FOREIGN UNITS OF LENGTH 243

244 FOREIGN UNITS OF LENGTH

Country/Region	Unit Name	Comments	Inches/64ths	Inches/100ths	Cm/20ths	Subdivisions
Denmark (continued)						
	Tomme		1-2/64	1.03	2.60	
	Fod		11-37/64	11.58	29.40	
	Alen		24-46/64	24.71	62.75	= 2 Fod
	Favn		74-9/64	74.14	188.30	= 3 Alen
Estonia						
Narva	"Ell"		23-35/64	23.55	59.80	
Parnu	"Ell"		21-38/64	21.60	54.85	
Tallin	"Foot"		10-34/64	10.53	26.75	
Tallin	"Ell"		21-5/64	21.08	53.55	= 2 "Foot"
France						
	Pouce		1-4/64	1.07	2.70	
	Paume	farrier's	3-13/64	3.20	8.10	
	Pied du Roi	til 1812	12-51/64	12.79	32.50	
	Pied metrique	1812 to 1840	13-8/64	13.12	33.35	= 1/3 meter
	Pas		31-47/64	31.73	80.60	
	Aune		46-54/64	46.85	118.85	
	Aune	usual, till 1837	47-16/64	47.25	120.00	
	Brasse		63-60/64	63.94	162.40	
	Pas	geometrical	63-61/64	63.95	162.40	
	Aune	in 1812	74-14/64	74.21	188.50	
	Toise		76-47/64	76.74	194.90	= 6 pied
Avignon	Pied	old measure	9-50/64	9.79	24.85	
Avignon	Aune	old measure	45-61/64	45.95	116.70	
Bayonne	Aune		34-51/64	34.80	88.40	
Besancon	Pied		12-11/64	12.17	30.90	
Bordeaux	Pied	old measure	14-3/64	14.05	35.65	
Bordeaux	Pas		35-7/64	35.11	89.20	
Bordeaux	Aune		46-60/64	46.93	119.60	
Burgundy	Pied	old measure	13-3/64	13.04	33.10	
Burgundy	Aune	old measure	31-53/65	31.82	80.85	
Carcassonne	Canne		70-17/64	70.26	178.45	
Dauphiny	Pied	old measure	13-27/64	13.43	34.10	
Dijon	Pied		12-22/64	12.35	31.35	
Dunkirk	Aune		26-40/64	26.62	67.60	
Franche-Comte	Pied	old measure	14-5/64	14.07	35.75	
Gironde	Compas		61-52/64	61.81	157.00	
Grenoble	Pied		13-26/64	13.41	34.05	
Guienne	Pied	old measure	13-41/64	13.64	34.65	
Lisle	Aune		27-45/64	27.70	70.35	
Lorraine	Pouce		1-8/64	1.13	2.85	
Lorraine	Pied	old measure	11-17/64	11.27	28.65	
Lorraine	Aune	old measure	25-11/64	25.18	63.95	
Lorraine	Toise	minimum	67-27/64	67.42	171.25	
Lyons	Pied	old measure	13-31/64	13.48	34.25	

Country/Region	Unit Name	Comments	Inches/64ths	Inches/100ths	Cm/20ths	Subdivisions
France (continued)						
Lyons	Aune	old measure	44-15/64	44.23	112.35	
Lyons	Aune		46-13/64	46.20	117.40	
Lyons	Elle		47-45/64	47.71	121.20	
Marseilles	Canne		79-15/64	79.23	201.25	
Montpellier	Palme		8-45/64	8.71	22.10	
Montpellier	Elle		31-17/64	31.26	79.40	
Montpellier	Canne		78-15/64	78.24	198.75	
Nantes	Aune		55-51/64	55.80	141.65	
Neufchatel	Pied		11-52/64	11.81	30.00	
Neufchatel	Elle		21-56/64	21.87	55.55	
Neufchatel	Aune		43-51/64	43.80	111.25	
Nice	Palmo		10-27/64	10.42	26.45	
Nice	Raso		21-39/64	21.61	54.90	
Nice	Aune		46-49/64	46.77	118.80	
Normandy	Pied	old measure	11-46/64	11.72	29.80	
Paris	Elle		46-57/64	46.89	119.10	
Picardy	Aune	old measure	31-32/64	31.50	80.00	
Provence	Aune		31-17/64	31.26	79.40	
Rennes	Aune	old measure	54-32/64	54.50	138.45	
Rouen	Aune	silk and woolen	45-51/64	45.80	116.40	
Rouen	Aune	linen	55	55.00	139.65	
Savoy	Pied		10-42/64	10.66	27.05	
Sedan	Pied		13-4/64	13.06	33.15	
St. Malo	Aune	old measure	53-03/64	53.04	134.70	
St. Omer	Aune	old measure	28-26/64	28.40	72.15	
Toulouse	Canne		70-45/64	70.71	179.60	
Valenciennes	Aune		25-60/64	25.93	65.85	
France/Germany						
Strassburg	Pied/Fuss		11-3/64	11.04	28.05	
Strassburg	Pied/Fuss/Schuh	city; old measure	11-25/64	11.39	28.95	
Strassburg	Schuh	rural; old measure	11-39/64	11.61	29.50	
Strassburg	Elle/Aune	old measure	21-12/64	21.19	53.85	
Germany						
	Zoll		61/64	0.95	2.40	
	Elle		26-16/64	26.25	66.70	
	Elle		27-22/64	27.35	69.50	
Aachen	Zoll	surveyor's	59/64	0.93	2.35	
Aachen	Zoll	builder's	61/64	0.95	2.40	
Aachen	Elle		26-21/64	26.33	66.85	
Altona	Elle	of Hamburg	22-36/64	22.56	57.30	
Amberg	Elle	maximum	32-56/64	32.88	83.50	
Anhalt	Elle		25-2/64	25.04	63.60	
Anspach	Fuss		11-46/64	11.72	29.80	
Augsburg	Fuss		11-42/64	11.66	29.60	

FOREIGN UNITS OF LENGTH

Country/Region	Unit Name	Comments	Inches/64ths	Inches/100ths	Cm/20ths	Subdivisions
Germany (continued)						
Augsburg	Elle	short	23-20/64	23.32	59.25	
Augsburg	Elle	long	24	24.00	60.95	
Baden	Fuss		11-52/64	11.81	30.00	
Baden	Elle	legal	23-40/64	23.62	60.00	
Baden	Klafter		70-55/64	70.87	180.00	
Bavaria	Schuh	builder's/surveyor's	11-24/64	11.37	28.85	
Bavaria	Fuss	since 1809	11-31/64	11.49	29.20	
Bavaria	Schuh		12-7/64	12.11	30.75	
Bavaria	Fuss	Rhineland	13-8/64	13.12	33.35	
Bavaria	Elle	Rhineland	47-16/64	47.25	120.00	
Bavaria	Elle	legal	68-51/64	68.8	174.75	
Bavaria	Klafter		68-60/64	68.94	175.20	
Berlin	Zoll	since 1816	1-2/64	1.03	2.60	
Berlin	Zoll	surveyor's	1-31/64	1.48	3.75	
Berlin	Fuss		12-12/64	12.19	30.95	
Berlin	Fuss	since 1816	12-23/64	12.36	31.40	
Berlin	Fuss	surveyor's	14-53/64	14.83	37.65	
Bonn	Fuss		14-29/64	14.45	36.70	
Bonn	Elle		25-49/64	25.76	65.45	
Bremen, Hamburg	Zoll	surveyor's	1-9/64	1.14	2.90	
Bremen	Fuss		11-24/64	11.38	28.90	
Bremen	Elle		22-49/64	22.76	57.80	
Bremen	Klafter		68-20/64	68.32	173.50	
Brunswick	Fuss		11-15/64	11.23	28.55	
Brunswick, Kasell	Elle		22-30/64	22.47	57.05	
Cleves	Fuss		11-41/64	11.63	29.55	
Coburg	Elle		23-4/64	23.07	58.55	
Cologne	Fuss		10-53/64	10.83	27.50	
Cologne	Fuss	old measure	11-21/64	11.32	28.75	
Cologne	"Ell"	long	27-21/64	27.33	69.40	
Darmstadt	Zoll	since 1818	55/64	0.86	2.00	
Darmstadt	Fuss	since 1818	9-54/64	9.84	25.00	
Darmstadt	Fuss	old measure	11-21/64	11.32	28.75	
Dresden	Elle		22-20/64	22.30	56.65	
Dusseldorf	Elle	old measure	23-16/64	23.25	59.05	
Dusseldorf	Elle		26-63/64	26.98	68.50	
Emden	Elle		26-26/64	26.40	67.05	
Erfurt	Elle	short	15-58/64	15.90	40.35	
Erfurt	Elle	long	23-24/64	23.38	59.40	
Frankfurt	Zoll	surveyor's	1-26/64	1.40	3.55	
Frankfurt	Fuss	surveyor's	14-1/64	14.01	35.60	
Frankfurt	Elle		21-35/64	21.55	54.75	
Frankfurt	Elle	of Brabant	27-34/64	27.53	69.90	
Frankfurt/Oder	Elle		26-8/64	26.12	66.35	

Country/Region	Unit Name	Comments	Inches/64ths	Inches/100ths	Cm/20ths	Subdivisions
Germany (continued)						
Freiburg	Fuss		11-35/64	11.55	29.35	
Freiburg	Stab		42-7/64	42.11	106.95	
Giessen	Elle		21-41/64	21.63	54.95	
Gotha	Elle		22-41/64	22.65	57.55	
Gotha	Klafter		67-60/64	67.94	172.60	
Hamburg, Altona	Palm	ship-building	3-49/64	3.76	9.55	
Hamburg, Altona	Fuss		11-18/64	11.28	28.65	
Hamburg	Fuss	surveyor's	12-23/64	12.36	31.40	
Hamburg	Elle	for silks, etc.	22-36/64	22.56	57.30	
Hamburg	Elle		22-55/64	22.86	58.05	
Hamburg	Elle	for woolens/prints	27-16/64	27.24	69.20	
Hamburg	Klafter		67-43/64	67.68	171.90	
Hanover	Fuss		11-32/64	11.50	29.20	
Hanover	Elle		22-63/64	22.99	58.40	
Heidelburg	Fuss	old measure	11	11.00	27.95	
Hildesheim	Fuss		11-3/64	11.05	28.05	
Hildesheim	Elle		22-6/64	22.10	56.15	= 2 Fuss
Holstein	Fuss		11-48/64	11.75	29.85	
Karlsruhe	Elle	old measure	21-52/64	21.81	55.40	
Kassel	Fuss	surveyor's	11-14/64	11.22	28.50	
Kassel	Fuss		11-21/64	11.32	28.75	
Kassel	Elle		22-7/64	22.11	56.15	
Koblentz	Fuss		11-28/64	11.44	29.05	
Koblentz	Elle		22	22.00	55.85	
Koblentz, Cologne	Elle		22-36/64	22.57	57.30	
Konigsberg	Fuss		12-7/64	12.11	30.75	
Konigsberg	Elle		22-40/64	22.62	57.45	
Leipsig	Zoll	builder's	1-7/64	1.11	3.00	
Leipsig	Fuss		11-7/64	11.11	28.20	
Leipsig	Elle		22-14/64	22.22	56.45	
Leipsig	Elle		27-8/64	27.12	68.90	
Leipsig	Stab		44-33/64	44.51	113.05	
Leipsig	Klafter		66-49/64	66.77	169.60	
Lindau	Fuss (long)		12-26/64	12.4	31.5	
Lubeck	Elle		22-43/64	22.67	57.60	
Luneburg	Fuss	old measure	11-29/64	11.46	29.10	
Mannheim	Fuss	old measure	11-27/64	11.43	29.05	
Mannheim	Elle		21-56/64	21.88	55.60	
Mecklenburg	Fuss		11-29/64	11.45	29.10	
Mecklenburg	Fuss		11-29/64	11.45	29.10	
Mecklenburg	Elle		22-36/64	22.56	57.30	
Munich	Elle		32-51/64	32.8	83.30	
Munster	Elle	old measure	31-53/64	31.83	80.85	
Nurnberg	Zoll	old measure	1	1	2.55	

Country/Region	Unit Name	Comments	Inches/64ths	Inches/100ths	Cm/20ths	Subdivisions
Germany (continued)						
Nurnberg	Fuss	old measure	11-61/64	11.96	30.40	= 12 Zoll
Nurnberg	Elle		25-54/64	25.84	65.65	
Nurnberg	Elle		26-46/64	26.72	67.90	
Oldenburg	Fuss		11-42/64	11.65	29.60	
Osnabruck	Elle	long	23-45/64	23.7	60.15	
Osnabruck	Elle		25-9/64	25.13	63.85	
Pomerania	Elle		25-40/64	25.62	65.10	
Prussia	Spanne	miner's	10-19/64	10.3	26.15	
Prussia	Fuss	Rhineland foot	12-22/64	12.35	31.40	
Regensburg	Fuss		11-27/64	11.42	29.00	
Regensburg	Elle		31-58/64	31.9	81.10	
Rhineland	Foot (Old)		12-23/64	12.36	31.40	
Rostock	Elle		22-49/64	22.76	57.85	
Saxony	Spanne	miner's	9-42/64	9.66	24.55	
Schaffhausen	Schuh	artisan's	11-47/64	11.73	29.80	
Silesia	Fuss		11-12/64	11.19	28.40	
Silesia	Elle	Prussian	22-43/64	22.67	57.60	
Stuttgart	Elle		24-5/64	24.08	61.15	
Trier	Schuh	artisan's/surveyor's	11-36/64	11.57	29.40	
Trier	Schuh	carpenter's	12-1/64	12.01	30.50	
Ulm	Elle	old measure	22-24/64	22.38	56.85	
Weimar	Zoll	surveyor's	1-50/64	1.78	4.50	
Weimar	Fuss		11-7/64	11.1	28.20	
Weimar	Fuss	surveyor's	17-49/64	17.76	45.10	
Weimar	Klafter		66-39/64	66.61	169.20	
Wesel	Fuss	old measure	9-16/64	9.25	23.50	
Wurttemberg	Elle		24-12/64	24.18	61.40	
Wurttemberg	Klafter		67-43/64	67.68	171.90	
Wurtzburg	Elle		22-51/64	22.8	57.90	
Zittau	Elle		22-27/64	22.43	57.00	
Great Britain						
	Inch		1	1.01	2.55	
	Nail		2-16/64	2.25	5.70	
	Palm		3	3	7.60	
	Hand		4	4	10.15	
	Link	surveyor's	7-59/64	7.92	20.10	
	Span		9	9	22.85	
	Foot		12	12	30.50	
	Yard	yard of Henry VII	35-62/64	35.96	91.35	
	Yard		36	36	91.45	
	Ell		40	40	101.60	
	Ell	cloth ell	45	45	114.30	
	Pace		60	60	152.40	
	Fathom		72	72	182.90	

FOREIGN UNITS OF LENGTH 249

Country/Region	Unit Name	Comments	Inches/64ths	Inches/100ths	Cm/20ths	Subdivisions
Great Britain (continued)						
Gibralter	Vare		33-8/64	33.12	84.15	
Jersey	Ell/Aune		48	48	121.90	
Scotland	Link	old measure	8-59/64	8.93	22.70	
Scotland	Ell		37-13/64	37.2	94.50	
Greece						
	"Foot"		12-5/64	12.08	30.70	
	"Cubit"		18-8/64	18.13	46.05	
	Pecheus (Old)		25-33/64	25.52	64.80	
Chios	Pic	short	25-63/64	25.98	66.00	
Chios	Pic	long	27	27	68.55	
Crete	Pic		25-7/64	25.11	63.75	
Ionian Islands	Pie		13-6/64	13.09	33.25	
Ionian Islands	Braccio	for woolens	27-12/64	27.19	69.05	
Ionian Islands	Passo		68-26/64	68.4	173.75	
Patras	Pic	for silks	25-1/64	25.01	63.55	
Patras	Pic	other fabrics	27	27	68.55	
Rhodes	Pic		29-49/64	29.76	75.60	
Zakinthos	Braccio	silk	25-24/64	25.37	64.45	
Hungary						
	Zoll	of Austria	1-2/64	1.04	2.65	
	Fuss		12-28/64	12.45	31.60	
	Elle		30-43/64	30.68	77.90	
	Elle		31-31/64	31.48	79.95	
Iceland						
	Alen		22-30/64	22.47	57.05	
Italy						
Ancona	Braccio		25-21/64	25.33	64.35	
Bergamo	"Foot"		17-11/64	17.17	43.60	
Bergamo	Braccio		25-51/64	25.8	65.55	
Bologna	Pie		14-62/64	14.96	38.00	
Bologna	Braccio	silk	23-29/64	23.46	59.55	
Bologna	Braccio		24-43/64	24.67	62.65	
Bologna	Braccio	woolen	25	25	63.50	
Bologna	Braccio		25-26/64	25.4	64.50	
Bolzano	Braccio		21-41/64	21.64	54.95	
Bolzano	Elle	Austrian	31-7/64	31.11	79.02	
Brescia	Braccio		18-26/64	18.4	46.75	
Cagliari	Palmo		9-50/64	9.78	24.85	
Carrara	Palmo	marble-work	9-38/64	9.59	24.35	
Carrara	Pie	surveyor's	11-35/64	11.55	29.35	
Carrara	Canna	woodland	24-38/64	24.59	62.45	
Casal	Pie		19-4/64	19.06	48.40	
Chamberi	Raso		22-40/64	22.62	57.45	
Cremona	Pie		18-58/64	18.91	48.05	

FOREIGN UNITS OF LENGTH

Country/Region	Unit Name	Comments	Inches/64ths	Inches/100ths	Cm/20ths	Subdivisions
Italy (continued)						
Cremona	Braccio		23-27/64	23.42	59.50	
Ferrara	Pie		15-58/64	15.9	40.40	
Ferrara	Braccio	for silk	24-62/64	24.98	63.45	
Florence	Soldo		1-9/64	1.15	29.15	
Florence	Pie	architect's	21-37/64	21.58	54.80	
Florence	Braccio		21-38/64	21.6	54.85	
Florence	Passo		64-48/64	64.75	164.45	
Genoa	Pie		11-44/64	11.68	29.70	
Genoa	Piede	Manual	13-31/64	13.49	34.25	= 8 Oncie
Genoa	Piede	Liprando	20-15/64	20.23	51.40	= 12 Oncie
Genoa	Braccio		22-57/64	22.88	58.10	
Genoa	Canna	canna "piccola"	87-33/64	87.52	222.30	
Genoa	Canna	"custom-house"	97-16/64	97.25	246.90	
Lodi	Braccio		17-61/64	17.96	45.60	
Lombardy	Palmo	since 1803	3-60/64	3.94	10.00	
Lombardy	Pie	mean	13-43/64	13.68	34.75	
Lucca	Braccio	silk	22-51/64	22.8	57.90	
Lucca	Pie		23-14/64	23.23	59.00	
Lucca	Braccio	woolen	23-51/64	23.8	60.40	
Mantua	Pie		18-24/64	18.38	46.70	
Messina	Palmo		10-26/64	10.4	26.40	
Milan	Pie	architect's/builder's	15-40/64	15.63	39.70	
Milan	Pie		17-9/64	17.13	43.50	
Milan	Braccio	for silk	20-38/64	20.6	52.35	
Milan	Braccio	old measure	23-06/64	23.09	58.65	
Modena	Braccio		22-47/64	22.74	57.75	
Modena	Braccio		24-20/64	24.31	61.75	
Naples	Palmo		10-24/64	10.38	26.35	
Naples	Braccio		25-13/64	25.2	64.00	
Naples	Braccio		27-33/64	27.51	69.85	
Naples	Canna		83-3/64	83.05	210.95	
Padua	Braccio	silk	25-19/64	25.3	64.30	
Padua	Braccio	woolen	26-52/64	26.82	68.10	
Palermo	Palmo		9-20/64	9.31	23.65	
Parma	Braccio di legno	surveyor's	21-22/64	21.34	54.20	
Parma	Cubit		22-25/64	22.39	56.90	
Parma	Braccio	for silk	23-10/64	23.15	58.80	
Parma	Braccio	for woolens	25-13/64	25.2	64.00	
Pavia	Pie		18-19/64	18.3	46.45	
Piacenza	Pie		18-32/64	18.5	47.00	
Piacenza	Braccio		26-37/64	26.57	67.50	
Piedmont	Pan		9-54/64	9.84	25.00	
Piedmont	Ras		23-29/64	23.45	59.55	
Pisa	Pie		11-48/64	11.75	29.85	

FOREIGN UNITS OF LENGTH

Country/Region	Unit Name	Comments	Inches/64ths	Inches/100ths	Cm/20ths	Subdivisions
Italy (continued)						
Ragusa	Braccio		20-13/64	20.21	51.30	
Ravenna	Pie		23-1/64	23.02	58.45	
Ravenna	Braccio		26-29/64	26.46	67.2	
Reggio	Braccio		20-55/64	20.86	53.00	
Rome	Palmo	clothier's	8-22/64	8.35	21.20	
Rome	Palmo	commercial	9-51/64	9.8	24.90	
Rome	Pie		11-35/64	11.55	29.35	
Rome	Pie	architect's	11-47/64	11.73	29.8	
Rome	"Cubit"		17-32/64	17.5	44.45	
Rome	Braccio	weaver's	25-03/64	25.04	63.60	
Rome	Braccio-d'ara		29-17/64	29.53	75.00	
Rome	Braccio		30-47/64	30.73	78.05	
Rome	Braccio	merchant's	33-25/64	33.39	84.80	
Rome	Passo		58-41/64	58.64	148.95	
Rome	Canna	canna "mercantile"	78-22/64	78.34	199	
Rome	Canna	"builders" canna	87-61/64	87.96	223.4	
Sardinia	Palmo		10-21/64	10.33	26.25	
Sardinia	Pie		18-54/64	18.85	47.85	
Sardinia	"Ell"		21-40/64	21.62	54.9	
Sicily	Palmo		9-34/64	9.53	24.2	
Sicily	Canna		76-16/64	76.25	193.6	
Sienna	Braccio	for cloth	14-55/64	14.86	37.75	
Sienna	Braccio	for linen	23-40/64	23.63	60	
Trent	Elle	for silk	24-6/64	24.09	61.2	
Trent	Elle	for cloth	26-41/64	26.64	67.65	
Treviso	Braccio	for woolens	26-39/64	26.62	67.60	
Trieste	Braccio	for silk	25-14/64	25.22	64.05	
Trieste	Braccio	for woolens	26-38/64	26.6	67.6	
Turin	Pie		13-31/64	13.48	34.25	
Turin	Pie liprando		20-15/64	20.23	51.40	
Turin	Raso		23-38/64	23.6	59.95	
Turin	Tesa		67-27/64	67.42	171.25	
Tuscany	Palmo		11-31/64	11.49	29.2	
Tuscany	Pie		11-60/64	11.94	30.35	
Tuscany	Braccio		22-63/64	22.98	58.35	
Tuscany	Passetto		45-61/64	45.95	116.7	
Tuscany	Passo		68-56/64	68.88	174.95	
Urbino	Pie		16-8/64	16.13	40.95	
Venice	Pie		12-45/64	12.7	32.25	
Venice	Palmo		13-43/64	13.68	34.75	
Venice	Pie		13-60/64	13.94	35.4	
Venice	Braccio	for silk	25-9/64	25.15	63.85	
Venice	Braccio	for woolens	26-58/64	26.91	68.35	
Venice	Passo	geometrical	54-45/64	54.71	138.95	

252 FOREIGN UNITS OF LENGTH

Country/Region	Unit Name	Comments	Inches/64ths	Inches/100ths	Cm/20ths	Subdivisions
Italy (continued)						
Venice	Chebbo		61-39/64	61.61	156.50	
Venice	Passo		68-25/64	68.39	173.70	
Verona	Pie		13-32/64	13.5	34.30	
Verona	Braccio	for woolens	25-36/64	25.57	64.95	
Vicenza	Pie		13-40/64	13.63	34.60	
Vicenza	Pie		14-5/64	14.07	35.75	
Vicenza	Braccio		26-61/64	26.96	68.5	
Latvia						
	Fuss	Rhineland foot	12-23/64	12.36	31.40	
	Elle		24-3/64	24.04	61.05	
Riga	Foot		10-51/64	10.79	27.4	
Riga	Elle		21-37/64	21.58	54.8	
Lithuania						
Konigsberg	Elle		22-40/64	22.63	57.50	
Malta						
	Span		7-13/64	7.2	18.3	
	Palmo		10-18/64	10.28	26.10	
	Pie		11-11/64	11.17	28.35	
	Canna		82-26/64	82.4	209.3	
Moravia						
	Fuss		11-42/64	11.65	29.60	
	Elle		31-8/64	31.13	79.05	
Netherlands						
	Voet	Rhineland	12-22/64	12.35	31.4	
	Duim	Netherlandic	25/64	0.39	1.00	
	Elle (Old)		27-5/64	27.08	68.80	
Amsterdam	Palm	ship-building	3-46/64	3.71	9.45	
Amsterdam	Palm	Netherlandic; legal	3-60/64	3.94	10.00	
Amsterdam	Voet		11-9/64	11.14	28.3	
Amsterdam	El		21-63/64	21.98	55.85	
Amsterdam	El	of Brabant	27-21/64	27.34	69.45	
Amsterdam	El	of Flanders	27-63/64	27.98	71.05	
Amsterdam	Faden		66-54/64	66.84	169.75	
Bergen-op-Zoom	El		27-17/64	27.27	69.25	
Breda	Voet		11-12/64	11.19	28.40	
Dendermonde	El	retail	27-26/64	27.4	69.60	
Dendermonde	El	wholesale	28-50/64	28.78	73.10	
Dordrecht	Voet		14-11/64	14.17	36.00	
Groningen	Voet		11-31/64	11.49	29.15	
Groningen	El		27-19/64	27.3	69.35	
Gueldres	El		26-8/64	26.12	66.35	
Haarlem	Voet		11-16/64	11.25	28.6	
Haarlem	El		28-42/64	28.65	72.75	
Hague	El		27-21/64	27.33	69.4	

Country/Region	Unit Name	Comments	Inches/64ths	Inches/100ths	Cm/20ths	Subdivisions
Netherlands (continued)						
Leyden	El		27-8/64	27.12	68.9	
Maastricht	Voet		11-3/64	11.06	28.05	
Middelburg	Voet		11-52/64	11.81	30	
Rotterdam	El		27-11/64	27.17	69	
Utrecht	Voet		10-47/64	10.74	27.3	
Zealand	El		27-11/64	27.18	69.05	
Norway						
	Fod		12-22/64	12.35	31.35	
Bergen	Palm	ship-building	3-31/64	3.49	8.85	
Bergen	El		24-45/64	24.71	62.75	
Poland						
	Calow	old measure	63/64	0.98	2.50	
	Lawek		1-45/64	1.7	4.30	
	Stopa	since 1819	11-22/64	11.34	28.80	
	Stopa	old measure	11-46/64	11.72	29.80	
	Trewice		14-2/64	14.03	35.65	
	Precikow		17	17	43.2	
	Lokiec	since 1819	22-43/64	22.68	57.60	
	El		24-19/64	24.3	61.70	
Cracow	Calow		1-11/64	1.17	3.00	
Cracow	Cwierzec		6-5/64	6.07	15.40	
Cracow	Lokiec		23-14/64	23.21	58.95	
Danzig	Zoll	old measure	60/64	0.94	2.40	
Danzig	Fuss	old measure	11-19/64	11.3	28.70	
Danzig	Elle, Arn	old measure	22-38/64	22.59	57.40	
Danzig	Elle, Arn	legal	26-16/64	26.26	66.70	
Elblag	Arn		22-19/64	22.3	56.65	
Stettin	Stopa		11-8/64	11.12	28.25	
Stettin	Arn		25-40/64	25.62	65.10	
Torun	Arn		22-27/64	22.42	56.95	
Warsaw	Cwierzec	since 1819	5-43/64	5.67	14.40	
Warsaw	Stopa		14-2/64	14.03	35.65	
Portugal						
	Dedo		46/64	0.72	1.85	
	Pollegada		1-5/64	1.08	2.75	
	Palmo		8-42/64	8.66	22.00	
	Pe		13	12.99	33.00	
	Pe (old)		13-8/64	13.12	33.35	
	Pe	architectural	13-21/64	13.33	33.85	
	Covado		25-63/64	25.98	66.00	= 3 Palmo
	Vara		39-11/64	39.17	99.50	
	Varre		43-45/64	43.7	111.00	
	Braca	marine	64-62/64	64.96	165.00	
	Braca		87-26/64	87.4	222.00	= 2 Vara

254 FOREIGN UNITS OF LENGTH

Country/Region		Unit Name	Comments	Inches/64ths	Inches/100ths	Cm/20ths	Subdivisions
Portugal (continued							
	Lisbon	Palmo da Junta	shipping	7-56/64	7.88	20.00	
	Lisbon	Palmo	commercial	8-62/64	8.98	22.80	
	Lisbon	Pe		12-60/64	12.94	32.85	
	Lisbon	Terca		14-22/64	14.34	36.45	
	Lisbon	Covado		25-52/64	25.82	65.60	
	Lisbon	Covado		26-45/64	26.7	67.80	
	Lisbon	Vara		33	33	83.8	
	Lisbon	Vara		43-20/64	43.31	110.00	
	Oporto	Passo	geometrical	64-34/64	64.54	163.95	
Rumania							
	Bucharest	Endesi		26-5/64	26.08	66.25	
	Bucharest	Halibin		27-39/64	27.61	70.15	
	Moldavia	"Foot"		8	8	20.3	
	Moldavia	Kot	for silk	24-55/64	24.86	63.15	
	Moldavia	Fathom		96	96	243.85	
	Transylvania	Elle		24-34/64	24.53	62.30	
Russia							
		Verschok		1-48/64	1.75	4.45	
		Foute		12	12	30.50	
		Fuss	Rhineland foot	12-23/64	12.36	31.4	
		"Foot"	old measure	13-11/64	13.17	33.45	
		Archin		28	28	71.10	
		Sajen		84	84	213.35	
	Crimea	Halebi		28-50/64	28.78	73.10	
	St. Petersburg	Paletz		32/64	0.50	1.27	
Spain							
		Dedo		44/64	0.68	1.75	
		Palmo		8-22/64	8.34	21.20	= 9 onza
		Pie		12-1/64	12.01	30.50	
		Codo		16-44/64	16.69	42.40	
		Codo	"of the River"	22-16/64	22.26	56.55	
		Braccio	commercial	26-24/64	26.38	67.00	
		Vara		33-23/65	33.36	84.75	
		Vare, Rod		36-1/64	36.01	91.45	
		Paso	geometrical	55-41/64	55.64	141.35	
		Braza	marine	65-48/64	65.75	167.00	
		Toesa		66-49/64	66.77	169.60	
	Alicante	Vara		29-61/64	29.95	76.05	
	Alicante	Braza		71-17/64	71.26	181.00	
	Aragon	Pulgada		54/64	0.84	2.15	
	Aragon	Palmo		7-38/64	7.59	19.25	
	Aragon	Pie		10-8/64	10.12	25.70	
	Aragon	Vara		30-23/64	30.35	77.10	
	Asturias	Vara		32-56/64	32.87	83.50	

Country/Region	Unit Name	Comments	Inches/64ths	Inches/100ths	Cm/20ths	Subdivisions
Spain (continued)						
Barcelona	Canna		21-4/64	21.06	53.50	
Barcelona	Media-Canna	for cloth	30-35/64	30.55	77.60	
Barcelona	Canna		61-7/64	61.10	155.20	= 2 M-Canna
Burgos	Pie		11-8/64	11.13	28.25	
Cadiz	Octava		4-7/64	4.11	10.45	
Canary Islands	Onza		59/64	0.93	2.35	
Canary Islands	Pie	Castilian	11-8/64	11.13	28.25	= 12 onza
Canary Islands	Vara		33-9/64	33.14	84.20	
Canary Islands	Braza/Brazada		71-52/64	71.81	182.40	
Castile	Pulgada	legal for Spain	58/64	0.91	2.30	
Castile	Palmo menor		2-47/64	2.74	6.95	
Castile	Palmo mayor		8-14/64	8.22	20.90	= 3 Palmo-m
Castile	Pie		10-61/64	10.96	27.85	
Castile	Vara		32-56/64	32.87	83.50	
Galicia	Vara	mercer's	43-6/64	43.10	109.45	
Granada	Vara	old measure	27-19/64	27.29	69.30	
Madrid	Tercia		10-61/64	10.96	27.85	
Madrid	Varo		39-11/64	39.17	99.50	
Majorca	Palmo		7-45/64	7.70	19.55	
Majorca	Media-canna		30-50/64	30.79	78.20	
Majorca	Canna		61-37/64	61.58	156.40	= 2 M-Canna
Majorca	Canna		67-32/64	67.5	171.40	
Minorca	Canna		63-10/64	63.15	160.40	
Oviedo	Vara		34-1/64	34.02	86.35	
Saragossa	Quarta		7-38/64	7.59	19.30	
Toledo	Pie		10-51/64	10.79	27.40	
Toledo	Vara		32-14/64	32.22	81.85	
Tortosa	Canna		62-43/64	62.67	159.15	
Valencia	Pulgada		63/64	0.99	2.50	
Valencia	Palmo menor		2-62/64	2.98	7.55	
Valencia	Palmo		8-59/64	8.93	22.65	= 3 Palmo-m
Valencia	Pie		11-58/64	11.9	30.25	
Valencia	Vara		35-45/64	35.71	90.70	
Sweden						
	Tum	"gamla"	62/64	0.97	2.45	= 1/12 Fot
	Tum	"nya"	1-11/64	1.17	2.95	= 1/10 Fot
	Fot		11-44/64	11.69	29.70	= 10/12 Tum
	Fot		12-56/64	12.88	32.70	
	Ala		23-23/64	23.36	59.35	
	Famn		69-60/64	69.94	177.65	
	Faden		70-9/64	70.14	178.15	
Switzerland						
	Zoll		62/64	0.96	2.45	
	Fuss, Vaud		11-52/64	11.81	30.00	

FOREIGN UNITS OF LENGTH

Country/Region	Unit Name	Comments	Inches/64ths	Inches/100ths	Cm/20ths	Subdivisions
Switzerland (continued)						
	Elle		23-40/64	23.62	60.00	= 2 Fuss
Aarau	Fuss		11-35/64	11.55	29.30	
Aarau	Elle		23-24/64	23.38	59.40	
Appenzell	Zoll		1-2/64	1.03	2.60	
Appenzell	Elle	for woolens	24-16/64	24.26	61.60	
Appenzell	Elle	for linen	31-36/64	31.56	80.15	
Basel	Zoll		1	1.00	2.55	
Basel	Fuss, Schuh		11-63/64	11.99	30.45	
Basel	Brasse, Braccio		21-26/64	21.41	54.40	
Basel	Elle		44-26/64	44.41	112.80	
Basel	Aune, Klafter		46-24/64	46.38	117.80	
Berne	Fuss	quarry	12-32/64	12.51	31.75	
Berne	Elle		21-23/64	21.36	54.25	
Berne	Klafter		69-18/64	69.27	175.95	
Geneva	Pied		19-13/64	19.21	48.80	
Geneva	Aune	retail	45-02/64	45.03	114.35	
Geneva	Aune	wholesale	46-51/64	46.79	118.85	
Lausanne	Aune	before 1823	42-23/64	42.36	107.60	
Lausanne	Aune	since 1823	47-16/64	47.25	120.00	
Lucerne	Schuh	builder's/surveyor's	11-12/64	11.19	28.45	
Lucerne	Aune, Elle		24-46/64	24.71	62.75	
Neuchatel	Fuss	surveyor's	11-20/64	11.31	28.70	
Schaffhausen	Elle		23-29/64	23.45	59.55	
Solothurn	Fuss		11-35/64	11.55	29.35	
St. Gall	Elle		28-61/64	28.95	73.55	
Thurgau	Fuss	Rhineland foot	12-23/64	12.36	31.40	
Zug	Schuh	stone-cutter's	10-37/64	10.58	26.85	
Zurich	Zoll	surveyor's	1-12/64	1.18	3.00	
Zurich	Klafter		70-56/64	70.87	180.00	
United States						
	Inch		1	1.00	2.55	
	Palm		3	3.00	7.60	
	Span		9	9.00	22.85	
	Foot		12	12.00	30.50	
	Yard		36	36.00	91.45	
	Pace		54	54.00	137.15	
	Pace		60	60.00	152.40	
	Fathom		72	72.00	193.05	

BIBLIOGRAPHY

(author unknown). "Weights and Measures". *The Encyclopedia Britannica*, 9th edition, volume 24. New York: Charles Scribner's Sons, 1888.

(author unknown). "Weights and Measures". *The Encyclopedia Britannica*, 11th edition, volume 27. New York: Encyclopedia Britannica, Inc., 1911.

Adams, John Quincy. *Report of the Secretary of State Upon Weights and Measures, Prepared in Obedience to a Resolution of the House of Representatives of the Fourteenth of December, 1819*, Executive Document No. 109. Washington: Gales & Seaton, 1821.

Alexander, John H. *Universal Dictionary of Weights and Measures, Ancient and Modern; Reduced to the Standards of the United States of America*. Baltimore: William Minifie and Co., 1850.

Bion, Nicholas. *The Construction and Principal Uses of Mathematical Instruments*, Paris, 1723. Translated by Edmond Stone; . . . *To Which Are Added, the Construction and Uses of Such Instruments As Are Omitted by M. Bion*, by Edmond Stone; 2nd edition. . . . *To Which is Added, a Supplement Containing a Further Account of Some of the Most Useful Mathematical Instruments As Now Improved*, Supplement by Edmond Stone. London: J. Richardson, 1758. (reprinted by the Astragal Press)

Clarke, F. W. *Weights, Measures, and Money, of All Nations*. New York: D. Appleton & Company, 1876.

Gilliland, William. "Some Measures of History". *The Bulletin of the Scientific Instrument Society*, No. 20 (1989), pp. 7-17.

Grier, William. *The Mechanic's Pocket Dictionary; Being a Note Book of Technical Terms, Rules, and Tables, in Mathematics and Mechanics*, 11th edition. Glasgow: Blackie & Son, 1851.

Harris, J. *Lexicon Technicum; or an Universal English Dictionary of Arts and Sciences*, 4th edition, volume 1. London: D. Browne et al, 1725.

Hallock, William, Ph.D. and Wade, Herbert T. *Outlines of the Evolution of Weights and Measures and the Metric System*. New York: The MacMillan Company, 1906.

Haswell, Charles H. *Engineers' and Mechanics' Pocket-Book*. New York: Harper & Brothers, 1844.

—.*Mechanic's and Engineer's Pocket Book of Tables, Rules, and Formulas pertaining to Mechanics, Mathematics, and Physics*, 74th edition. New York: Harper & Brothers, 1909.

Huntar, Alexander. *A Treatise of Weights, Mets, and Measures of Scotland*. Edinburgh: John Wreittown, 1624. (reprinted by Theatrum Orbis Terrarum, Ltd.)

Kelly, Patrick. *The Universal Cambist and Commercial Instructor; Being a Full and Accurate Treatise on the Exchange, Monies, Weights & Measures, of All Trading Nations and their Colonies, with an Account of their Banks, Public Funds, and Paper Currencies*, 2nd edition, volumes 1 and 2. London: Lackington & Co. et al for the author, 1821.

Kennelly, Arthur E. *Vestiges of Pre-Metric Weights and Measures Persisting in Metric-System Europe, 1926-1927*. New York: The MacMillan Company, 1928.

Woolhouse, W. S. B. *The Measures, Weights, & Moneys of All Nations; and an Analysis of the Christian, Hebrew, and Mahometan Calenders*, 2nd edition, corrected. London: John Weale, 1859.

Zupko, Ronald E. *British Weights and Measures; A History from Antiquity to the Seventeenth Century*. Wisconsin: The University of Wisconsin Press, 1977.

—.*French Weights and Measures Before the Revolution: A Dictionary of Provincial and Local Units*. Bloomington: The University of Indiana Press, 1978.

—.*Italian Weights and Measures From the Middle Ages to the Nineteenth Century*. Philadelphia: The American Philosophical Society, 1981.

—.*A Dictionary of Weights and Measures for the British Isles: the Middle Ages to the Twentieth Century*. Philadelphia: The American Philosophi-cal Society, 1985.

—.*Revolution in Measurement: Western European Weights and Measures Since the Age of Science*. Philadelphia: The American Philosophi-cal Society, 1990.

IX
Rule Accessories

During the heyday of the carpenters' folding rule (approximately 1800-1930) a large number of accessories were invented and marketed for use with rules to extend their functionality and allow them to replace other items in the tool kit. Some adapted a rule for use as a square or bevel, or as a plumb or level. Others used the rule as the beam for a set of trammel points. Still others were concerned with convenience, protecting the rule and/or securing it in the carpenter's pocket for ready availability.

This chapter, an expansion of the the author's 1983 article on this subject (Ref. 1), describes a number of these rule accessories, from the most elementary to the extremely complex. It does not pretend to be comprehensive, however, and there are undoubtedly others which have not yet turned up. (This is one of the exciting thing about collecting; one is always encountering new objects while searching for nice examples of objects already known).

Only true rule accessories are described here. In order to so qualify an accessory must have: a) been intended for use on a common two foot, four fold carpenters' rule (attachments intended for use on a steel square, such as the Stanley No. 44 Bit & Square Level, have been omitted, for instance); b) been offered for sale separately from the rule, and be capable of being attached or removed at will (attachments which were mounted permanently on the rule by third parties prior to sale, such as Nichols' patent rule dividers or Cowgill's patent bevel attachment, are considered to be modifications, and not accessories); c) been actually manufactured and offered for sale. (The large number of accessories which were patented but never manufactured are interesting, but by their very nature uncollectible.)

RULE FENCE ATTACHMENTS

Probably the most common type of accessory was the attachable fence (sometimes called a "gauge"). This was a device which could be attached to the rule to form one or more fixed fences on one surface of a rule at an angle to an edge and thus allow it to be used as a square or bevel. Six attachments of this type are shown here. The simplest of these provided a single fence; the most complex (the Arrow "Angler" and the Stanley 3-Angle Tool) each had five.

These fence attachments could perform other functions besides squaring and mitering. By mounting them a specific distance from the end of the rule it was possible to then use the rule as a depth gauge or a marking gauge (this latter by resting the pencil in the notch between the bulge of the joint and the end of the stick). Additionally, by using two rules sliding against one another and employing the fence attachment as a clamp, it was possible to form a crude extension stick.

SINGLE-FENCE RULE GAUGES

The most basic type of rule gauge provided only a single fence at right angles to the edge of the rule. A rule with one of these gauges attached could be used as a square, or as a depth or marking gauge, but could not mark miters.

A single-fence rule gauge by an unknown maker is shown in Figure 9-1 below. This gauge is made of folded sheet steel, and locks to the rule by a cam lever which bears against one edge.

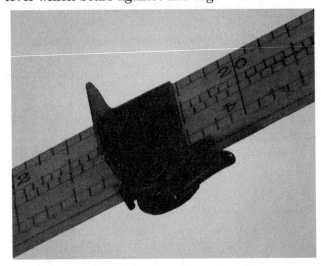

Figure 9-1: Rule gauge by unknown maker

The inventor of this gauge also is unknown. The marking on it: "PAT. APL'D FOR" would indicate that it was a patented design, but the exact patent is not indicated, either by date or number.

This rule gauge is obviously factory made, but the name and location of the maker is unknown.

THREE-FENCE RULE GAUGES

The next level of sophistication in rule fence attachments were gauges which provided three fences, a right angle fence at one end, and a V-shaped fence (which could be used to mark miters at ±45°) at the other.

A number of examples of this type of gauge are known, varying in their material, construction, and method of attaching to the rule. Three of them are shown in Figure 9-2. They are, respectively, the Campbell Combination Tool, the Robinson Rule Tool, and the Crescent Rule Gauge.

The Campbell Combination Tool

The Campbell Combination Tool was made of machined cast brass, and used a cam lever to clamp onto the rule. Later examples were made of aluminum instead of brass, but the tool was otherwise identical.

This tool was invented by Robert Campbell, of Elizabeth, New Jersey, and patented by him on October 1, 1907 (U.S. Patent No. 867,556, entitled "Combination Tool"). This patent describes a tool equipped with a level vial parallel to the right-angle fence as one possible configuration, but no examples including such a level have been found.

Campbell's tool was offered by at least two dealers in woodworking tools, being listed in the 1923 catalog of the Rubelman Hardware Co. (Ref. 2), and the 1914 catalog of the Orr & Lockett Company (Ref. 3). Judging from the number of surviving examples, it must have been a popular item in the carpenters' kit of tools.

Subsequent to his 1907 patent Campbell continued to develop his tool, replacing the cam lever clamp with a thumb screw. He received a patent for this improvement on May 6, 1913 (U.S. Patent No. 1,061,045, entitled "Rule Attachment"). No examples of this improved version are known, however, and it is probable that it never made it into production.

The Robinson Rule Tool

The Robinson Rule Tool was made of cast iron, and fastened to the rule with a round head machine screw which bore against its edge. Construction was crude, with only the fence faces machined and nickel plated. (It is interesting to note that the back step against which the rule rested was *not* machined; this must have adversely affected the accuracy of the three fences). It is not known if the Robinson Rule Tool was ever patented; it is not marked in any way, and patent searches have not found any device which corresponds to it in construction. This tool was offered in a ca. 1914 catalog of the Sears, Roebuck & Co. (Ref. 4), and in the 1923 catalog of the Rubelman Hardware Co. (Ref. 2), but cannot have been too popular; the one shown here is the only example known to the author.

The Crescent Rule Gauge

The Crescent Rule Gauge, was made of a single piece of nickel-plated sheet steel, with bent-out projections forming a right-angle and two miter fences on one side, and a pair of projections which gripped the body of the rule with a sliding friction fit on the other.

This tool was invented by Thomas W. Southard, of Williamsport, Pennsylvania, and

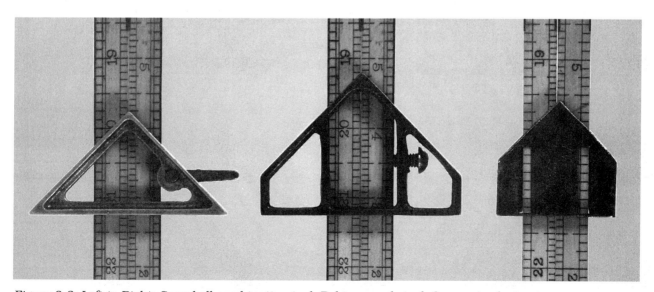

Figure 9-2: Left to Right: Campbell combination tool, Robinson rule tool, Crescent rule gauge

patented by him on September 16, 1914 (U.S. Patent No. 1,110,968, entitled "Rule Attachment"). It is not known which, if any, dealers offered it for sale, but it could not have been many; examples are quite scarce.

FIVE-FENCE RULE GAUGES

The next level of complexity in rule fence attachments was to increase the number of fences from three to five. This was done by giving the two miter fences a double angle, at both ±45° and ±60°, to allow the marking of miters for both squares and hexagons.

Two examples of this configuration are shown below, the Stanley No. 2 Three-Angle Tool and the Arrow "Angler.

The Stanley No. 2 3-Angle Tool

The No. 2 3-Angle Tool, shown in Figure 9-3 below, was made of nickel-plated cast iron, and was held on the rule by a bow-shaped spring which pressed against one side of the rule leg. In addition to the five fences, it also had a spirit vial parallel to the right-angle fence, allowing it to be used as a level.

The 3-Angle Tool was invented by William H. Stanley, of Alameda, California, and patented by him on August 8, 1911 (U.S. Patent No. 999,899, entitled "Combination Tool"). It was first offered in the Stanley catalog that same year, and continued to be so listed through 1935. It must have been a popular attachment; it was also offered in numerous catalogs published by various tool retailers throughout that period.

> **WILLIAM H. STANLEY**
>
> When one considers the similarity in name between the patentee and the company which made the 3-angle tool, it is tempting to assume that William Stanley was some connection of the Stanley family of New Britain. No concrete evidence other than the name can be found to support such an assumption, however, and when one considers the distance between Alameda and New Britain, any such connection seems highly unlikely.

The Arrow "Angler"

The Arrow "Angler," shown in Figure 9-4, was made of machined, nickel-plated cast iron. In basic configuration it was similar to the Stanley No. 2 3-Angle Tool, having the same arrangement of five fences: a right angle fence, and miter fences at ±45° and ±60°. It did not incorporate a level, however, as did the Three-Angle Tool, but

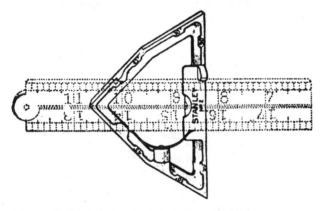

Figure 9-3: The Stanley No. 2 3-angle tool

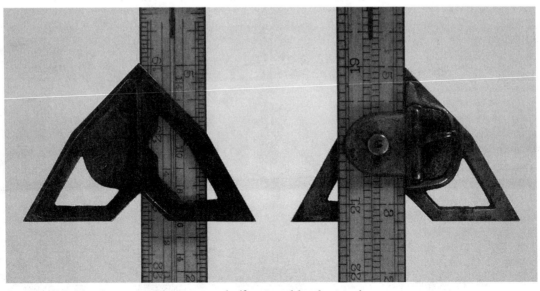

Figure 9-4: The Arrow "Angler" on a rule (front and back views)

had a V-shaped notch in the right-angle fence, in line with the edge of the rule, to allow it to mark the centers of squares and circles, a feature which the 3-Angle tool did not have.

The most unusual feature of the "Angler," however, was its method of clamping to the rule. Instead of using pressure on the edge of the rule to clamp it sideways against a stop on the back of the tool, it used pressure on the flat of the rule to clamp it vertically against the flat back surface. This was done by means of an overhanging bracket, equipped with a cam-lever pressure pad, on the back of the tool. A stop was provided on the back of the tool to align the rule at right angles to the primary fence, allowing the four miter fences to function at their designed angles. However, the user was able to ignore this stop, and clamp the rule to the tool in a range of positions, making it possible to mark or gauge a line at any desired angle.

The Arrow "Angler" was invented by Philip F. Freytag, of Utica, New York, and patented by him on May 12, 1914 (U.S. Patent No. 1,096,782, entitled "Combination Tool"). The maker, according to advertisements, was Barnes & Irving, Inc., of Syracuse, New York.

PROTRACTOR FENCE ATTACHMENTS

The most versatile type of rule fence attachments are those in which the fence can be set and clamped at any angle, providing not only square and miter angles, but any other desired angle, as with a protractor or bevel.

In addition to its utility in measuring/marking angles, a protractor fence attachment could also function as a compass to draw semicircles centered on the edge. The radius of the semicircle would be determined by the position of the rule in the clamp fixture, and the corner formed by the main joint and the end of the rule would guide the pencil.

Two examples of this type of accessory are shown below, the Perfection Rule Gauge and Miller's Multiple Gauge.

THE PERFECTION RULE GAUGE

The Perfection Rule Gauge (Figures 9-5 and 9-6) was a two-piece assembly of machined cast iron, one piece being the fence, and having a protractor scale marked on it, and the other being a rotating bracket to hold the rule, with an index to indicate the angle on the protractor scale. The two were connected by a pivot pin which allowed the bracket to be set to any angle relative to the fence, with the angle indicated on the protractor scale; the bracket could then be clamped in position by tightening a threaded thumb nut on the end of the pivot pin.

The protractor scale was calibrated to ±45° in increments of 5°, and was also marked with letters at certain key angles, to wit: M (Miter) at ±45°, P (Pentagon) at ±36°, H (Hexagon) at ±30°,

Figure 9-5: The Perfection Rule Gauge

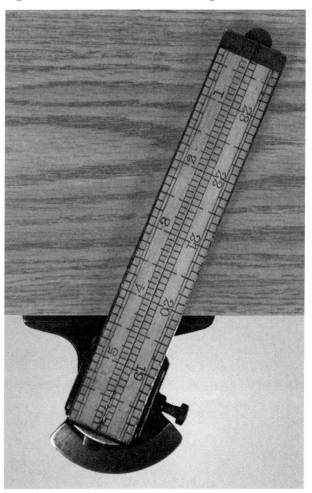

Figure 9-6: The Perfection Rule Gauge in use

O (Octagon) at ±22½°, and W (the proper pitch angle for a window or door sill) at ±9°.

The Perfection Rule Gauge was invented by Frederick Reissman, of West Point, New York, and patented by him on November 10, 1896 (U.S. Patent No. 571,052, entitled "Miter-Gage Attachment for Rules"). The gauge was offered in at least one dealer's catalog, that of Tower & Lyon Company, in New York City, in 1904 (Ref. 5).

MILLER'S MULTIPLE GAUGE

Miller's Multiple Gauge (Figure 9-7) was generally similar to the Perfection Rule Gauge, differing primarily in the material used in its construction. Where the Perfection Rule Gauge was made of machined cast iron, Miller's Multiple Gauge was made of stamped and formed steel.

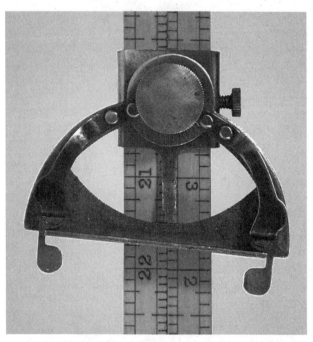

Figure 9-7: Miller's multiple gauge (shown with curved edge followers extended)

There were minor variations between the two. In the Perfection gauge, the protractor scale was on the same side as the pivoting bracket; in Miller's gauge it was on the opposite side, where it could not be obscured by the body of the rule. The major functional difference was the presence on Miller's gauge of two round bearing points which could be extended beyond the fence to allow the gauge to follow a curved edge when using it as a marking gauge. These curved edge followers were pivoted on the body of the gauge at each end of the fence, and were held in position, either extended or retracted, by lock springs.

Miller's Multiple Gauge was invented by Arlington H. J. Miller, of Easton, Pennsylvania, and patented by him on February 23, 1915 (U.S. Patent No. 1,129,707, entitled "Gage"). It is not known who manufactured it.

RULE TRAMMEL POINTS

Another class of rule accessory was that of trammel points which used the rule as their stick or beam. These rule trammel points were not much different from ordinary trammel points, the primary difference being that they are designed to fit a stick not more than ¼ inch thick, and from 1 to 1½ inches wide, the usual maximum dimension of most rules.

Two types of rule trammel points have been noted, those that used the rule on edge as their beam, and those that used the rule lying flat as their beam. Two examples of the first type, the Stanley No. 99 and the S&J Mfg. Co. rule trammel points, and one example of the second type, Devoe's rule trammel points, are shown below and on the following pages.

THE STANLEY NO. 99 RULE TRAMMEL POINTS

The No. 99 rule trammel points (Figure 9-8), made by the Stanley Rule & Level Co., consisted

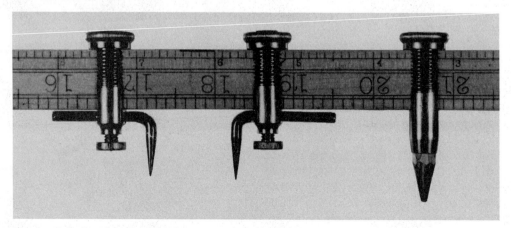

Figure 9-8: Stanley No 99 rule trammel points

of three cylindrical brass heads, two with steel points and a third with a pencil socket. These heads were slotted to slip onto a rule edgewise, and threaded on the top to accept thumb nuts which clamped them into place. They fit rules of any width up to 3/4 of an inch, and were thus usable with all ordinary two and four fold rules.

The No. 99 rule trammel points were the invention of Justus A. Traut, of New Britain, Connecticut, the well-known Stanley employee and inventor, and were patented by him on June 14, 1887 (U.S. Patent No. 365,031, entitled "Trammel Point"). They were first offered in the Stanley catalog the next year, and continued to be so listed through 1935.

S&J Mfg. Co. Rule Trammel Points

The S&J rule trammel points (Figure 9-9) are later and considerably cruder than the rather elegant Stanley product. Made of stamped and formed sheet steel, they probably date from the 1920s or 1930s. They are also unusual in that they are obviously intended to use a common household yardstick as their beam, instead of a four fold rule. They exactly fit the shape of the common yardstick, $1\frac{1}{8}$ x $\frac{5}{32}$ inches, sliding onto the stick from the end, and being held in place by pressure springs on the back.

Unlike the No. 99 rule trammel points, the S&J trammel points did not have a separate fixture to hold the pencil for beam compass work. Instead, the clamping bar which held the steel point in place could be loosened and a pencil slid in under its other end in place of the point.

The only marking on these trammel points is the name of the makers, THE S&J MFG. CO., and their location, Detroit, Michigan. They were presumably patented, but neither patent number nor patent date appear anywhere on them, and patent searches have not turned up any patent which appears to describe them.

Devoe's (and other similar) Trammel Points

Devoe's trammel points (Figure 9-10) are multi-purpose devices, described in the 1904 Tower & Lyon catalog (Ref. 5) as "for the use of carpenters and joiners, stair builders, cabinet makers, ship carpenters, and wood workers in general," and also called an "improved drafting tool." They consisted of a pair of notched cylindrical metal blocks, each equipped with a thumbscrew permitting it to be clamped onto the edge of a thin stick, and with a removable sharp steel point. They were about three inches long, including the point.

With the steel points installed, they could be clamped onto a steel square or open rule and used as trammel points. With the points removed, they could be attached to a steel framing square to convert it into a stair gauge to mark the rise and run of treads on stair joists.

Devoe's trammel points were invented by C. D. Janssen and William Devoe, of Torrington,

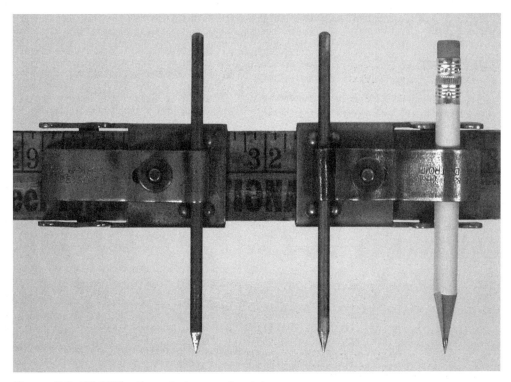

Figure 9-9: S&J Mfg. Co. rule trammel points

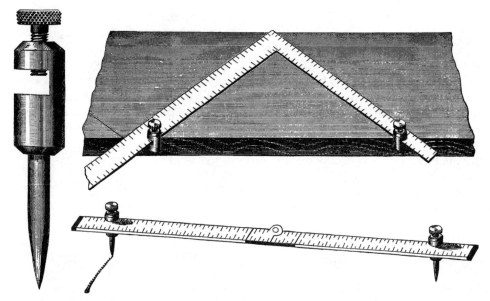

Figure 9-10: Devoe's trammel points and their uses

Connecticut, and patented by them on February 15, 1887 (U.S. Patent No. 357,771, entitled "Attachment for Squares"). These points were offered in at least two dealers' catalogs, those of Tower & Lyon Company, of New York City, in 1904 (Ref. 5), and the T. B. Rayl Company, of Detroit, in 1905 (Ref. 6).

At least two other variations of this general pattern are known, and are shown with a pair of Devoe's trammel points in Figure 9-11. Devoe's points are on the left. The center pair is by the Tower & Lyon Co., and is marked "T&L CO N.Y." on the thumb screws. The small pair on the right is by an unknown maker.

COMPLEX RULE TOOLS

Two other rule tools deserve special mention because of their exceptional complexity and versatility. These were the No. 1 Odd-Jobs Tool and the Boye Rule Tool. Fairly similar in design, each could perform the functions of the simpler attachments described above, plus other functions as well. Each had right angle and V fences, one or more spirit vials, and a steel point near the angle of the V fence which served as one point of a beam compass.

It is difficult to decide which was the more versatile of the two. Both could function as a square, miter gauge, depth gauge, beam compass, level, plumb, and miter level. However, the Boye Rule Tool had a built-in protractor, while the Odd Jobs had a built in scratch awl/pencil holder, and could thus be used as a mortise gauge.

THE NO. 1 ODD-JOBS TOOL

The No. 1 Odd-Jobs Tool, made by the Stanley Rule & Level Co., was a remarkable piece of multi-

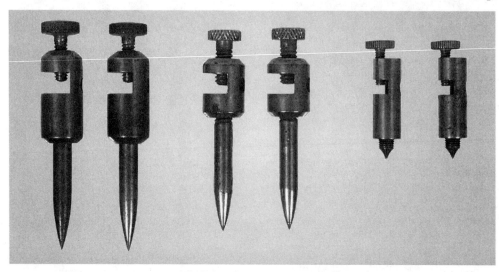

Figure 9-11: Devoe's (left), and two other similar sets of rule trammel points

function design, differing from any other pattern of rule attachments in at least two respects. First, the Odd-Jobs was longer than other rule accessories; the distance between the right angle fence and the tip of the V fence being more than three inches. This made the sides long enough so that the single spirit vial could function both as a level and as a plumb. Second, the Odd-Jobs was equipped with a scratch awl which could be used independently, or clamped to the tool and, with a pencil attached in the holder at its other end, act as the scribe of a mortise or marking gauge.

The Odd-Jobs was made of nickel-plated cast iron, and attached to a common one inch wide rule by means of a thumbscrew. After 1898 a separate rule was not required; a special one foot rule with an embedded scribe point was furnished with the tool. (Technically, at this time, it ceased to be a rule "accessory", but became instead a complete tool in its own right.)

The design of the Odd-Jobs was based on a patent issued on January 25, 1887 to George F. Hall of Long Branch, New Jersey (U.S. Patent No. 356,533, entitled "T-square and Gage Attachment for Rules") Stanley must have purchased the patent from Hall very shortly thereafter, because the Odd-Jobs first appeared in its 1888 catalog, only a year after Hall's patent was issued.

Not all of the features of Halls patent were retained. The scribe, sliding in a hole in the right angle fence was retained, but Hall's complex scheme for turning the rule into a miter gauge was abandoned and the V fence substituted. The resulting Odd-Jobs tool as finally produced and sold (shown below in Figure 9-12), was as described in a subsequent joint patent granted to George F. Hall, of Newark, New Jersey, and Justus A. Traut, of New Britain. Connecticut, on September 18, 1888 (U.S. Patent No. 389,647, entitled "Attachment for Carpenter's Rules"). Stanley described the Odd-Jobs tool as "Ten Tools in One," and included elaborate diagrams and pictures in its catalogs (Figure 9-13) to illustrate the many uses of this tool to the prospective purchaser.

THE BOYE RULE TOOL

The Boye Rule Tool (Figure 9-14) was made of machined, die-cast aluminum. In general form it was similar to the Robinson Rule Tool, but the

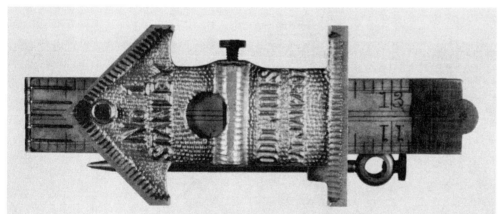

Figure 9-12: The Stanley No. 1 Odd-Jobs tool

Figure 9-13: Applications of the Odd-Jobs tool

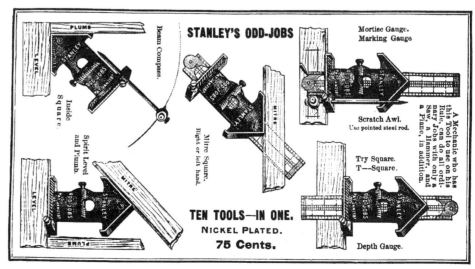

266 RULE ACCESSORIES

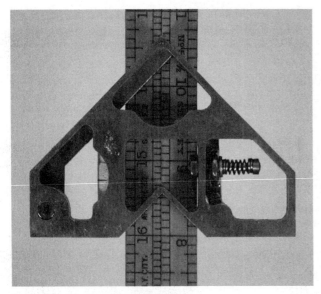

Figure 9-14: The Boye rule tool

Boye had not one, but two level vials. Additionally, the Boye had a swing-out protractor arm, a small protruding point in its side near the apex of the V fence, and a small V-shaped notch in the right-angle fence. The rule was attached to the tool by a spring-loaded pressure foot which pressed against its edge.

The instruction sheet which was packed with the rule (Figure 9-15) illustrated all of the operations which it could perform. The two spirit vials allowed it to be used as both a plumb and level. The small point near the apex of the V fence would act as the pivot when using it as a beam compass, and the protractor (its arm graduated in inches and eighths) for measuring, gauging, and setting off angles. The two 45° fences formed a right angle which could be used as a square, and the notch in the center of the right angle fence made it possible also to use it as a center gauge.

The Boye rule tool was invented by George F. Hall, of Newark, New Jersey, and patented by him on February 10, 1925 (U.S. Patent No. 1,526,143, entitled "Combination Square, Level, and Rule Attachment"). The patent was assigned at the time of issue to John T. Uhl, also of Newark. The manufacturer was the Boye Needle Company, of

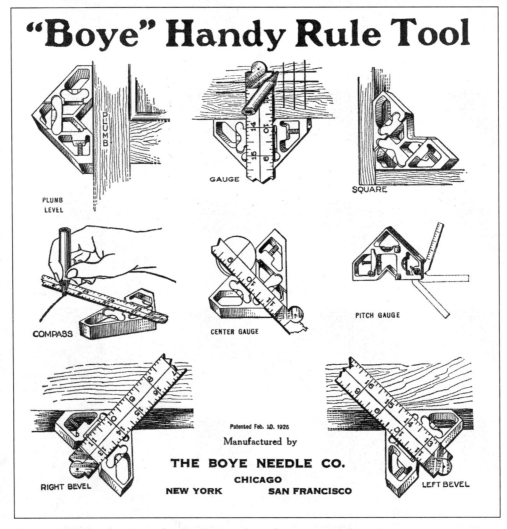

Figure 9-15: Instructions for the Boye rule tool

Chicago, Illinois (hence the name), better known today as a maker of sewing, knitting and crocheting needles.

It is not clear to the author how well this tool would perform when used as a compass, swinging on the point at its tip. When used in this way, the rule was about 3/16 inch above the surface of the work, and thus could not accurately guide a pencil or scribe in marking that surface.

> ### GEORGE F. HALL
>
> It is interesting to note that the Odd-Jobs Tool, patented in 1887, and the Boye Rule Tool, patented in 1925, 38 years later, were both invented by the same individual, George F. Hall, of Newark, New Jersey. Although Hall's original 1887 patent for the Odd-Jobs Tool was granted while he was residing in Long Branch, New Jersey, he had removed to Newark, 25 miles away, shortly thereafter, and was residing there when his and Traut's joint patent for the modified tool was issued a year later.
>
> Hall continued to reside in Newark throughout his working life, being listed as a resident and a patternmaker in the 1890 and subsequent U.S. census returns, and in Newark city directories, until 1922. He was still residing in Newark and (presumably) still working as a patternmaker (by that time close to retirement) when he invented and patented the Boye Rule Tool in 1925.

OTHER RULE ACCESSORIES

RULE POCKET CASES

It was all too easy to drop or lose a folding rule while on the job, especially since the usual place to carry one was in the shirt or overall front pocket, where it could fall out the first time the workman leaned over. If the pocket was deep enough to discourage this type of accident, then another problem manifested itself; a rule placed in a deep pocket would slide down in that pocket to the point where it would be hard to grasp. The cases shown in Fig. 9-16 were another type of rule accessory, one designed to keep the rule conveniently at hand while at the same time preventing its falling out and being lost or damaged.

Two examples are shown, very similar in design. They are rectangular boxes made of folded sheet metal, approximately 1 1/4 inches by 3/8 of an inch by 5 inches, closed at the bottom and open at the top with a slightly flared edge. A spring clip, mounted on the outside of the flat side of the case, holds the case in place when placed in the pocket.

A one inch wide two foot four fold rule will just fit in the case with the top inch protruding from the pocket, and a flat spring riveted to one side inside the case presses against the edge of the rule to prevent it from coming out accidentally.

One of the cases in the figure is made of aluminum, folded and riveted at the seams, and is marked with the patent date just below the clip. This case design was invented by Percy W. Tooth, of Leastalk, California and patented by him on May 13, 1913 (U.S. Patent No. 1,061,886, entitled "Rule-Pocket"). The case as produced differs from the patent in two respects: the edge is flared, rather than rolled as shown in the drawing, and the construction of the clip is slightly different.

Figure 9-16: Two rule pocket cases

Figure 9-17: Rule pocket case in use

The second case is of the same basic design as the first, although differing in several respects. It is made of black enameled sheet steel, instead of aluminum, and has its seams tacked with spot welds. Also, the clip mechanism differs from both that on the first case and the clip shown in the patent drawing. It is not marked in any way.

Both cases are obviously factory made, but the maker and date of manufacture of either is unknown. They could represent the product of a single maker at different periods of production, or the similar products of two different makers, one copying the other. The lack of a patent date on the enameled steel case would seem to support this second scenario, but we cannot be certain.

The Gould Tri-Mytre Rule Case

The Gould Tri-Mytre (*sic*) Rule Case was also intended to hold and protect the rule from wear and damage when not in use. It differed significantly from the pocket cases described above, however. It did not clip to the top edge of the pocket to enhance accessibility, as they did, but was simply a container to enclose and protect the rule. It had a second function which they did not, however: it was designed to allow its use as a rule fence attachment when desired.

This case was a U-shaped trough of nickel-plated steel 6 inches long, 1 inch wide, and 3/8 of an inch deep, which would just hold a folded common four fold rule and protect three of its four surface from damage while in the pocket or a tool box (see Figure 9-18). It did not afford complete protection; one surface of the folded rule was still exposed and could be scratched or dented, but apparently the protection it offered was deemed acceptable.

The back of this case was cut with three slots 1/2 of an inch wide and 3/16 of an inch deep, just large enough to hold one leg of a rule. One of these slots was at right angles to the length of the case; the other two were at 45° angles, left and right. When the stick of a rule was placed in one of these slots the case then became in effect a square or miter fence (see Figure 9-19). Each slot had a semicircular cutout on one of its edges, to accommodate the rule main joint and center the rule in the fence.

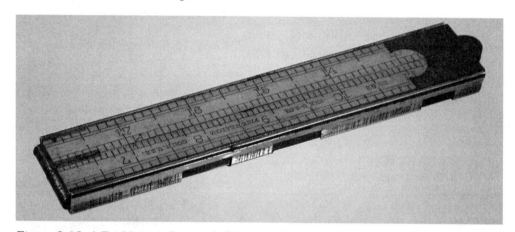

Figure 9-18: A Tri-Mytre rule case holding a rule

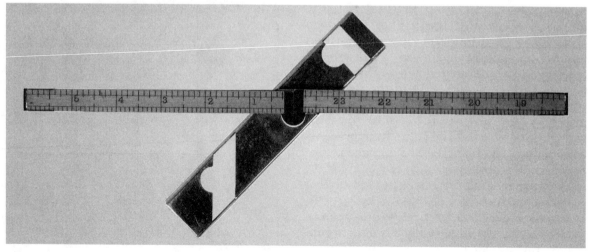

Figure 9-19: The Tri-Mytre rule case functioning as a fence

The Gould Tri-Mytre Rule Case was invented by Henry M. Gould, of Portland, Maine, and patented by him on April 9, 1918 (U.S. Patent No. 1,262,038, entitled "Combined Rule and Case"). This patent was assigned at the time of issue to the Tri-Mytre Rule Case Company, of Portland Maine, a corporation specifically formed to manufacture this attachment.

RULE END HOOK

This rule accessory (Figure 9-20) is a hook-shaped fitting designed to be attached to the end of a measuring or board stick to facilitate its one-handed use in measuring wide stock. With this hook installed, the rule became essentially a hook rule, and could be held in one hand and hooked onto the far edge of the work while taking the dimension.

This example is in no way a precision product, and would not be suitable for fine work at the bench. It is made of cast iron, nickel plated, and the only machining done to it was to drill and tap it for the attaching screw. The inventor and/or maker are unknown, and it is unlikely that it was ever patented; except for the nickel plating it could easily be craftsman made.

Figure 9-20: End hook installed on yardstick

RULE ACCESSORIES YET TO BE FOUND

Some rule accessories have no known example, and we are aware of them only through advertising and catalog listings. Three such devices are listed here, the Caldwell Rule Gauge, the Sunset Rule Gauge, and the Eden Specialty Company Rule Attachment Set.

THE CALDWELL RULE GAUGE

The Caldwell Rule Gauge (Figure 9-21) was a simple rule gauge providing only a single, right angle fence. It was made of sheet brass, with a clamping mechanism consisting of a pair of cam-operated jaws which gripped the edges of the rule. We know of the existence of the Caldwell Rule Gauge only through its listing in the 1923 catalog of the Rubelman Hardware Co. (Ref. 2). It is not known who the manufacturer was.

THE SUNSET RULE GAUGE

The Sunset Rule Gauge (Figure 9-22, next page) was also a simple rule gauge providing only a single, right angle fence. It differed from all other simple rule gauges in that the fence pivoted, and could be either latched up, for marking gauge use, or unlatched and laid down, allowing the gauge to be left on the rule while using it as an ordinary rule or carrying it in the pocket (the gauge is shown twice on the rule in the figure, on the left with the fence up and on the right with the fence down). This gauge, made of sheet brass, was a spring-loaded sliding fit on the two tip sticks of a common one inch folding rule (fitting on just the two tip sticks allowed the rule to be opened half way without removing the gauge).

The makers of the Sunset Rule Gauge made one claim for their device that was made for no other accessory known to the author—that their gauge would act as a finger protector, to avoid splinters and cuts while sliding the rule along the edge of a board when scribing a line parallel to that edge. One wonders that this claim was made by no other rule gauge maker. Virtually all gauges would, in effect, perform this same function when scribing lines, and it would have been a worth-while attribute to point out to prospective users.

Another unusual feature of this gauge was the manner in which it was adapted for use as a compass. This was accomplished by making the two

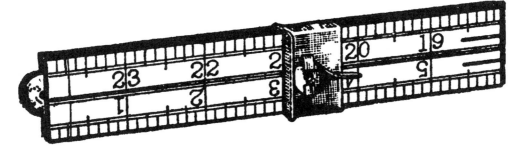

Figure 9-21: The Caldwell rule gauge

This simple device, made of brass is applicable to any ordinary two foot folding pocket rule, and is easily adjusted to be carried on the rule in the pocket when not in use.
It embraces in itself when on the rule the following

1st **Finger Protector** as it protects the finger from splinters or a rough surface when scribing a line parallel to the edge. It is worth its price ten times over for this purpose alone to any mechanic.

2nd **Adjustable Gauge** as it can be placed at any required position on the rule, simply by sliding.

3rd **Marking Gauge** with a pencil or scratch awl all work as done with a marking gauge, mortise gauge or panel gauge can be performed.

4th **Hinge Gauge** it can be set in an instant for marking the cuts on doors and castings in fitting butts or hinges.

5th **Compass or Divider** by sliding the rivet on either side of gauge from end of rule to measurement of radius of circle you get center or revolving point, your pencil or awl gives you the marking point.

Price $0 15

Figure 9-22: The Sunset rule gauge

rivets which fastened the fence to the body prominent and pointed enough so that they could be imbedded in the wood to form a pivot for swinging a circle. This seems to be a practical way to accomplish this function; it is surprising that no other maker adopted the same scheme.

The Sunset Rule Gauge was invented by John J. McManus, of San Francisco, California, and patented by him on February 27, 1894 (U.S. Patent No. 515,550, entitled "Gage for Rules"). It is not known who the manufacturer was. We know of its existence only through its listing in the industrial supply catalog of the Osborn & Alexander Company, of Pittsburgh, Pennsylvania.

THE EDEN RULE ATTACHMENT SET

This was a set of four devices, sold together, which could be attached to any one inch wide two foot, four fold rule to perform a variety of functions. This set consisted of: a pie-shaped 2 inch by 2 inch right-angle clamp attachment, a pair of trammel points, one with a steel point and the other with a pencil clip, and a piece of flat metal ½ inch wide and 6 inches long, with a 45° sideways bend 2 inches from one end.

The clamp attachment and the pair of trammel points were both the invention of John Kralund of New York City, and had been patented by him. The clamp attachment was patented on January 30, 1906 (U.S. Patent No. 811,070, entitled "Rule Attachment"); the trammel points on December 25, 1906 (U.S. Patent No. 839,604, entitled "Rule Attachment") (see Figure 9-23). This second patent was assigned at the time of issue to James A. Eden, Sr. The fourth element in the set, the flat piece of metal, was not described in either patent.

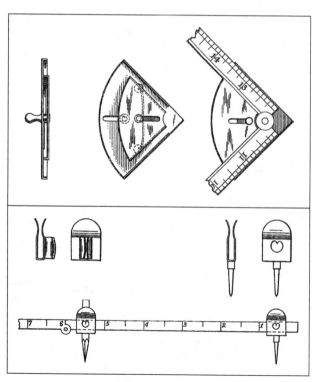

Figure 9-23: The Eden clamp attachment and trammel points (from the patent drawings)

The right-angle clamp attachment was made of sheet steel, with the two flat sides formed into fences, and was provided with a spring-loaded sliding clamp which would hold the legs of a rule firmly against the insides of these fences. If the two legs of a rule on either side of the main joint were placed in this clamp, the rule would be held open at exactly 90°, and could be used as a square (see figure). If only one leg was placed in the clamp, leaving the other side empty, then the clamp could be used as the sliding fence of a depth gauge.

The trammel points were a cheap pair of spring steel clips which just snapped onto the rule. They had no fastening screw, but were held in place by friction. The user was instructed to "clamp them with the finger and thumb" while marking or scribing a circle.

The fourth element in the set, the flat piece of metal with the 45° sideways bend near one end, was included to allow the clamp to be used as an inside or outside miter square (see Figure 9-24). If the short section of the piece was placed in one side of the clamp, then the other side of the clamp would form either a 45° or a 135° angle with the long section (depending on whether the clamp was located on the inside or the outside of the bend) and could be used to gauge and mark miter angles. (It might be observed that in this configuration the clamp was not functioning as a rule attachment, since no rule was involved).

The Eden Rule Attachment set was made by the Eden Specialty Company, of Brooklyn, New York, presumably owned by the same James A. Eden, Sr. who was the assignee of Kralund's rule trammel point patent. We know of the existence of the set only through an advertisement in the August 1908 issue of Dayton's *Carpenter and Builders Index*.

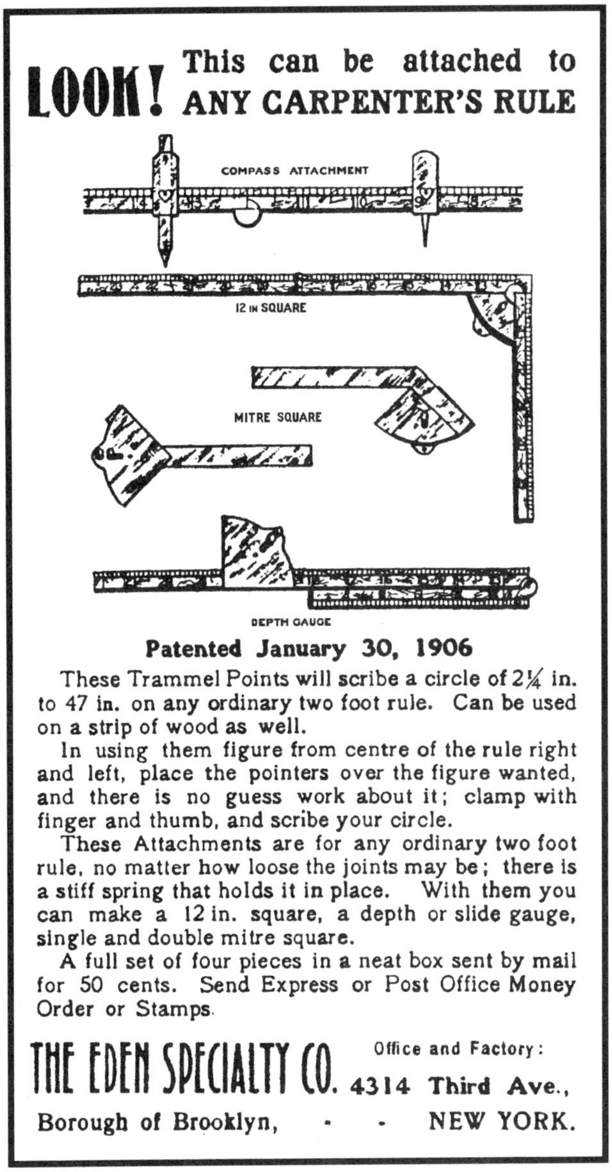

Figure 9-24: Advertisement for the Eden rule attachment set

CONCLUSION

None of the rule accessories described here are sold today. When the boxwood rule fell into disuse, supplanted by the "Zig-Zag" and push-pull tape rules, the demand for such accessories declined as well. One by one they ceased being made, and the few which remained in production during the 1920s were eventually discontinued during the depression, or, at the latest, with the advent of World War II.

REFERENCES

1. —, "A Sampling of Rule Accessories," *The Fine Tool Journal*, vol. 27, no. 1 (September, 1983), 4-7.
2. *Mechanics Tool Catalogue.* (Catalogue No. 43) (St. Louis: The George A Rubelman Hardware Co., 1923).
3. *Catalogue of Woodworkers Tools* (Chicago: Orr & Lockett Hardware Co., ca. 1914).
4. *Tools—Machinery—Blacksmiths' Supplies* (Chicago: Sears, Roebuck & Co., ca. 1914).
5. *Tower & Lyon Co., Manufacturers of Hard-ware Specialties, Fine Tools, and Police Specialties* (Catalogue No. 12) (New York: Tower & Lyon Co., 1904).
6. *Rayl's, Detroit Michigan - Woodworker's Tools* (Catalogue No. 21) (Detroit: The T. B. Rayl Co., ca. 1905).

X
A Rule Bibliography

A great deal of printed material relating to rules and measuring instruments has been published during the last 50 years. Unfortunately, most collectors are only aware of a fraction of this material. Many who have recently entered the field of rule collecting do not have complete files of the journals where such material has been published. Many catalog reprints are out of print and not readily available. Additionally, most libraries, except those concerned with the history and development of tools and technology, do not have large collections in this area.

This bibliography is an attempt to inform collectors as to what information is available and where it can be found. It comprises 1) a list of current or recent catalog reprints dealing with measuring instruments, 2) a bibliography of books and articles, grouped by subject, 3) a list of encyclopedias and encyclopedic dictionaries that contain useful information and references, 4) a discussion of the available indexes of U.S. patents, and 5) a list of other bibliographies and directories which will possibly be of use as well.

MANUFACTURERS' AND WHOLESALERS' CATALOGS

In 1972 Herman Maddocks, Jr., of West Boylston, Mass. reprinted the 1892 catalog of the Stanley Rule & Level Co., to our knowledge the first rule manufacturer's catalog to be so reproduced. Since that time a large number of other such catalogs have been reprinted, mostly by two individuals, Ken Roberts and Roger K. Smith, and two organizations, The Mid-West Tool Collectors Association and the Astragal Press. Their efforts have made information available that we could not otherwise have obtained, and we are in their debt.

We have listed here all such reprinted catalogs which bear on rules and measuring instruments. Catalogs not related to measurement have been omitted, as have catalogs which have not been reprinted, and thus are not readily available to the average collector.

Many more original catalogs of interest to rule collectors exist (see Barlow and Romaine, cited in the BIBLIOGRAPHIES & DIRECTORIES section), but have not been (and probably never will be) reprinted. The best places to view such catalogs is in museums, historical societies, and private collections. The best place to acquire them is at book, ephemera, and paper shows, and from tool, scientific instrument, and ephemera dealers.

Belcher Brothers & Co.'s Price List of Boxwood & Ivory Rules. New York: Belcher Brothers & Co., 1860. [reprinted by Ken Roberts Publishing Co.]

Brown Brothers & Co., Manufacturers of Shaw's U.S. Standard Ring Travelers, Belt Hooks, Wire Goods, Etc., and dealers in ... General Mill Furnishings. Providence: Brown Brothers & Co., 1890. [This catalogue is concerned almost exclusively with supplies for textile mills. It is included here solely because it contains the only known advertisement for Hogg's Improved Slide Rule; see *Stanley Tool Collector News* 4, (Winter, 1991): 10-12]

Carpenters & Mechanics Tools, Catalogue No. 102. New Britain: The Stanley Rule and Level Co., 1909 [reprinted by Roger K. Smith]

Catalogue and Invoice Prices of Rules, Planes, Gauges, &c. Manufactured by Hermon Chapin, Union Factory, Pine Meadow, Conn. Pine Meadow: Hermon Chapin, 1853. [reprinted by Ken Roberts Publishing Co.]

Catalogue and Price List of Fine Tools for Mechanics. Athol: The L. S. Starrett Co., 1895. [reprinted by the Bud Brown Publishing Co.]

Catalogue and Price Lists of Brown & Sharpe Mfg. Co., Providence, Rhode Island, United States. Providence: Brown & Sharpe Mfg. Co., January 1, 1887. [reprinted by the Astragal Press as part of *A Brown & Sharpe Catalogue Collection*]

Catalogue and Price Lists of Darling, Brown & Sharpe, Providence, R.I., United States. Providence: Darling, Brown and Sharpe,

1868. [reprinted by the Astragal Press as part of *A Brown & Sharpe Catalogue Collection*]

Catalogue No. 101 and Price Lists of Machinist's Tools, Rules, Squares, Micrometer Calipers, Gauges, and Accurate Test Tools, made by Brown & Sharpe Mfg. Co., Providence, Rhode Island, U.S.A. Providence: Brown & Sharpe Mfg. Co., 1899. [reprinted by the Astragal Press as part of *A Brown & Sharpe Catalogue Collection*]

Catalogue of Measuring Rules Tapes, Straight Edges, and Steel Band Chains, Spirit Levels, &c. Birmingham: John Rabone & Sons, July 1892. [reprinted by Ken Roberts Publishing Co.] [contains a good history of the company by Ken Roberts, plus much good information about rulemaking in Birmingham]

Illustrated and Priced Catalogue of American Tools, Twist Drills, Chucks, Vices, &c., &c. London: Buck & Hickman Ltd. ("Tool Makers and American Merchants"), August 1902 [reprinted by the Mid West Tool Collectors Association]

Illustrated Catalogue and Price List of Carpenters' Tools, Catalogue No. 7. Philadelphia: J. B. Shannon, August 1873. [reprinted by William Cavallini]

Illustrated Catalogue of Rules, Levels, Plumbs & Levels, Thermometers, Planes, Improved Woodworkers' and Mechanics' Tools, &c., Catalogue No. 18. Birmingham: Edward Preston & Sons, Ltd., May, 1909. [Reprinted by the Astragal Press (now out of print); contains an extensive history of the company by Mark Rees.]

Illustrated Price List of Rules, Spirit Levels, Planes and Tools, &c. Birmingham: Edward Preston & Sons, July, 1901 [Reprinted by Ken Roberts Publishing Co.; contains a good history of the company by Ken Roberts.]

Illustrated Price List of Wood Working Tools Manufactured by Alexander Mathieson & Son, Ltd., Glasgow (8th edition). Glasgow: Alexander Mathieson & Sons, Ltd., 1899 [Reprinted by Ken Roberts Publishing Co.; contains a short history of the company by Ken Roberts.]

Leonard Bailey & Co.'s Illustrated Catalogue and Price List. Hartford: Case, Lockwood and Brainard Co., 1876. [Includes a list of Stephens & Co.'s boxwood and ivory rules; reprinted by the Astragal Press as part of *The Stanley Catalog Collection, Volume II.*]

List Prices Per Dozen of Box Wood & Ivory Rules Manufactured by E. A. Stearns & Co., Brattleboro, Vt. Brattleboro: E. A. Stearns & Co., ca. 1848. [reprinted in *The ACTIVE Scrapbook* 10 (March, 1976)]

Price List and Illustrated Catalogue of Rules, Planes, Gauges ... Manufactured by H. Chapin's Son (E. M. Chapin), Union Factory, Pine Meadow, Conn., USA. Hartford: Case, Lockwood, and Brainard, January 1890. [reprinted by the Mid-West Tool Collectors Association.]

Price List of Box Wood and Ivory Rules Manufactured by Charles L. Mead, Successor to E. A. Stearns & Co., Brattleboro, Vt. Brattleboro: E. A. Stearns & Co., January 1, 1861. [reprinted in *The ACTIVE Scrapbook* 35 (January, 1981)]

Price List of Boxwood and Ivory Rules, Levels, Try Squares, Sliding T Bevels, Gauges, &c. New Britain: The Stanley Rule and Level Co., February 1859. [reprinted by the Astragal Press as part of *The Stanley Catalog Collection*]

Price List of Boxwood and Ivory Rules, Manufactured by The Stanley Rule and Level Co. (Successors to E. A. Stearns & Co., Brattleboro, Vt.). New Britain: The Stanley Rule and Level Co., April 1, 1863. (reprinted by Roger K. Smith)

Price List of Boxwood and Ivory Rules. New Britain: A. Stanley & Co., January 1855. [reprinted by the Astragal Press as part of *The Stanley Catalog Collection*]

Price List of Rules, Planes, Gauges, Hand Screws, Bench Screws, Levels, &c., manufactured by Hermon Chapin, Pine Meadow, Conn. Hartford: Press of Case, Lockwood and Company, July 1859. [reprinted by Ken Roberts Publishing Co.]

Price List Of U. S. Standard Boxwood and Ivory Rules, Levels, Try Squares, Gauges, Handles, Mallets, Hand Screws, &c. New Britain: Stanley Rule and Level Co., January 1, 1867. [reprinted by the Astragal Press as part of *The Stanley Catalog Collection*]

Price List Of U. S. Standard Boxwood and Ivory Rules, Levels, Try Squares, Gauges, Iron and Wood Bench Planes, Mallets, Hand Screws, Spoke Shaves, &c. New Britain: The Stanley Rule and Level Co., January 1870. [reprinted by the Astragal Press as part of *The Stanley Catalog Collection*]

Price List Of U. S. Standard Boxwood and Ivory Rules, Levels, Try Squares, Gauges, Iron and Wood Bench Planes, Mallets, Hand Screws, Spoke Shaves, Screw Drivers, Etc. New Britain: The Stanley Rule and Level Co., January 1872. [reprinted by the Astragal

Press as part of *The Stanley Catalog Collection, Volume II*]

Price List Of U. S. Standard Boxwood and Ivory Rules, Levels, Try Squares, Gauges, Mallets, Screw Drivers, Hand Screws, Iron and Wood Planes, Spoke Shaves, Etc. New Britain: The Stanley Rule and Level Co., January 1874. [reprinted by the Astragal Press as part of *The Stanley Catalog Collection, Volume II*]

Price List Of U. S. Standard Boxwood and Ivory Rules, Plumbs and Levels, Try Squares, Bevels, Gauges, Mallets, Handles, Awl Hafts, Screw Drivers, Hand Screws, Iron and Wood Planes, Spoke Shaves, Etc. New Britain: The Stanley Rule and Level Co., January 1877. [reprinted by the Astragal Press as part of *The Stanley Catalog Collection, Volume II*]

Price List Of U. S. Standard Boxwood and Ivory Rules, Plumbs and Levels, Try Squares, Bevels, Gauges, Mallets, Iron and Wood Adjustable Planes, Spoke Shaves, Screw Drivers, Awl Hafts, Handles, Etc. New Britain: The Stanley Rule and Level Co., January 1879. [reprinted by the Astragal Press as part of *The Stanley Catalog Collection*]

Price List of U. S. Standard Boxwood and Ivory Rules, Plumbs and Levels, Try Squares, Bevels, Gauges, Mallets, Iron and Wood Adjustable Planes, Spoke Shaves, Screw Drivers, Awl Hafts, Handles, Etc. New Britain: The Stanley Rule and Level Co., January 1884. [reprinted by the Astragal Press as part of *The Stanley Catalog Collection, Volume II*]

Price List of U. S. Standard Boxwood and Ivory Rules, Plumbs and Levels, Try Squares, Bevels, Gauges, Mallets, Iron and Wood Adjustable Planes, Spoke Shaves, Screw Drivers, Awl Hafts, Handles, Etc. New Britain: The Stanley Rule and Level Co., January 1888. [reprinted by the Astragal Press as part of *The Stanley Catalog Collection*]

Price List of U. S. Standard Boxwood and Ivory Rules, Plumbs and Levels, Try Squares, Bevels, Gauges, Mallets, Iron and Wood Adjustable Planes, Spoke Shaves, Screw Drivers, Awl Hafts, Handles, &c. New Britain: The Stanley Rule and Level Co., January 1892. [reprinted by Herman C. Maddocks, Jr.]

Price List Of U. S. Standard Boxwood and Ivory Rules, Plumbs and Levels, Try Squares, Bevels, Gauges, Mallets, Iron and Wood Adjustable Planes, Spoke Shaves, Screw Drivers, Awl Hafts, Handles, &c. New Britain: The Stanley Rule and Level Co., January 1898. [reprinted by the Astragal Press as part of *The Stanley Catalog Collection*]

Rules, Planes, Gauges, Plumbs and Levels, Hand Screws, Handles, Spoke Shaves, Box Scrapers, Etc.. Catalogue No. 114. Pine Meadow: The Chapin Stephens Co., 1914 [reprinted by Ken Roberts Publishing Co.]

Stanley Rule and Level Company, Catalogue No. 28. New Britain: The Stanley Rule and Level Co., January 1902. [reprinted by Alvin Sellens]

Stanley Tools for Carpenters and Mechanics, Catalogue No. 129. The Stanley Rule & Level Plant of The Stanley Works, 1929. [reprinted by Roger K. Smith]

Stanley Tools, Catalogue No. 110. New Britain: The Stanley Rule and Level Co., 1911. [reprinted by Roger K. Smith]

Stanley Tools, Catalogue No. 139. New Britain: Stanley Tools Division of the Stanley Works, September 1, 1939. [reprinted by Roger K. Smith]

Stanley Tools, Catalogue No. 34. New Britain: The Stanley Rule & Level Co., 1915 [reprinted by the Stanley Publishing Co.]

Steel Carpenters' Squares. Shaftsbury: The Eagle Square Manufacturing Co., ca. 1910. [reprinted by Antique Crafts and Tools in Vermont.]

Tools for Coopers and Gaugers (and) Produce Triers, Catalogue No. 20. New York: A. F. Brombacher & Co., Inc., 1922. [Reprinter unknown]

BOOKS & ARTICLES

DRAFTING INSTRUMENTS

Dickinson, Henry W. "A Brief History of Draughtsmen's Instruments." *Transactions of the Newcomen Society* 27 (1949-1951): 73-84.

Hambly, Maya. *Drawing Instruments 1580-1980.* London: Sotheby's Publications, 1988.

Heather, J. F. *A Treatise on Mathematical Instruments, Including Most of the Instruments Employed in Drawing, for Assisting the Vision, in Surveying and Levelling, in Practical Astronomy, and for Measuring the Angles of Crystals: in which Their Construction, and the Methods of Testing, Adjusting, and Using Them, are Concisely Explained.* Third edition. London: John Weale, 1856.

Hulme, F. Edward. *Mathematical Drawing Instruments and How to Use Them.* London: Trübner & Co., 1880.

Kentish, Thomas. *Treatise on a Box of Instruments.* London: publisher unknown, 1849.

Lyles, John. *Catalogue of Drawing Instruments* (in the *John Lyles Collection*—ed) Santa Ana: Published by the author, 1993.

Scott-Scott, Michael. *Drawing Instruments, 1850-1950* (Shire Albem No. 180). Princes Risborough: Shire Publications, Ltd., 1986.

Sutcliffe, G. Lister. "Compendium of Drawing and Drawing Instruments." (Sections IV and V of *The Modern Carpenter Joiner and Cabinet-Maker*, G. L. Sutcliffe, editor.) London: Publisher unknown, 1902. [reprinted in 1990 by the National Historical Society]

Townsend, Raymond R. "Parallel Rules - Then and Now." *The Fine Tool Journal* 36, no.1 (March/April, 1988): 5-6.

—. "Railroad Drawing Instruments." *The Fine Tool Journal* 40, no. 1 (Summer, 1990): 17-20.

Warren, S. Edward, C.E. *A Manual of Drafting Instruments and Operations*. New York: John Wiley & Son, 1874

Makers

(Author Unknown). "New Britain Manufactures, No. 7 The Stanley Rule and Level Co." *The New Britain Record* (May 18, 1866).

(Author Unknown). "Rules." *The Great Industries of the United States*. Horace Greely, et al, editors, 739-743. Hartford: J. B. Burr & Hyde, 1873.

Courser, Fred W., and Roberts, Kenneth D. "The New Hampshire Rule Business of R. B. and H. R. Haselton." *The ACTIVE Scrapbook* 42 (March, 1983): 1-2

Brundage, Larry. "Making Tools in Bangor, Maine File, Saw, and Rule Makers." *The Gristmill* 39 (March, 1985): 14-15

Drake, H. E. "Rule Manufacturing." (private papers), 1910. [reprinted in *Wood Planes in 19th Century America, Vol. II*; see ROBERTS, below]

Fales, Clifford. "The Hedge Clark Connection, or An Unusual Rule Joint." *The Gristmill* 47 (March, 1987): 10-11.

Farnham, Alexander. *Early Tools of New Jersey and the Men Who Made Them*. Stockton NJ: Kingwood Studio Publications, 1984.

—. *Search For Early New Jersey Toolmakers* Stockton NJ: Kingwood Studio Publications, 1992.

Hill, James. "E. A. Stearns & Co." *Mensuration; The Newsletter of the Rule Collector's Association* 1 (Summer, 1987): 1-6.

Holden, William. "Wm. Greenleaf: Talented Eccentric." *The Gristmill* 28 (June, 1982): 8-9.

Jacob, Walter W. "The Stanley Rule & Level Company: Its Historical Beginning." *The Chronicle of the Early American Industries Association* 51, no. 3 (September, 1998): 80-84

—. "The Stanley Rule & Level Company: Charles L. Mead and the Acquisition of E. A. Stearns." *The Chronicle of the Early American Industries Association* 51, no. 4 (December, 1998): 120-122.

Kebabian, John. "Early American Factories: The Eagle Square Company." *The Chronicle of the Early American Industries Association* 23, no. 1 (March, 1970): 7-8.

—. "Willis Thrall." *The Chronicle of the Early American Industries Association* 29, no. 1 (March, 1976): 12-13.

—. "Stearns Rules." *The ACTIVE Scrapbook* 35 (January, 1981).

Kebabian, Paul. "Early Vermont Square Makers and the Eagle Square Company." *The Chronicle of the Early American Industries Association* 36, no. 4 (December, 1983): 65-71.

—. "The Eagle Square Company, 1817-1874." *Stanley Tool Collector News* 17 (Spring, 1996): 31-39.

Keller, David. "Lufkin." *Cooper Industries, 1833-1983*. Athens: The Ohio University Press, 1983. [The best available history of the Lufkin Rule Company.]

Leavitt, Robert K. *History of The Stanley Works*. New Britain: The Stanley Works, 1951.

Lynk, Scott. *Stanley "Special" and Custom Rules*. Vergennes: Published by the author, 2002. [An exhaustive study of the many nonstandard rules made by this company.]

Roberts, Kenneth D., editor. *Lufkin Measuring Instruments; Excerpts From Trade Catalogues, 1888 to 1940*. Fitzwilliam: Ken Roberts Publishing Co., 1983.

— *Wood Planes in 19th Century America, Vol. II: Planemaking by the Chapins at Union Factory, 1826-1929*. Fitzwilliam: Ken Roberts Publishing Co., 1983.

Smith, Philip Chadwick Foster. "Notes On a Visit to the Buff & Buff Surveying Instrument Factory, Jamaica Plain, Massachusetts, Monday, 3 June 1968." *The Chronicle of the Early American Industries Association* 53, no. 2 (June, 2000): 76, 80.

Stanley, Philip E. *Boxwood & Ivory; Stanley Traditional Rules, 1855 1975*. Westborough: The Stanley Publishing Co., 1984.

Walter, H. S. "History of the Rule Shop." *The Rule Shop Seismograph* I, no. 5 (June 16, 1919).

Walter, John. "The Stanley Rule & Level Company." (in two parts) *The Tool Shed* 97 (June, 1997): 1, 4-6, & *The Tool Shed* 98 (September, 1997): 1, 4-5.

Wing, Donald, and Wing, Anne. "A Rural Rulemaker: Captain Anthony Gifford." *The Chronicle of the Early American Industries Association* 55, no. 2 (June, 2002): 46-51.

PATTERNMAKING

Chase, I. McKim, ME. *The Art of Pattern Making*. New York: John Wiley & Sons, 1903. [Has an interesting section on how a patternmaker can make a shrinkage rule to any special allowance he requires.]

RULES - MAKING

(Author Unknown). *Graduating, Engraving, and Etching*. New York: The Industrial Press, 1921. [reprinted by Lindsay Publications, Inc.]

(Author Unknown). "I Check Stanley Rules." *Tool Talks No. 7*. New Britain: The Stanley Tools Division of The Stanley Works, 1938. [reprinted in *Stanley Tool Collector News* 10 (Winter, 1993): 12-15.]

(Author Unknown). "I Supervise the Making of Boxwood Rules." *Tool Talks No. 11*. New Britain: The Stanley Tools Division of The Stanley Works, 1938. [reprinted in *Stanley Tool Collector News* 14 (Spring, 1995): 31-34].

Aldinger, Hanry. "Common Graduation and Early U.S. Slide Rules." *Journal of the Oughtred Society* 7, no. 2 (Fall, 1998): 42.

Barclay, R. L.. "The Metals of the Scientific Instrument Maker: Part I: Brass." *The Bulletin of the Scientific Instrument Society* 39 (December, 1993): 32-36.

—. "The Metals of the Scientific Instrument Maker: Part II: Steel." *The Bulletin of the Scientific Instrument Society* 40 (March, 1994): 13-14.

Battison, Edwin A. "The Development of Accurate Scales." *Tools and Technology* 1, no. 2 (Summer, 1977): 1-2.

Goodson, David. "Crafting Scale Rules by Hand - An Historical Note." *Journal of the Oughtred Society* 7, no. 2 (Fall, 1998): 36-41.

Hopp, Peter M. "Ivory, Bone, or Plastic?" *Journal of the Oughtred Society* 9, no. 2 (Fall, 2000): 21-23.

Hubbard, Guy. "Development of Machine Tools in New England (Part 4)." *American Machinist* 59, no. 9 (August 30, 1923): 311 315. [Includes a description & photograph of Hedge's dividing machine.]

Kean, Herb. "Tool Woods." *The Tool Shed* 76 (April, 1993): 1, 4-6.

Ramsden, Jesse. *Description of an Engine for Dividing Strait Lines on Mathematical Instruments*. London: William Richardson, 1779.

Stubbs, Alan. "A Linear Dividing Engine - Joseph Brown, 1850." *Tools and Technology* 1, no. 1 (Spring, 1977): 1-2.

Welsh, Greg. "Who Made That Rule?" *The Tool Shed* 77 (June, 1993): 8.

Whelan, John M. "Steel." *The Tool Shed* 36 (April, 1985): 6-7, 12.

RULES - CARPENTERS' & ENGINEERS'

(Author Unknown). *Instructions for the Engineer's Sliding Rule, With Examples of Its Application*. Pine Meadow: Hermon Chapin, 1858.

(Author Unknown). *The Carpenter's Slide Rule, Its History and Use*. Third edition. Birmingham: John Rabone & Sons, 1880. [reprinted by Ken Roberts Publishing Co.]

Babcock, Bruce. "An Error on a Slide Rule for 50 Years?" *Journal of the Oughtred Society* 2, no. 2 (October, 1993): 15-17.

Jillson, Arnold. *A Treatise on Instrumental Arithmetic: or Utility of the Slide Rule*. Hartford: Case, Lockwood & Brainard, 1874. [This is the set of instructions on the use of the Carpenters' and Engineers' sliding rules which was offered by the Stanley Rule & Level Co. to their customers.]

Kebabian, Paul. "The Engineer's Rule." *The ACTIVE Scrapbook* 30 (July, 1979).

Knott, John V. "Joshua Routledge, 1775-1829." *Journal of the Oughtred Society* 4, no. 2 (October, 1995): 25.

Roberts, Kenneth D. "Carpenter's and Engineer's Slide Rules (Part I, History)." *The Chronicle of the Early American Industries Association* 36, no. 1 (March, 1983): 1-5.

Routledge, Josiah. *Instructions for the Use of the Practical Engineers' & Mechanics' Improved Slide Rule, as Arranged By J. Routledge, Engineer*. Birmingham: John Rabone and Son, Ca. 1867. [Reprinted by Ken Roberts Publishing Co.]

Stanley, Philip E. "Carpenters' and Engineers' Slide Rules (Part II, Routledge's Rule)." *The Chronicle of the Early American Industries Association* 37, no. 2 (June, 1984): 25-27.

—. "Carpenters' and Engineers' Slide Rules (Part III, Errors in the Data Tables)." *The Chronicle of the Early American Industries Association* 40, no. 1 (March, 1987): 7-8.

—. "Full and Complete Instructions . . ." *Mensuration; The Newsletter of the Rule Collector's Association* I, no. 3 (Fall, 1989): 7-8.

—. "Commentary on 'An Error on a Slide Rule for 50 Years?'" (See BABCOCK, above). *Journal of*

the *Oughtred Society* 3, no. 1 (March, 1994): 35-38.
—. "A Stanley/Jillson Engineer's Rule" *Stanley Tool Collector News* 19 (Winter, 1996-1997): 29-31.
Wyman, Thomas. "Slide Rules of the Stanley Rule & Level Company and Other American Makers." *The Chronicle of the Early American Industries Association* 54, no. 3 (September, 2001): 114-117.

RULES - CLEANING

Evans, John W. "Thoughts on the Cleaning and Restoration of Antique Woodworking Tools." *The Fine Tool Journal* 26, no. 1 (March, 1983): 3-5.
Hill, James, and Stanley, Philip. "Cleaning Rules ...and Storing Rules." *Mensuration; The Newsletter of the Rule Collector's Association* I, no. 2 (Fall, 1988): 6-8.
Kean, Herbert P. *Restoring Antique Tools*. Mendham: Astragal Press, 2002.

RULES - EARLY

Babcock, Bruce. "A Guided Tour of an 18th Century Carpenter's Rule." *Journal of the Oughtred Society* 3, no. 1 (March, 1994): 26-34.
Bernard, Baader, Sayward, Townsend, et al. [Replies to a previously published enquiry by Roberts regarding early rules.] *The Fine Tool Journal* 27, no. 2 (October, 1983): 23-26.
Bedwell, Thomas. *Mesolabium Archtectonicum; that is, A Most Rare and Singular Instrument, for ... Measuring of Planes and Solids*. London: Wilhelm Bedwell, Publisher, 1631. [reprinted by Theatrum Orbis Terrarum, Ltd.]
Browne, John. *The Description and Use of an Ordinary Joint Rule Fitted With Lines, for the Ready Finding the Lengths and Angles of Rafters and Hips, and Collar Beams in any Square or Bevelling Roofs at any Pitch, and the Ready Drawing of the Architrave, Friese, and Cornice of any Order*. London: William Fisher, 1686. [bound as an appendix into Fisher's English edition of *The Mirror of Architecture: or the Ground Rules of the Art of Building*, by Vicenzo Scamozzi.]
Evans, John W. "Phineas Pett and His Folding Rule." *The Fine Tool Journal* 27, no. 4 (January, 1984): 68-69, and *The Fine Tool Journal* 27, no. 5 (February, 1984): 89.
Hodge, Al. "Birmingham (England) Carpenter's Rules." *The Tool Shed* 81 (April, 1994): 6.
Kebabian, Paul. "The English Carpenter's Rule Notes on its Origin." *The Chronicle of the Early American Industries Association* 41, no. 2 (June, 1988): 24-27.

—. "Further Notes on the Early English Three Fold Ship Carpenter's Rule." *The Chronicle of the Early American Industries Association* 42, no. 1 (March, 1989): 3-5.
Knight, Richard. "A Carpenter's Rule from the Mary Rose." *Tools & Trades, the Journal of the Tool and Trades History Society* 6 (August, 1990): 43-55.
—. "An Important 17th Century Rule." *Tools & Trades, the Journal of the Tool and Trades History Society* 10 (August, 1997): 74-75.
—. "A Postscript to 'An Important 17th Century Rule'." *Tools & Trades, the Journal of the Tool and Trades History Society* 11 (June, 1999): 45.
—. "The Carpenter's Rule". *TATHS, Newsletter of the Tool and Trades History Society* 20 (Winter, 1988): 12-19.
MacKay, Andrew. *The Description and Use of the Sliding Rule, in Arithmetic and in the Mensuration of Surfaces and Solids. Also the Description of the Ship Carpenter's Sliding Rule, Its Use Applied to the Construction of Masts, Yards, &c., Together With the Description and Use of the Gauging Rule, Gauging Rod, & Ullage Rule*. Second edition, ... Improved, Enlarged, and Illustrated With an Accurate Engraving of the Different Rules. Edinburgh: Oliphant, Waugh, and Innes, 1811.
More, Richard. *The Carpenters Rule*. Second edition. London: Published by the Author, 1602. [reprinted by Theatrum Orbis Terrarum, Ltd.]
Sharman, Ann, & Sharman, Peter. "A Gunnery Inclinometer & Rule?" *TATHS, Newsletter of the Tool and Trades History Society* 56 (Spring, 1997): 15-20.
—. "A Gunner's Rule - continued." (correspondence) *TATHS, Newsletter of the Tool and Trades History Society* 57 (Summer, 1997): 31-34.
—. "A Gunner's Rule and Quadrant - ad nauseum." (correspondence) *TATHS, Newsletter of the Tool and Trades History Society* 60 (Spring, 1998): 28-31.
Stanley, Philip E. "Benjamin Dearborn and His 'Anglet'." *The Chronicle of the Early American Industries Association* 55, no. 2 (June, 2002): 52-54.
Townsend, Ray. "Early Measuring and Leveling Instruments." *The Fine Tool Journal* 29, no. 3 (November, 1984): 36-39.

RULES - GAUGING

(Author Unknown) *Gauging and Wantage Rods*. New Britain: The Stanley Rule and Level Co.,

1911. [reprinted in *Stanley Tool Collector News* 13 (Winter, 1994): 27-28.]

Gandenberger, Gil and Gandenberger, Mary. "The Stanley Cask Gauging Rod." *Stanley Tool Collector News* 16 (Winter, 1995): 14-17.

Lynk, Scott. "Stanley Milk Sticks." *The Fine Tool Journal* 50, no. 1 (Summer, 2000): 6-7.

Nelson, Robert E. "Barrel Mensuration & Ullage." *Mensuration; The Newsletter of the Rule Collector's Association* I, no. 3 (Fall, 1989): 1-5.

Packham, Jim. "Barrel Gauging." *The Chronicle of the Early American Industries Association* 50, no. 4 (December, 1997): 121-124.

—. "H. S. Pearson Gauging Rod: A Tool Analysis." *The Chronicle of the Early American Industries Association* 53, no. 3 (September, 2000): 93-95.

Sauer, Robert J. "Is 'Gauging' an Archaic Practice?" *Journal of the Oughtred Society* 8, no. 1 (Spring, 1999): 5-7.

Walsh, James I. "Capacity and Gauge Standards for Barrels and Casks of Early America." *The Chronicle of the Early American Industries Association* 52:4 (December, 1999): 151-154.

RULES - GENERAL

Fales, Clifford. "How Thick is 'Extra Thick'?" *The Gristmill* 54 (March, 1989): 8-9.

Gauntlett, James. (To Karl West, 27 March 1999). "In Consideration of Measurement." *The Chronicle of the Early American Industries Association* 53, no. 2 (June, 2000): 78-80.

Lynk, Scott. "A Rare New Discovery?" *The Chronicle of the Early American Industries Association* 53, no. 1 (March, 2000): 36.

Rabone, John, Jr. "Measuring Rules." *Birmingham and the Midland Hardware District, the Resources, Products, and Industrial History*. Samuel Timmins, ed., 628-632. London: (publisher unknown), 1866. [This article is included as an addendum to the reprinted 1892 catalogue of John Rabone & Sons.]

Roberts, Kenneth D. *Introduction to Rule Collecting*. Fitzwilliam: Ken Roberts Publishing Co., 1982.

Smith, Roger K. "Tools Made in the Midwest — Measuring Devices." *The Gristmill* 38 (December, 1984): 14.

Stanley, Philip E. "A Concordance of Major American Rule Makers." *The Fine Tool Journal* 31, no. 1 (September, 1985): 7-18.

—. "A Concordance of Major American Rule Makers." 2nd ed. *Fundamentals of Rule Collecting*. Kenneth D. Roberts, ed., 43-61. Fitzwilliam: Ken Roberts Publishing Co., 1997.

—. "Six Fold Rules" *Mensuration; The Newsletter of the Rule Collector's Association* I, no. 1 (Summer, 1987): 10-11.

—. "Measuring Instruments A Diversity." *The Chronicle of the Early American Industries Association* 55, no. 2 (June, 2002): 45,58.

Stanley, Philip E., and Hill, James. "A Proposed System for Classifying Rules as to Rarity" *Mensuration; The Newsletter of the Rule Collector's Association* I, no. 1 (Summer, 1987): 9,11.

Welsh, Greg. "History of Measurement." *The Tool Shed* 104 (November, 1998): 8-9.

West, Karl H., Jr. "Caliper Rule." *The Chronicle of the Early American Industries Association* 52:1 (March, 1999): 23.

RULES - GUNTER'S SCALES

Babcock, Bruce. "Some Notes on the History and Use of Gunter's Rule." *Journal of the Oughtred Society* 3, no. 2 (September, 1994): 14-20.

Graham, Ralph. "Gunter's Scale." *The Gristmill* 55 (June, 1989): 17.

Jezierski, Dieter von. "Further Notes on the Operation of Gunter's Rule." *Journal of the Oughtred Society* 6, no. 2 (Fall, 1997): 7-8.

RULES - LUMBER

(Author Unknown). "Board and Log Rules." *The Chronicle of the Early American Industries Association* 18, no. 4 (December, 1965): 63-64.

Baader, William. "Board Rules...How to Use." *The Gristmill* 42 (December, 1985): 20.

Freese, Frank. *A Collection of Log Rules*. USDA Forest Service General Technical Report No. FPL1. Madison: US Dept. of Agriculture Forest Service, 1974. [A comprehensive survey of all of the different rules, scales and tables for estimating sawn lumber quantities from log size. A MUST for anyone interested in log calipers of log sticks. Only its length (51 pages) prevented its inclusion as one of the reprints in Chapter III.]

Heggie, Andrew. "Timber Measurement." (correspondence) *TATHS, Newsletter of the Tool and Trades History Society* 49 (Summer, 1995): 19.

—. "More Timber Offerings." (correspondence). *TATHS, Newsletter of the Tool and Trades History Society* 55 (Winter, 1996): 29-33.

Hodge, Al. "Hoppus' Tables and the System of Girt Measure." *The Tool Shed* 64 (November, 1990): 1,7.

Knight, Richard. "The Case of the Missing Tool." *TATHS, Newsletter of the Tool and Trades History Society* 55 (Winter, 1996): 24-28.

Major, Ken. "Timber Measurement." *TATHS, Newsletter of the Tool and Trades History Society* 65 (Summer, 1999): 23-24.

Nelson, Robert. "Missing Tool - Case Solved." (correspondence) *TATHS, Newsletter of the Tool and Trades History Society* 56 (Spring, 1997): 34-37.

RULES - SCALES

Finch, Bob. "Another Curious Measuring Device." *The Chronicle of the Early American Industries Association* 50, no. 3 (September, 1997): 105.

Roberts, Kenneth D. "The E & M Scales." *Stanley Tool Collector News* 14 (Spring, 1995): 17-19.

Sear, T. N. "Instrument Profile: Airy's Dry Measure Guage for Checking Trade Volume Measures." *The Bulletin of the Scientific Instrument Society* 45 (June, 1995): 28-29.

Stanley, Philip E. "The E & M Scales." *Mensuration; The Newsletter of the Rule Collector's Association* I, no. 1 (Summer, 1987): 6 9.

RULES - SECTOR

Galilei, Galileo. *Operations of the Geometric and Military Compass*, 1606. Translated, with an introduction, by Stillman Drake. Washington: The Smithsonian Institution Press, 1978.

Gunter, Edmund. *The Description and Use of the Sector, the Crosse Staffe, and Other Instruments*. London: Published by the Author, 1624. [reprinted by Theatrum Orbis Terrarum, Ltd.]

Hood, Thomas. *The Making and Use of the Geometricall Instrument, Called a Sector*. London: John Windet, publisher, 1598. [reprinted by Theatrum Orbis Terrarum, Ltd.]

Pagnini, Giovanni. *Costruzione ed Uso del Compasso di Proporzione*. Naples: Ignazio Russo, 1753. [This work, in Italian, is a detailed description of the construction and use of the sector, copiously illustrated. We only recently obtained a copy of it, and it is still in the process of being translated.]

Stanley, Philip E. "William Cox on the Sector." *Journal of the Oughtred Society* 6, no. 2 (Fall, 1997): 13-18.

Waters, David. *The Art of Navigation in England in Elizabethan and Early Stuart Times*. Second edition. Greenwich: National Maritime Museum, 1978. [Contains excellent descriptions of the theory and use of Gunter's scale and the sector.]

RULES - SPECIFIC

(Author Unknown). "A Combination Pocket Rule." From "New and Interesting Inventions" in *Scientific American* magazine, (December 28, 1912).

(Author Unknown). "Hogg's Improved Slide Rule." *Stanley Tool Collector News* 4 (Winter, 1991): 10-12.

(Author Unknown). "The Stanley No. 133 Rule." *Mensuration; The Newsletter of the Rule Collector's Association* I, no. 2 (Fall, 1988): 4.

Baader, William. "A Most Unusual Rule — More For Your Money." *The Gristmill* 40 (June, 1985): 8-9, 18

—. "A Metric Zig Zag Rule." *The Gristmill* 41 (September, 1985): 20.

—. "Master Framing Rule." *The Gristmill* 43 (March, 1986): 17.

—. "The Marsh Dry Measure Gauge." *The Gristmill* 45 (September, 1986): 20.

—. "Ell Rules." *The Gristmill* 46 (December, 1986): 7.

—. "Extension Tape Rule." *The Gristmill* 47 (March, 1987): 9.

—. "Wykoff's Combination Rule." *The Gristmill* 53 (December, 1988): 16.

—. "Cutting Gage for Packing." *The Gristmill* 56 (September, 1989): 18-19.

—. "Protractor and Finger Traveler." *The Gristmill* 58 (March, 1990): 7.

—. "Extension Rule." (Jewel's patent) *The Gristmill* 61 (December, 1990): 21.

—. "Telescopic Measuring Rule." *The Gristmill* 59 (June, 1990): 16.

Crowe, David B. "A Chinese Two-Fold Rule." *The Gristmill* 71 (June, 1993): 19.

Fales, Cliff. "Whatsit Rule." *The Chronicle of the Early American Industries Association* 47, no. 4 (December, 1994): 118.

Farnham, Alexander. "A New Jersey Folding Advertising Rule." *The Tool Shed* 35 (February, 1985): 118-119.

Hogg, James. *A Practical Course of Instruction With Hogg's Improved Slide Rule in Arithmetic and Mensuration; (and) in the Calculations of Machinery for the Superintendent and Overseer in the Textile Manufacture, With Rules, Calculations, and Tables*. Boston: Rand, Avery Company, 1887.

Jacob, Walter W. "Stanley Advertising Architect's Rule." *The Fine Tool Journal* 47, no. 1 (Summer, 1997): 14.

—. "Stanley No. 83 Metric/English Carpenter's Rule With Slide." *The Fine Tool Journal* 47, no. 2 (Fall, 1997): 17.

Lynk, Scott. "Stanley Two Fold/Four Foot English/Metric Rule." *The Fine Tool Journal* 46, no. 4 (Spring, 1997): 15.

—. "Stanley Advertising Architect's Rule." *The Fine Tool Journal* 47, no. 1 (Spring, 1997): 14.

—. "Stanley No. 87(C) Ivory Caliper Rule." *The Fine Tool Journal* 47, no. 1 (Summer, 1997): 15.
—. "Stanley Number 1 Rule With a Brass Protractor." *The Fine Tool Journal* 47, no. 2 (Fall, 1997): 16.
—. "Stanley Metric Rules." *The Fine Tool Journal* 47, no. 4 (Spring, 1998): 11.
—. "English, Metric and Spanish." *The Fine Tool Journal* 48, no. 2 (Fall, 1998): 16.
—. "In-House Shop Tools at the Stanley Mills: Log Calipers." *The Fine Tool Journal* 48, no. 4 (Spring, 1999): 20.
—. "A Rare New Discovery." *The Fine Tool Journal* 49, no. 2 (Fall, 1999): 19-20.
—. "No. 036 Combination Rule." *The Fine Tool Journal* 49, no. 3 (Winter, 2000): 16-17, 19.
—. "Type Study of the No. 036 Combination Rule." *The Chronicle of the Early American Industries Association* 55, no. 2 (June, 2002): 55-58.
Lynk, Scott and Kebabian, Paul. "C. D. Cowgill's Bevel Attachment for Folding Rules." *Stanley Tool Collector News* 13 (Winter, 1994): 10-11.
Rees, Mark. "A Coachmaker's Rule?" (correspondence) *TATHS, Newsletter of the Tool and Trades History Society* 16 (Winter, 1987): 43-45.
—. "Not So Much a Rule - More a Vade Mecum." *TATHS, Newsletter of the Tool and Trades History Society* 45 (Spring, 1994): 35-38.
Richardson, Max. "Some Intriguing Rules." *The Tool Shed* 80 (February, 1994): 11.
—. "An Ironmonger's Rule." *The Tool Shed* 108 (September, 1999): 9-10.
Roberts, Kenneth D. "The L. C. Stephens Patent Combination Rule, Square, Level, and Bevel." *The Chronicle of the Early American Industries Association* 35, no. 2 (June, 1982): 29, C4.
Smith, Roger K. "Tools Made in the Midwest - William E. Owen, Teacher and Inventor." *The Gristmill* 59 (June, 1990): 14-15. [Describes Owen's patented "Combined Compass Protrac-tor and Ruler."]
Stanley, Philip E. "No. 036 Combination Rule Instruction Sheet." *Stanley Tool Collector News* 1 (Winter, 1990): 15.
—. "The No. 58-1/2 Rule." *Stanley Tool Collector News* 5 (Spring, 1992): 14-20
—. "A 'Take Down' Extension Stick, or Somebody Goofed!" *Mensuration; The Newsletter of the Rule Collector's Association* I, no. 3 (Fall, 1989): 5-7.
Tontz, Clay. "Re: Brass Folding Rule." *The Gristmill* 73 (December, 1993): 27; *The Gristmill* 74 (March, 1994): 27.
Walter, John. "Long's Patent Improved Gear or Cog Wheel Calculating Rule." *Stanley Tool Collector News* 17 (Spring, 1996): 25-30.

Walter, John. "Stowell Patent Rule." *Stanley Tool Collector News* 13 (Winter, 1994): 8.

RULE-RELATED TOOLS

(Author Unknown). "Those Odd Jobs." *The Fine Tool Journal* 46, no. 1 (Summer, 1996); 10-12.
Jacob, Walter. "Stanley's Odd Jobs." *The Tool Shed* 105 (February, 1999): 1, 4-6.
Kidwell, Peggy Aldrich. "American Parallel Rules: Invention on the Fringes of Industry." *Rittenhouse; The Journal of American Scientific Enterprise* 10, no. 39 (May, 1996): 90-96.
Stanley, Philip E. "A Sampling of Rule Accessories." *The Fine Tool Journal* 27, no. 1 (September, 1983): 4-8.
Townsend, Raymond R. "Parallel Rules ... Then and Now." *The Fine Tool Journal* 36, no. 1 (March/April, 1988): 5-6, 9.

SCIENTIFIC INSTRUMENTS

Bion, Nicholas. *The Construction and Principal Uses of Mathematical Instruments.* Paris: 1723. Translated by Edmond Stone; ... *To Which Are Added, the Construction and Uses of Such Instruments As Are Omitted by M. Bion*, by Edmond Stone. Second edition; ... *To Which is Added, a Supplement Containing a Further Account of Some of the Most Useful Mathematical Instruments As Now Improved*, Supplement by Edmond Stone. London: J. Richardson, 1758. [reprinted by the Astragal Press]
Turner, Anthony. *Early Scientific Instruments; Europe, 1400 1800.* London: Sotheby's Publica-tions, 1987.

SLIDE RULES

(Author Unknown). *Mensuration Made Perfectly Easy, by the Assistance of a New Improved Sliding Rule; Which at One Operation, Solves Questions into Superficies and Solids of all Denominations. by an Entire New and Concise Method Never Before Put Into Practise. To Which is Added a Description and Use of Mr. Scamozzi's Lines , in Finding the Lengths and Angles of Rafters, Hips, Collar Beams, &c. also Chords, Hours, Latitudes, and Inclinations of Meridians, With their Use in Mechanic.* Birmingham: A. & M. M. Chapman, 1823.
(Author Unknown). "Hogg's Improved Slide Rule." *Stanley Tool Collector News* 4 (Winter, 1991): 10-12.
Baader, William J. "Dialling." *The Fine Tool Journal* 37, no. 2 (November/December, 1988): 20-21.
Cajori, Florian. *A History of the Logarithmic Slide Rule.* New York: The Engineering News

Publishing Co., 1909. [reprinted by the Astragal Press]

Coggeshall, Henry. *Timber Measure by a Line of More Ease, Dispatch, and Exactness Than Any Other Way Now in Use, By a Double Scale* London: (publisher unknown), 1677.

—. *The Art of Practical Measuring, by the Slide Rule: Shewing How to Measure Round, Square, or Other Timber ... Whereunto Is Added, In a Short Method, the Use of Scamozzi's Lines for Finding the Lengths and Angles of Hips, Rafters, &c ... by John Ham*. Seventh edition. London: Edward and Charles Dilly, 1767.

Feely, Wayne. "K & E Slide Rules." *The Chronicle of the Early American Industries Association* 49, no. 2 (June, 1996): 50-52.

—. "The Paisley Slide Rule." *The Chronicle of the Early American Industries Association* 49, no. 4 (December, 1996): 113.

—. "The Low-Tech Caliputer Still Has Its Uses." *The Gristmill* 85 (December, 1996): 20.

—. "The Cal-Tape Slide Rule." *The Chronicle of the Early American Industries Association* 50, no. 1 (March, 1997): 22.

—. "Chemical Slide Rules." *The Chronicle of the Early American Industries Association* 50, no. 2 (June, 1997): 44-48.

—. "The Fuller Spiral Scale Slide Rule." *The Chronicle of the Early American Industries Association* 50, no. 3 (September, 1997): 93-98.

—. "Thacher Cylindrical Slide Rules." *The Chronicle of the Early American Industries Association* 50, no. 4 (December, 1997): 125-127.

—. "The Last Days of K & E Slide Rules." *The Chronicle of the Early American Industries Association* 52, no. 2 (June, 1999): 62-65.

Hoare, Charles. *The Slide Rule and How to Use It*. London: (Publisher Unknown), 1890.

Hodgson, Fred T., ed. *The Mechanics' Slide Rule, and How to Use It*, (Work Manuals, No. II). New York: The Industrial Publication Co., 1881. [The author also recommends the books on this subject by Heather, Hoare, Jillson, Kentish, Riddell, and Tonkes, and the article "Slide Rule" in the *Penny Cyclopedia*.]

Hogg, James. *A Practical Course of Instruction With Hogg's Improved Slide Rule in Arithmetic and Mensuration; (and) in the Calculations of Machinery for the Superintendent and Overseer in the Textile Manufacture, With Rules, Calculations, and Tables*. Boston: Rand, Avery Company, 1887.

Nesbit, Anthony. "Description and Use of the Carpenter's Rule." *A Treatise on Practical Mensuration*. Twelfth ed., 149-155. London: Longman, Brown, Green, and Longmans, 1855.

Palmer, Aaron. *A Key to the Endless, Self Computing Scale*. Boston: Smith & Palmer, 1844. [This is the book of instructions for the earliest known American circular slide rule.]

Pickworth, Charles N. *Instructions for the Use of A. W. Faber's Improved Calculating Rule*. Newark: A. W. Faber, Ca. 1900.

Rees, Jane. "A 'Vade Mecum Slide Rule' Designed Expressly for the Timber Trade by George Bousfield." *TATHS, Newsletter of the Tool and Trades History Society* 66 (Autumn, 1999): 10-23.

Riddell, Robert. *Lessons in Carpentry by the Slide Rule*. Publication data not known.

Tonkes, William. *The Engineers' Slide Rule*. Publication data not known.

Zoller, Paul. "The Soho Slide Rule: Genesis and Archeology." *The Bulletin of the Scientific Instrument Society* 57 (June, 1998): 5-13.

SQUARES - FRAMING

(Author Unknown). "The Story of the Steel Square." *The Rule Shop Seismograph* I, no. 9 (October 15, 1919).

Kebabian, Paul. "The Bridge Builder's Square and E. H. Robinson's Ruling Device." *The Gristmill* 47 (March, 1987): 14-18.

Klingler, Gene. "The Carpenter's Square, a Neglected Tool." *The Chronicle of the Early American Industries Association* 38, no. 2 (June, 1985): 23-24.

Nelson, Bob. "The Van Namee Framing Square." *The Gristmill* 93 (December, 1998): 18-19.

—. "The Klinglesmith/Farner Bevel." *The Gristmill* 95 (June, 1999): 16-17.

Pozzato, Ronald. "The Scale on the Square." *The Chronicle of the Early American Industries Association* 47, no. 4 (December, 1994): 128.

SURVEYING

Blagrave, John. *A Book of the Making and Use of a Staffe, Newly Invented by the Author, Called the Familiar Staffe*. London: Published by the Author, 1590. [reprinted by Theatrum Orbis Terrarum, Ltd.]

Dow, Melvin C. "The Surveyor's Chain." *The Chronicle of the Early American Industries Association* 16, no. 4 (December, 1963): 37-38.

Kamrass, Murray. "On Rods, Roods, Perches and Poles: Linear Measuring With Chains and Tapes." *Rittenhouse; The Journal of American Scientific Enterprise* 1, no. 3 (May, 1987): 85-89.

Kebabian, John. "Surveying." *The Gristmill* 56 (September, 1989): 15.

Townsend, Raymond R. "Early Measuring and Leveling Instruments." *The Fine Tool Journal* 29, no. 3 (November, 1984): 36-39.

Vogel, Robert, and Hands, Edmund. "The Pools of Easton, Massachusetts." *The Chronicle of the Early American Industries Association* 50, no. 1 (March, 1997): 1-11.

TAPE MEASURES

Jacob, Walter. "Stanley Tapes Measure the Worls, Parts I through V." *The Chronicle of the Early American Industries Association* 53, no. 4 (Dec-ember, 2000) through 54, no. 4 (December, 2001).

TOOLS - GENERAL

Garvin, James L., and Garvin, Donna-Belle. *Instruments of Change — New Hampshire Hand Tools and Their Makers.* Canaan, New Hampshire: the New Hampshire Historical Society, 1985. (This catalogue of an exhibition at the New Hampshire Historical Society in 1984-1985 is very well prepared and extremely informative.)

Gaynor, James M., and Hagedorn, Nancy L. *Tools; Working Wood in Eighteenth-Century America.* Williamsburg, Virginia: The Colonial Williamsburg Foundation, 1994.

Gaynor, Jay, ed. *Eighteenth-Century Woodworking Tools.* Williamsburg, Virginia: The Colonial Williamsburg Foundation, 1997.

Goodman, W. L. *The History of Woodworking Tools.* New York, New York: David McKay Company, 1964.

Salaman, Raphael A. *Dictionary of Woodworking Tools, ca. 1700-1970, and Tools of Allied Trades.* Second edition, revised by Philip Walker, F.S.A. London: Unwin Hyman, 1989.

—. *Dictionary of Leather Working Tools, 1700-1950, and Tools of Allied Trades..* New York, New York: Macmillan Publishing Co., 1986.

Sellens, Alvin. *Dictionary of American Hand Tools.* Augusta, Kansas: Published by the Author, 1990.

Singer, Charles, Holmyard, E. J., Hall, A. R., and Williams, Trevor I. *A History of Technology* (5 volumes). Oxford: The Clarendon Press, 1956.

Walter, John. *Antique & Collectible Stanley Tools — A Guide to Identity and Value.* Marietta, Ohio: The Tool Merchant, 1996.

WEIGHTS & MEASURES

Dickinson, Henry W., and Rogers, Henry. "Origin of Gauges for Wire, Sheets and Strip." *Transactions of the Newcomen Society* 21 (1940-1941): 87-98.

Gilliland, William. "Why 12?" *Mensuration; The Newsletter of the Rule Collector's Association* I, no. 3 (Fall, 1989): 6.

—. "Some Measures of History." *The Bulletin of the Scientific Instrument Society* 20 (March, 1989): 7 17. [reprinted in *Mensuration; The Newsletter of the Rule Collector's Association* II, no. 1]

Hunter, Alexander. *A Treatise of Weights, Mets and Measures of Scotland.* Edinburgh: John Wreittoun, Publisher, 1624. [reprinted by Theatrum Orbis Terrarum]

Kidwell, Peggy Aldrich. "Publicising the Metric System in America: From F. R. Haswell to the American Metric Bureau." *Rittenhouse; The Journal of American Scientific Enterprise* 5, no. 20 (August, 1991): 111-117.

Ott, Max E. "Generous Scotch Measure?" (correspondence) *TATHS, Newsletter of the Tool and Trades History Society* 39 (Autumn, 1992): 46-47.

Roberts, Kenneth D. "A. & T. W. Stanleys' Meter-Diagram." *Stanley Tool Collector News* 2 (Spring, 1991): 16-21.

Stanley, Philip E. "Rules: Obsolete Units of Length." *Tools & Trades, the Journal of the Tool and Trades History Society* 13 (August, 2002): 97-114.

Zupko, Ronald E. *British Weights and Measures; A History from Antiquity to the Seventeenth Century.* Madison, Wisconsin: The University of Wisconsin Press, 1977.

—. *French Weights & Measures before the Revolution; A Dictionary of Provincial & Local Units.* Bloomington, Indiana: The Indiana University Press, 1979.

—. *Italian Weights and Measures from the Middle Ages to the Nineteenth Century.* Philadelphia, Pennsylvania: The American Philosophical Society, 1981.

—. *A Dictionary of Weights and Measures for the British Isles: the Middle Ages to the Twentieth Century.* Philadelphia, Pennsylvania: The American Philosophical Society, 1985.

MISCELLANEOUS

Blanchard, Clarence. "Condition, Condition, Condition." *The Fine Tool Journal* 45, no. 4 (Spring, 1996): 7-9.

Cowan, M. J. "Monetary Values, Past and Present." *The Bulletin of the Scientific Instrument Society* 21 (June, 1989): 7-8 [Useful for evaluating historical costs in terms of present-day dollars or pounds.]

Pollak, Emil, ed. *The Stanley Catalog Collection.* Morristown, New Jersey: The Astragal Press,

1989. [Collected reprints of seven A. Stanley and Stanley Rule & Level Co. catalogs from 1855-1898.]

Pollak, Martyl, ed. *The Stanley Catalog Collection, Volume II*. Morristown, New Jersey: The Astragal Press, 1998. [Collected reprints of five Stanley Rule & Level Co. catalogs and two Leonard Baily & Co. catalogs from 1872- 1892.]

Rubin, Jack. "Updating Historical $ and £ Data to Current Values." *Rittenhouse; The Journal of American Scientific Enterprise* 1, no. 3 (May, 1987): 82-84 [Useful for evaluating historical costs in terms of present-day dollars or pounds.]

Ward, Vernon U. "Statements of Condition." *The Fine Tool Journal* 26:5 (July, 1983): 1-2.

—. "The Classification of Condition for Antique Tools, Revisited." *The Fine Tool Journal* 41:4 (Winter, 1991): 8-10.

PERIODICALS

A number of periodical publications will be found useful as a source of information for those interested in measuring instruments. Some, such as *The Gristmill* or *The Tool Shed*, are the organs of various tool collecting organizations. Others, such as *The Compass* or *The Rule Shop Seismograph* were published by the makers of measuring instruments, and describe and discuss their product lines. Still others, such as *The Fine Tool Journal* (*) and *Tesseract*, are the catalogs of dealers in antique tools and scientific instruments; such catalogs frequently contain valuable information in their descriptions of items for sale.

(*) *The Fine Tool Journal* is much more than just a catalog; each issue includes significant articles on antique tools and other editorial material.

The ACTIVE Scrapbook. Paul Kebabian, ed. Published 3-4 times a year from 1972 to 1987 by Antique Crafts & Tools in Vermont, 11 Scottsdale Rd., South Burlington VT 05401

The Bulletin of the Scientific Instrument Society. Published quarterly by the Scientific Instrument Society, Howard Dawes, executive secretary, P.O. Box 15, Pershore, Worcestershire WR10 2RD, United Kingdom.

The Chronicle of the Early American Industries Association. Patty MacLeish, ed. Published quarterly by The Early American Industries Association, Elton W. Hall, Treasurer, 167 Bakerville Rd., South Dartmouth MA 02748.

The Compass. William Cox, ed. Published monthly from 1891 to 1894 by the Keuffel & Esser Co., Hoboken, New Jersey.

The Fine Tool Journal. Clarence Blanchard, ed. Published quarterly by Antique and Collectible Tools, Inc., 27 Fickett Road., Pownal ME 04069

The Gristmill. Mary Lou Stover, ed. Published quarterly by The Mid-West Tool Collector's Association, Willy Royal, President, 215 Anthony Circle, Charlotte NC 28211-1417.

Transactions of the Newcomen Society. Dr. Robert Otter, ed. Published semiannually since 1921 by the Newcomen Society, The Science Museum, London SW7 2DD, England.

Mensuration; The Newsletter of the Rule Collector's Association. Philip E. Stanley, ed. Published from 1987 to 1990 by the Rule Collectors Association, 40 Harvey Lane, Westborough MA 01581-3005.

Journal of the Oughtred Society. Robert Otnes, ed. Published semiannually by the Oughtred Society, 2160 Middlefield Rd., Palo Alto CA 94301

Rittenhouse; The Journal of American Scientific Enterprise. Randall C. Brooks, ed. Published semiannually by David & Yola Coffeen and Raymond Giordano., P.O. Box 9724, Station T, Ottawa ON K1G 5A3 Canada.

The Rule Shop Seismograph. Published monthly from 1915 to 1932 by the Stanley Rule & Level Plant of the Stanley Works, New Britain CT 06051.

Stanley Tool Collector News. John Walter, ed. Published in 19 editions from 1990 to 1996 by John Walter, P.O. Box 227, Marietta OH 45750.

Tesseract. Published quarterly by David & Yola Coffeen, Box 152, Hastings-on-Hudson, NY 10706 [a quarterly catalog of scientific instruments]

TATHS, Newsletter of the Tool and Trades History Society. Published quarterly by the Tool and Trades History Society, 60 Swanley Lane, Swanley, Kent BR8 7JG, United Kingdom.

Tools and Technology. Mary-Anne Ahearn, ed. Published quarterly by the American Precision Museum, Windsor VT.

The ATTIC Tool Chest. Robert Dunn, acting ed. Published semiannually by Antique Tools and

Trades in Connecticut, 34 Thomas Ave., Uncasville CT 06382.

The Tool Shed. Robert Garay, ed. Published five times a year by CRAFTS of New Jersey, 147 Dupont Ave., Hopatcong NJ 07843.

Tools & Trades, the Journal of the Tool and Trades History Society. Published more or less annually by the Tools and Trades History Society, 60 Swanley Lane, Swanley, Kent BR8 7JG, United Kingdom.

ENCYCLOPEDIAS

Beginning with Diderot's Encyclopedia and the Encyclopedia Britannica, technical and general encyclopedias have become a fixture in reference libraries, and many contain information which is of interest to students of measuring instruments. While it is true that 99+% of any encyclopedia will be unrelated to the readers specific interest, an occasional nugget will be encountered by the persistent reader.

Even if you are not looking for specific information, casual browsing can be rewarding. Serendipity works. A description of the technique used by early rulemakers to make the joint plates for folding rules was found by the author while idly turning the pages of the section on FILING in Appleton's Dictionary of Machines, Mechanics, Engine-Work and Engineering.

The following is a list of encyclopedias and encyclopedic dictionaries which the author has found useful and interesting. Some are only available in the original, but it is surprising how many have been reprinted in whole or in part. Where an encyclopedia has been reprinted, we have tried to indicate where and when.

Appleton's Dictionary of Machines, Mechanics, Engine-Work and Engineering. 2 volumes. New York: D. Appleton & Co., 1865. [the 1880 edition of this dictionary was reprinted jointly in 1982 by the Mid-West Tool Collectors Association and the Early American Industries Association].

The Cyclopedia, or Universal Dictionary of the Arts, Sciences, and Literature. 39 volumes of text; 6 volumes of plates. Abraham Rees, ed. London: 1819. [Far and away the best work in English on the technology of the period. The sections of this encyclopedia dealing with the manufacturing industries were reprinted in five volumes in 1972 by David & Charles Reprints].

Cyclopedia of Useful Arts & Manufactures. 3 volumes, Charles Tomlinson, ed. London: George Virtue, Ca. 1854. [Contains an excellent description of hand graduation methods; see GRADUATION]

The Encyclopedia Britannica. 11th edition, 29 volumes. New York: Encyclopedia Britannica, Inc., 1910-1911. ["reliable, scholarly, and amazingly comprehensive," Ferguson, see below. Contains a comprehensive list of foreign length standards; see WEIGHTS & MEASURES]

Encyclopédie, ou Dictionnaire raisonné des sciences, des arts et des métiers (Encyclopedia, or Classified Dictionary of Sciences, Arts and Trades]. 36 volumes, Denis Diderot and Jean d'Alembert, eds. Paris: Le Breton, 1751. [In French, but an incredible compendium of information on science and trades in the 18th century. The plates are magnificent!] [Reprinted in 1969 in a six-volume reduced page size edition by the Readex Microprint Corporation].

Knight's American Mechanical Dictionary. 4 volumes. Edward H. Knight, ed. New York: Hurd & Houghton, 1876. [reprinted jointly in 1979 by the Mid-West Tool Collectors Association and the Early American Industries Association]

Machinery's Encyclopedia. 8 volumes. Erik & Franklin D. Jones, eds. New York: The Industrial Press, 1917. [contains a good description of machine graduation methods; see GRADUATION]

Modern Machine Shop Practice. 2 volumes. Joshua Rose, ed. New York: Charles Scribner's Sons, 1887. [An extremely comprehensive description of all aspects of machine shop operation in the late 19th century. Not properly an encyclopedia, but listed here because it is in the same category as *Machinery's Encyclopedia*.]

The Penny Cyclopedia. 29 volumes. William Nicholson, ed. New York: 1833-1846. [The article SLIDE RULE in this work was listed as a reference in *The Mechanics' Slide Rule and How to Use It*; see Hodgson, Fred T., above.]

GUIDES TO OTHER SOURCES

PATENT INDEXES

During the first 100 years of the United States no fewer than four indexes of all Patents issued by the United States Government were compiled and published by the U.S. Secretary of State (in 1831), and by the U.S. Patent Office (in 1840, 1849, and 1874). Of these four the first and last are the most useful.

1831: Letter from the Secretary of State Transmitting a List of All Patents Granted by the United States, the Acts of Congress Relating Thereto, and the Decisions of the Courts of the United States Under the Same. 21st Congress, 2nd Session, Doc. No. 50. Washington: Duff Green, January 13, 1831.

This was a subject index, cross-indexed by patentee, of all U.S. Patents from 1790 up to the end of 1829.)

1840 & 1849: These two indexes, the first covering 1790-1838 and the second 1790-1848, were subject indexes only, with no cross index.

1874: Leggett, Mortimer D., editor. Subject-Matter Index of Patents for Inventions Issued by the United States Patent Office from 1790 to 1873, Inclusive. Washington: US Government Printing Office, 1874.

This last index, which was also subject-only, was three volumes in length, covering as it did the entire period from 1790 up to the end of 1873. [This index was reprinted in 1976 and again in 2001 by the Arno Press; sets of this reprint can occasionally be found at used book shows or in used book dealer lists.]

The most significant of these indexes is probably the first. It is particularly valuable because it is cross-indexed by patentee, as well as subject. Equally importantly, it was compiled from original materials, materials subsequently lost in the Patent Office fire of 1836, and thus is the closest to the original materials. All indexes prepared after that fire were prepared without access to the pre-1836 patents, which had all been destroyed, and are thus one step further removed from the documents they are attempting to catalog and describe.

The index of 1874 is the next most valuable, completely superseding as it does those of 1840 and 1849, and except for the above-mentioned lack of access to the early patents, and for its sometimes irregular subject name assignments, would be the only index needed.

COPIES OF PATENTS

It has recently become possible to get copies of U.S. Patents using a computer connected to the internet, instead of the more cumbersome process of sending in a fee and waiting for return mail. The U.S. Patent and Trademark Office has a Patent website (http://www.uspto.gov/patft/) permitting access to the text and drawings of all patents issued after 1836 and almost all "restored" patents from 1790 to 1836. If the patent number is known, select Patent Number Search; if the class & subclass are known, select Advanced Search. Then enter the patent number(*) or class and subclass in the appropriate location. For the patents of most interest to collectors (pre-1976) it is not possible at this time to search by any other parameter than the two just mentioned, such as date of issue or name of inventor, etc. It is to be hoped that in the future the functionality of this web site will be enhanced so that full search capability will be possible for the early patents as well.

(*)Patents issued prior to July, 1836 were not originally assigned patent numbers, and are thus sometimes called "name&date" patents. Numbers were later assigned to these pre-1836 patents, based on their listings in the annual Secretary of State reports to the Congress. These ex post facto numbers were prefixed with the letter X, and they are thus sometimes called "X-number" patents.

When searching for a pre-1836 patent by patent number, be sure to include the letter "X" at the beginning of the number (e.g.: X2369) so that the search engine can know that a pre-1836 patent is being referenced.

BIBLIOGRAPHIES & DIRECTORIES

In recent years efforts supported by the Early American Industries Association and the Scientific Instrument Society have led to the publication of two very useful directories, one dealing with American tool makers, and the other with British scientific instrument makers. Both will be found valuable to rule collectors in identifying makers and dating rules.

In addition to these two directories, a number of bibliographies have been published relating to the literature of tools and scientific instruments. Two (Barlow and Romaine) are not tool related, but are primarily concerned with trade catalogues; they will still be found useful in identifying maker's catalogues to be sought.

Anderson, R. G. W., Burnett, J., and Gee, B. *Handlist of Scientific Instrument-Makers' Trade Catalogues, 1600-1914.* Edinburgh: National Museums of Scotland, 1990. [includes rule makers]

Barlow, Ron, and Reynolds, Ray. *The Insider's Guide to Old Books, Magazines, Newspapers and Trade Catalogs.* El Cajon: Windmill Publishing Co., 1995.

Clifton, Gloria. *Directory of British Scientific Instrument Makers 1550-1851.* London: Philip Wilson Publishers, 1995. [includes rule makers]

Ferguson, Eugene S. *Bibliography of the History of Technology.* Boston: The MIT Press, 1968. [Particularly valuable in pointing out available information resources.]

Hindle, Brooke. *Technology in Early America; Needs and Opportunities for Study.* Williamsburg: The University of North Carolina Press, 1966. [Not properly a bibliography, but a good reference nonetheless.]

Letocha, Michael, Hesse, Peter, and Sauer, Siegfried. *Bibliographie Geschichte der Technik* (Bibliography of the History of Technology). Dresden: State Library of Saxony, 1986. [In German, but a good reference on European publications.]

Nelson, Robert E., ed. Directory of American Toolmakers. (Location unknown): Early American Industries Association, 1999.

Nicole, George. *The Woodworking Trades — A Select Bibliography.* Plymouth: The Twybill Press, 1993.

Romaine, Lawrence B. *A Guide to American Trade Catalogs, 1744-1900.* New York: R. R. Bowker, 1960. [reprinted by Dover Books]

Turner, G. L'E., and Bryden, D. J. *A Classified Bibliography on the History of Scientific Instruments.* Oxford: The Scientific Instrument Commission of the International Union on the History and Philosophy of Science, 1997.

Simcock, A. V. *A Supplement to 'A Classified Bibliography on the History of Scientific Instruments'.* Oxford: The Scientific Instrument Commission of the International Union on the History and Philosophy of Science, 1998.